# THEORETICAL PHYSICS AND BIOLOGY
# PHYSIQUE THÉORIQUE ET BIOLOGIE

# INSTITUT DE LA VIE

# PHYSIQUE THÉORIQUE ET BIOLOGIE

Comptes rendus de la Première Conférence Internationale de
Physique Théorique et Biologie,

Palais des Congrès, Versailles, 26–30 juin 1967

Edité par

M. MAROIS

## COMITÉ D'ORGANISATION

P. AUGER (Paris) – A. FESSARD (Paris) – H. FRÖHLICH (Liverpool) –
P. P. GRASSÉ (Paris) – A. LICHNÉROWICZ (Paris) – I. PRIGOGINE (Bruxelles) –
L. ROSENFELD (Copenhague) – M. MAROIS, Secrétaire Général de la Conférence

## PATRONAGES

M. le Ministre d'Etat Chargé de la Recherche Scientifique et des Questions
Atomiques et Spatiales (France)

Union Internationale de Physique Pure et Appliquée

1969

NORTH-HOLLAND PUBLISHING COMPANY
AMSTERDAM • LONDRES

*INSTITUT DE LA VIE*

# THEORETICAL PHYSICS
# AND BIOLOGY

*Proceedings of The First International Conference on
Theoretical Physics and Biology,*

*Palais des Congrès, Versailles, 26–30 June 1967*

*Edited by*

M. MAROIS

### ORGANIZING COMMITTEE

P. AUGER (Paris) – A. FESSARD (Paris) – H. FRÖHLICH (Liverpool) –
P. P. GRASSÉ (Paris) – A. LICHNEROWICZ (Paris) – I. PRIGOGINE (Bruxelles) –
L. ROSENFELD (Copenhagen) – M. MAROIS, Conference General Secretary

### SPONSORS

*M. le Ministre d'Etat Chargé de la Recherche Scientifique et des Questions
Atomiques et Spatiales (France)*

*International Union of Pure and Applied Physics*

1969

NORTH-HOLLAND PUBLISHING COMPANY

AMSTERDAM • LONDON

PHYSICS

© 1969 NORTH-HOLLAND PUBLISHING COMPANY - AMSTERDAM and L'INSTITUT DE LA VIE

*Library of Congress Catalog Card Number 76–89938*
*Standard Book Number 7204 4051 3*

PUBLISHERS:
NORTH-HOLLAND PUBLISHING CO. - AMSTERDAM
NORTH-HOLLAND PUBLISHING CO. LTD. - LONDON

SOLE DISTRIBUTORS FOR THE WESTERN HEMISPHERE:
WILEY INTERSCIENCE DIVISION
JOHN WILEY & SONS, INC. - NEW YORK

PRINTED IN THE NETHERLANDS

# THEORETICAL PHYSICS AND BIOLOGY

## ERRATA

| pages | ligne | | | |
|---|---|---|---|---|
| 2 | 29 | étrange | au lieu de | étrangers |
| 53 | 30/31 | de M. Prigogine | | du M. Prigogine |
| 100 | 3 | ce point | | se point |
| 113 | 10 | de M. Fröhlich | | du M. Fröhlich |
| 114 | 22 | polymeres | | polymers |
| | 28 | amassé | | ammassé |
| 115 | 37 | qu'aujourd'hui | | queaujourd'hui |
| | 39 | facteur | | factuer |
| 117 | 40 | raisonnable | | raisonable |
| 118 | 18 | ordinateurs | | ordinatuers |
| 119 | 10 | j'obtiendrais | | j'obtiendrai |
| 123 | 15 | a posteriori | | postériori |
| 124 | 25 | glycosidiques | | glycosyliques |
| | 29 | uracile | | uracil |
| | 30 | cocristallisations | | cocrystallisations |
| | 33 | idem | | idem |
| | 40 | lorsqu'au lieu | | lorsqueau lieu |
| | 40 | glycosidique | | glycosylique |
| 129 | 17 | glycosidique | | glycosylique |
| | 20 | glycosidique | | glycosylique |
| 132 | 1 | d'Albert | | de Albert |
| 133 | 15 | d'ionisation | | d'ionization |
| 133 | dernière | qu'un certain | | qu'on certain |
| 139 | 39 | naphtopyrène | | naphtopygène |
| 145 | 16 | quelle que | | quelque |
| 154 | 10 | ontogenesis | | onthogenesis |
| 164 | 11 | M. Bergmann | | le M. Bergmann |
| | 38 | M. Bergmann | | le M. Bergmann |
| 165 | 3 | toute | | route |
| 175 | 20 | de M. Pullman | | du M. Pullman |
| | 21/22 | M. Pullman | | le M. Pullman |
| | 28 | M. Pullman | | le M. Pullman |
| 215 | 19 | séparément | | séparemment |
| | 26 | chacune d'elles | | chacune d'elle |
| 216 | 15 | ponctuation | | punctuation |
| 217 | 6 | nucléotidique | | nuclétidique |

| 221 | 9 | par groupes | par groupe |
|---|---|---|---|
| 250 | 16 | requise | requisée |
| 252 | 36 | déplacements | déplacemenss |
| | 39 | et qui vont | et yui voot |
| 253 | 36 | je serais | je serait |
| 254 | 1 | M. Monod | le M. Monod |
| | 10 | M. Gros | le M. Gros |
| 260 | 31 | réunis | réunies |
| 314 | 20 | M. Wolff | le M. Wolff |
| 317 | 29 | processus que dans | processus dans |
| 340 | 8 | à M. Lindauer | au M. Lindauer |
| 342 | 15 | de M. Lloyd | du M. Lloyd |
| 347 | 37 | es sciences | des sciences |
| 351 | 3 | mesurer | |
| | 4 | temporelles | |
| | 5 | comme | |
| | 6 | illustrée | |
| | 7 | les | |
| 358 | 18 | Ce dernier | Le dernier |
| 359 | 34 | désynchronisation | désynchronization |
| 371 | 21 | diagnostique | diagnostic |
| 383 | 9 | aléatoires | alléatoires |
| | 12 | allongées | alongées |
| 387 | 30 | dangereux | dangéreux |
| 400 | 24 | l'ADN | l'DNA |
| | 26 | rythmes | rythme |
| 403 | 4 | relation | rélation |
| 406 | 33/34 | d'ambiance | d'ambiante |
| | 13 | unes de ces | uns de ces |
| 411 | 7 | M. Fessard | le M. Fessard |
| 413 | 2 | générales | générale |
| 423 | 34 | effectuer | effecturer |
| 424 | 32 | reviendrai | reviendra |
| | 34 | de M. Fessard | du M. Fessard |
| | | de M. Halberg | du M. Halberg |
| 426 | 40 | M. Prigogine | le M. Prigogine |

# Contents

Preface . . . . . . . . . . . . . . . . . . . . . . . . . . . . . . VII

Official Addresses . . . . . . . . . . . . . . . . . . . . . . . . . 1

*Journée du 26 juin 1967, 1ère séance*

H. Fröhlich (Liverpool): Quantum Mechanical Concepts in Biology    13
I. Prigogine (Bruxelles): Structure, Dissipation and Life . . . . . . . 23
Discussions . . . . . . . . . . . . . . . . . . . . . . . . . . . . . 53

*Journée du 26 juin 1967, 2ème séance*

Jack D. Cowan (Teddington): Some Remarks on Neurocybernetics . .    65
Walter M. Elsasser (Princeton): What is Theoretical Biology . . . .    74
General Discussions . . . . . . . . . . . . . . . . . . . . . . . . . 77

*Journée du 27 juin 1967, 1ère séance*

Bernard Pullman (Paris): Les Calculs Quantiques en Biologie . . . .   113
S. Bresler (Leningrad): Molecular Forces and Molecular Recognition   152
Britton Chance (Philadelphia): The Time Domains of Biochemical Reactions in Living Cells . . . . . . . . . . . . . . . . . . . . . . . . 156
Discussions . . . . . . . . . . . . . . . . . . . . . . . . . . . . . 159

*Journée du 27 juin 1967, 2ème séance*

Hans H. Ussing (Copenhagen): Active Transport . . . . . . . . . . . 179
A. Katchalsky (Israel): Non-equilibrium Thermodynamics of Bio-Membrane Processes . . . . . . . . . . . . . . . . . . . . . . . . . 188
Albert Szent-Györgyi (Woods-Hole): Charge, Charge Transfer and Cellular Activity . . . . . . . . . . . . . . . . . . . . . . . . . . 192
Discussions . . . . . . . . . . . . . . . . . . . . . . . . . . . . . 194

*Journée du 29 juin 1967, 1ère séance*

S. Ochoa (New York): Genetic Coding . . . . . . . . . . . . . . . . 211
François Gros (Paris): Remarques sur le Code Génétique . . . . . . . 215
Alfred Fessard (Paris): Les problèmes du Code Nerveux . . . . . . . 230
Maarten A. Bouman (Soesterberg): Quantum Noise and Vision . . . 246
Discussions . . . . . . . . . . . . . . . . . . . . . . . . . . . . . 250

*Journée du 29 juin 1967, 2ème séance*

J. MONOD, J. WYMAN and J. P. CHANGEUX: On the Nature of Allosteric
Transitions: a Plausible Model . . . . . . . . . . . . 267
ETIENNE WOLFF: L'Origine de la Symétrie Bilatérale chez l'Embryon . . 302
Discussions . . . . . . . . . . . . . . . . . . . . . . . 308

*Journée du 30 juin 1967, 1ère séance*

B. B. LLOYD (Oxford): The Concept of Regulation in Physiology . . 321
M. LINDAUER and H. MARTIN (Frankfort): Special Sensory Performances
in the Orientation of the Honey Bee. . . . . . . . . . 332
Discussions . . . . . . . . . . . . . . . . . . . . . . . 339

*Journée du 30 juin 1967, 2ème séance*

FRANZ HALBERG (Minneapolis): Chronobiologie; Rythmes et Physiologie
Statistique . . . . . . . . . . . . . . . . . . . . 347
Discussions . . . . . . . . . . . . . . . . . . . . . . . 394

*Journée du 30 juin 1967, 3ème séance*

General discussions and conclusions . . . . . . . . . . . 415
I. SEGAL (M.I.T.): Remark on the Parametrization of the States of Complex
Systems . . . . . . . . . . . . . . . . . . . . . . 418
List of Participants . . . . . . . . . . . . . . . . . . . 439

# *Preface*

Une interrogation fondamentale sur la vie et le destin de l'homme, tel est l'objet premier de l'Institut de la Vie. Fidèle à sa vocation et à son style, l'Institut de la Vie a pris l'initiative d'organiser une conférence internationale de Physique théorique et de Biologie. Cette manifestation fut placée sous le patronage de Monsieur le Ministre d'État chargé de la recherche scientifique et technique et de l'Union internationale de Physique pure et appliquée.

Ouverte par le Professeur P. Grassé, Président de l'Académie des Sciences, marquée par un discours de Monsieur Maurice Schumann, Ministre d'Etat chargé de la recherche scientifique et des questions atomiques et spatiales, cette conférence fut caractérisée par:

- son thème: la vie
- les multiples disciplines confrontées: de la physique théorique à la biologie
- la très haute qualité des participants: plus de soixante dix savants, dont plusieurs Prix Nobel, appartenant à quatorze nations,
- la volonté de dialogue d'abord, de coopération ensuite, malgré la diversité des langages et des méthodes.

Au cours des huit séances réparties en quatre journées, furent traités quelques aspects des problèmes de la vie depuis les conceptions théoriques jusqu'aux mécanismes physiologiques. Les titres de ces quatre journées montrent bien l'itinéraire choisi:

- concepts théoriques: concepts quantiques en biologie, physique statistique et thermodynamique
- physicochimie de la vie: données et problèmes: molécules biologiques et effets quantiques, aspects biochimiques
- l'information en biologie: information et codes, les structures formelles et la symétrie
- mécanismes physiologiques: régulations, rythmes biologiques.

Pour tempérer la rigueur logique de ce programme, un vaste champ libre fut laissé aux discussions, après la présentation de chaque rapport et de chaque communication sollicitée. Ainsi la réunion a-t-elle atteint son premier but: l'intercommunication entre disciplines très différentes.

Une telle conférence n'est pas une manifestation isolée. Elle marque le début d'une série dont nous n'entrevoyons pas le terme. Une deuxième conférence se tiendra à Versailles du 30 juin au 4 juillet 1969. Et l'entreprise de l'Institut de la Vie s'inscrit dans un ample dessein: conjuguer les ressources de la science contemporaine pour l'exploration d'un sujet fascinant, la vie.

Nous exprimons notre gratitude:

– aux membres du Comité Scientifique d'Organisation pour leur concours intellectuel: les Professeurs P. Auger, A. Fessard, H. Fröhlich, P. P. Grassé, A. Lichnérowicz, I. Prigogine et L. Rosenfeld ainsi qu'à tous les participants de la conférence.

– aux organismes publics et privés, aux entreprises et personnalités privées pour leur soutien matériel: le Ministère des Affaires Sociales, la Délégation Générale à la Recherche Scientifique et Technique, la Mutuelle Générale de l'Education Nationale, la Banque de France, la Banque Nationale pour le Commerce et l'Industrie, la Banque Neuflize, Schlumberger, Malet et Compagnie, la Banque Rothschild, la Banque Varin et Bernier, la Caisse des Dépôts et Consignations, la Caisse des Marchés de l'Etat, Monsieur André Chausson, les Ciments Lafarge, la Compagnie Minière, la Compagnie de Saint-Gobain, le Conseil National du Patronat Français, le Crédit Foncier, le Crédit Lyonnais, le Crédit National, la Discount Bank, la Fondation Synthesis (Eindhoven), Mademoiselle Joliot, la Société André Citroën, la Société Française des Pétroles B.P., la Société Générale, et les nombreux cotisants de l'Institut de la Vie, des plus prestigieux aux plus modestes.

– au Comité féminin d'accueil pour l'harmonie et la grâce qu'il a créées.

– à Madame P. Wissmer pour son inlassable activité.

– à Monsieur Frank de la North-Holland Publishing Company pour la conscience et le soin avec lesquels il a réalisé la présente édition.

M. MAROIS

OFFICIAL ADDRESSES

ALLOCUTIONS OFFICIELLES

Pierre AUGER
Professeur à la Faculté des Sciences de Paris
Président de la Commission des Sciences de l'UNESCO

Maurice SCHUMANN
Ministre d'Etat Chargé de la Recherche Scientifique et des Questions Atomiques
et Spatiales (France)

S. L. SOBOLEV
Academy of Sciences of the U.R.S.S.

R. S. MULLIKEN, Prix Nobel
Department of Physics, University of Chicago

ALLOCUTION D'OUVERTURE DE LA CONFÉRENCE

Pierre P. GRASSÉ
Président de l'Academie des Sciences, Paris

ALLOCUTION D'OUVERTURE DE LA PREMIÈRE SÉANCE

L. ROSENFELD
Nordisk Institut for Teoretisk Atomfysik, Copenhague

## Allocution de PIERRE AUGER

Monsieur le Ministre, Messieurs les Présidents, Messieurs les Directeurs généraux, Messieurs les Professeurs, Messieurs les Elèves, Au nom du Comité d'organisation, je vous remercie vivement, Monsieur le Ministre, d'avoir bien voulu honorer de votre présence cette réunion. Nous remercions tous les participants d'avoir répondu si nombreux à notre appel et d'avoir donné à ce colloque, dès son début, un niveau scientifique très élevé. Mais il n'y a pas que la substance intellectuelle proprement dite qui compte, dans un colloque, il y a cet impondérable que l'on appelle l'atmosphère, et il me semble que vous avez créé une bonne atmosphère, — ce qui par parenthèse nous console un peu de l'évidente mauvaise volonté de l'atmosphère extérieure. Recevez à ce point de vue les excuses du Comité d'organisation, qui n'a pu faire mieux. Il faudrait peut-être organiser quelques colloques sur la physique théorique et la météorologie, et on rencontrerait d'ailleurs bien des problèmes semblables à ceux que vous traitez: systèmes à *n* corps, structures dynamiques; nous avons réagi collectivement par l'instauration de transports actifs entre le Congrès et le Trianon et individuellement par l'usage de membranes à perméabilité sélective.

Ces considérations sur le temps qu'il fait pourraient paraître un peu marginales, à propos d'un colloque qui traite, sous une forme assez particulière sans doute, du problème de la Vie. Oui, scientifiquement, mais non du point de vue humain, du point de vue, disons, populaire. Le public s'intéresse passionnément aux progrès de la science, il admire les succès des savants, et attend avec espoir les résultats de colloques comme celui-ci, mais disons-le franchement, il est très souvent déçu.

Aux explorateurs de la haute atmosphère et de l'espace, constructeurs de satellites, il adressera ses félicitations pour la découverte des zones de Van Allen ou du vent solaire. Mais ce n'était pas exactement cela qu'il demandait, ce qu'il aurait voulu c'est savoir quel temps il fera cet été. Aux membres de ce colloque il adressera ses félicitations pour la meilleure compréhension de la pompe à sodium, mais ce qu'il aurait voulu, c'est savoir ce qu'est la vie, ou plus modestement ce qu'est le cancer. En cela, ce public montre qu'il n'a pas oublié la magie de jadis et ses mirages séduisants où on lui promettait la jouvence, la pluie et le soleil à volonté, le plomb en or. Nous lui donnons des particules étrangers, le zéro absolu (ou presque), des RNA messagers, des isotopes plus ou moins explosifs. Alors il faut lui expliquer, lui faire comprendre que c'est par cette voie complexe, détournée, inattendue et rude que ses grands problèmes seront résolus — ou peut-être que leur solution apparaîtra soudain comme inutile parce que dépassée.

Au fond cela provient en grande partie de ce que la science se présente au public comme constituée d'une multitude de spécialités: on n'est plus physicien, mais physicien nucléaire, et même physicien nucléaire basses énergies, ce terme pouvant d'ailleurs prêter à équivoque car ces basses énergies correspondent à des potentiels cent fois supérieurs à ceux de nos lignes à très haute tension! L'idée de la science comme une mosaïque de spécialités est un cadeau, un mauvais cadeau de ce dix-neuvième siècle que Léon Daudet appelait stupide. Mon père, qui était chimiste, me racontait comment sa carrière avait été rendue difficile à l'Université à cause de son obstination à vouloir faire de la chimie, et non pas seulement de la chimie organique, et plus spécialement de la chimie organique aromatique, et plus spécialement de la chimie organique aromatique hétéro-cyclique. S'il avait fait un tel choix et s'y était tenu, il aurait fait partie d'un cénacle qui l'aurait poussé en avant, alors qu'il était considéré comme un traître pour s'être intéressé aux acides phosphoriques, bassement inorganiques...

Alors comment voulez-vous que le public comprenne ce que font ensemble des physiciens, — pire, des physiciens théoriciens! — et des biologistes, qui pour lui sont encore soit des médecins, soit des naturalistes. Il faut dire, en passant, que la Biologie si elle a pu pénétrer dans les Facultés, n'a pas encore pu pénétrer dans l'Académie où elle ferait sans doute figure de "parvenue" ou de nouveau riche devant les titres de noblesse des valeurs traditionnelles comme l'Economie Rurale par exemple, qui possède une section entière à elle toute seule. Le public doit être informé du grand mouvement actuel de la science vers une rupture des barrières entre spécialités et même entre les disciplines générales. Il doit savoir que les grands principes, les grandes lois traversent tout le champ des sciences, qu'elles soient exactes ou naturelles, et à condition qu'elles ne soient pas inexactes ou surnaturelles. Conservation de l'énergie, du moment cinétique, croissance de l'entropie, principe de symétrie, de relativité, lois quantiques, bien d'autres encore permettent de revenir à une conception unitaire de la science, conception qui avait été celle de l'antiquité et qui n'avait cédé que sous la pression de connais-sances nouvelles venues en foule et en ordre dispersé.

Il y a pourtant une différence fondamentale entre le genre d'unité qui se présente à nous actuellement, et le genre d'unité recherché par l'antiquité—et il faut bien le dire, par la Renaissance elle-même, et jusqu'à l'aube de la révolution scientifique du 17e siècle. Cette première unité était basée sur l'Homme, mesure de toutes choses, et individualité type. L'Homme devant se projeter sur l'Univers entier et se retrouver partout, dans le macroscopique du cosmos, dans le microscopique de l'atome. Ces atomes qui avaient les uns pour les autres des amitiés ou des haines. Maintenant, nous assignons à l'Homme une place plus limitée, parmi les organismes vivants, et c'est la Nature que nous projetons dans cet Homme, dans sa vie, dans sa pensée. L'antiquité tentait une biologie de la physique, de la chimie, de l'astronomie, tandis que nous tentons, vous tentez ici, une physique de la biologie—au moins d'une part aussi grande que possible de cette biologie. La jonction entre l'univers cosmique et atomique et le microcosme humain paraît devoir se faire en partant d'individualisations physico-chimiques moléculaires, caractérisées par des nombres précis, si complexes qu'ils puissent être, et en allant vers l'être vivant, virus, bactérie, cellule ou homme, à l'inverse de la recherche traditionnelle.

Cette marche décidée vers des généralisations précises et fécondes—au lieu des généralisations vagues et inhibitrices d'autrefois—cette ouverture sur des perspectives d'ensemble, ne signifie en aucune manière la disparition des spécialités et des spécialistes. Mais la spécialité est réservée à l'action, c'est-à-dire, à l'avance pas à pas dans la recherche. Nous sommes tous des spécialistes dès que nous agissons: spécialistes du volant sur la route, spécialistes de la pédagogie quand nous répondons aux nombreux "pourquoi" de nos enfants—ou petits enfants hélas pour moi,—ou même de la docimologie quand ces enfants nous présentent leur carnet scolaire. La docimologie est paraît-il la science de la notation des élèves, en particulier aux examens et concours. Nous sommes parfois de médiocres spécialistes, je le concède, dans ces activités, mais des spécialistes quand même. Or le physicien qui monte un appareil est électricien ou électronicien—et un bon spécialiste, s'il vous plaît, et plus tard un spécialiste des plaques nucléaires, des chambres à bulles, etc. La même chose vaut pour les chimistes et les biologistes, et je pense aussi pour les mathématiciens. Mais de plus en plus, dans la réflexion et la communication tout d'abord, dans l'expérimentation ensuite, nous trouvons des savants qui s'intéressent à la fois aux sciences dites exactes, mathématiques ou physiques, et aux sciences dites naturelles, c'est-à-dire aux aspects divers de la biologie. Mieux encore, des physiciens deviennent des biologistes, comme nous avons vu des mathématiciens devenir psychologues, ou des philosophes devenir ethnologues et apporter dans ces sciences de véritables révolutions.

En réalité, il ne s'agit pas de nouvelles rencontres, mais de véritables retrouvailles. N'oublions pas que Galvani était médecin et physicien, ce qui lui permit de mettre la biologie des cuisses de grenouille au service de la physique de jonction métallique et de découvrir le principe des piles électriques. Que Mayer, médecin, découvrit le principe de conservation de l'énergie, que Pasteur, chimiste, fonda la microbiologie et l'immunologie. Plus près de nous Schrödinger, Gamow, physiciens, s'intéressèrent à la génétique.

Mes chers collègues, tout ceci me fait penser à une remarque très pénétrante faite par un de mes amis, diplomate, qui se trouva placé professionnellement en contact avec des savants atomistes. "Je n'ai aperçu le véritable sens social de la science," m'a-t-il dit, "que le jour où j'ai compris que les scientifiques de par le monde étaient tous membres d'un même club." Un club sans réglementation écrite, sans formalisme à l'entrée ni cotisations annuelles, un club tacite en quelque sorte, mais cependant très fermé à ceux qui ne sont pas qualifiés pour en faire partie. Un club d'extension mondiale et dont l'influence, pour ne pas être souvent très apparente, n'en est pas moins considérable. Eh bien, vous formez ici un brillant échantillonnage des membres du club en question, venus ici à l'invitation de l'Institut de la Vie, pour confronter et échanger vos idées sur cette vie, justement, ou au moins sur certains de ses aspects qui paraissent les plus accesibles.

Au nom des membres du club mondial des scientifiques réunis ici, Monsieur le Ministre, je vous remercie encore une fois de nous avoir fait l'honneur de prendre place parmi nous et aussi d'avoir bien voulu m'écouter avec tant de patience.

## *Allocution de* MAURICE SCHUMANN

Si l'on a vu des naturalistes devenir des chimistes, on a vu aussi des littéraires devenir ministres de la Science! Aussi ai-je eu tout à l'heure le sentiment d'être visé, quand mon ami, le Pr. Auger, s'est adressé à certains d'entre nous en les appelant "Messieurs les élèves". Nous allons voir si je suis un assez bon élève pour avoir compris le raisonnement que le Pr. Auger a tenu tout à l'heure et que je me suis efforcé de saisir au vol. D'une part, la biologie exprime, ou commence à exprimer les phénomènes vitaux en langage moléculaire. Et d'ailleurs Pr. Auger nous a lui-même cité un brillant exemple, en évoquant les acides ribonucléiques, le rôle qu'ils jouent dans la production des cellules. D'autre part, la physique théorique progresse, et progresse même rapidement dans la représentation des molécules: elle interprète les propriétés des molécules en langage mathématique. Par conséquent, le lien entre la biologie et la physique théorique est le lien entre le phénomène vital et le langage mathématique. C'est alors que surgit la grande question: va-t-on mettre, allez-vous mettre ou tenter de mettre la vie en équations? Allez-vous confier ces équations aux machines à calculer, qui effectuent plusieurs millions d'opérations en une seconde? Allez-vous, allons-nous ramener les phénomènes physiques à des déplacements d'électrons?

Et aussitôt surgit, à peine modifiée, la question que Jean Rostand posait, avec angoisse, lorsqu'il se demandait si l'homme était capable de prendre en main les commandes chimiques de son destin. Nous pourrions dire: l'homme est-il capable de prendre en main les commandes physiques de son destin? Y a-t-il une désintégration de l'homme parallèle à la désintégration de la matière? Dès que je pose cette question trois réponses me rassurent. Ces trois réponses se réfèrent à trois thèmes auxquels il m'arrive maintenant de penser chaque jour.

Le premier de ces trois thèmes—puisque je viens de parler de machines électroniques—est ce que l'on appelle désormais, d'un mot international, le "software". Je veux dire (nous voulons dire) l'ensemble des connaissances qu'il faut posséder pour appliquer les dispositifs techniques au traitement de l'informatique. Quelque chose d'infiniment rassurant à constater est que le "software", c'est-à-dire la part irréductible de l'intelligence humaine et du cerveau humain, entre pour 60%, d'après les estimations les plus récentes, dans le coût total de ces machines, dont certains se demandaient si elles n'avaient pas pour objet de rendre, à la limite, le cerveau humain inutile.

Le deuxième élément qui me rassure, c'est ce que je tends à appeler, de plus en plus, "la prédominance de l'application sur l'explication". A mon tour, je m'explique. Et je m'explique en raisonnant, si vous le voulez bien, par analogie.
... En fait, nous ne savons pas désintégrer, c'est-à-dire faire disparaître, en la transformant en énergie, n'importe quelle masse, ou n'importe quelle matière.

Nous savons obtenir de l'énergie avec une petite partie de la masse d'un élément naturel. En réalité, on ne fait pas disparaître de la matière, mais on modifie la composition des noyaux et en conséquence l'énergie de liaison qui soude entre elles les particules des noyaux. De même, on peut bien me dire qu'un individu qui pèse 100 kilos contient 63 kilos d'oxygène, 19 de carbone et 9 d'hydrogène. Mais on me dit pas comment les cellules peuvent constituer un coeur qui bat, une main qui écrit, une bouche qui parle, ou un cerveau qui pense. Ce qui revient à dire que le vrai mystère ne recule pas. Me dira-t-on qu'il en ira autrement si l'on se situe au niveau ou au sein même des molécules? Assistera-t-on alors à des expériences qui permettront d'assimiler aux phénomènes physicochimiques le mécanisme intime de la vie? Mon sentiment est que le vrai triomphe est ailleurs, et que le titre même d'Institut de la VIE, qui nous rassemble ce soir, l'exprime. Le vrai triomphe, vous le remportez (pardonnez-moi, mes chers Maîtres cet apparent pragmatisme) quand vous déchiffrez le code génétique, quand vous prévoyez les modifications de l'hérédité, la guérison de certaines maladies, peut-être même du cancer, quand vous donnez les formules des acides nucléiques. Je songe, par exemple, à ce laboratoire de Chimie moléculaire, où peut-être, à l'heure où nous parlons, on est en train de choisir les types de molécules, qui, une fois marquées et chargées de radioisotopes, permettront non seulement de dépister les cancers, mais encore, fixées sélectivement sur les tumeurs, d'en accélérer la guérison. Vous allez en somme plus loin sur la voie de l'action bienfaisante que sur celle de l'explication totale. Et la Mathématique est l'outil avec lequel vous transformez le monde des réalités vivantes, comme elle fut l'outil avec lequel la Science a transformé le monde des réalités inertes.

Enfin, ma troisième et dernière réponse, donc ma troisième et dernière consolation, est ce que M. Auger appelait tout à l'heure le "Club". Je veux dire cette universalité de la recherche dont vous êtes les témoins et dont la découverte est si réconfortante pour un homme public, qui, quand le bonheur lui échoit d'être ministre de la Science, se dit qu'il peut et qu'il doit aller dans les capitales les plus diverses, pour y tenir et y entendre le même langage, pour y asseoir la coopération technique et scientifique sur les mêmes bases. En ce sens, le temps revient où Mersenne correspondait avec Torricelli, Leibniz avec toute l'Europe, le temps même un peu plus proche où, en pleine guerre napoléonienne, le physicien anglais Davy venait tout près d'ici pour recevoir un prix de l'Institut de France. Mettre "non la vie en équation mais l'équation au service de la vie": voilà l'enseignement de vos travaux, voilà peut-être aussi votre devise! Nul ne l'a mieux illustrée que Teilhard de Chardin, quand, dans une belle phrase qui fut mise symboliquement en musique, transformée en cantate par André Jolivet, il disait: "Jusqu'au dernier moment des siècles, la matière sera jeune, exubérante, étincelante et nouvelle . . . pour qui voudra!"

## Allocution de S. L. SOBOLEV

Après d'aussi brillants discours, le modeste mathématicien que je suis n'ose plus même évoquer d'aussi vastes questions. Participant à un si important Colloque, je parlerai cependant, pour dire toute ma reconnaissance à M. Marois, à M. le Ministre, à tous ceux qui ont fait, de notre séjour à Versailles, une détente si agréable.

Comme mathématicien, j'ajoute que certaines questions ne peuvent être résolues dès maintenant par la physique théorique ou expérimentale, ni par la biologie; elles nécessitent l'apport de nouvelles idées, qui ne peuvent venir que des mathématiques. L'histoire nous en donne de nombreux exemples. Aucune grande découverte, dans les sciences, techniques ou naturelles, et autres, qui n'ait été précédée d'un demi-siècle de recherche mathématique! Tout le monde se moquait, disant que cette recherche était inutile et sotte. C'était l'exemple d'une géométrie non euclidienne, qui "ne pouvait être sérieuse". Mais quand, après soixante ou soixante-dix ans, fut élaborée la théorie de la relativité, tout le monde comprit que c'était vrai, et très intéressant.

Je ne suis qu'un modeste mathématicien, mais je ne suis pas très modeste quand je parle mathématiques!


## Address by R. S. MULLIKEN

After the preceeding speakers, there is very little I could say, even if I were good at after dinner speaking. However I would like to express that this is the most interesting conference I have ever attended, with such a remarkable group of people of which some are old friends and some of them new friends. I congratulate the Comittee and I thank Professor Marois and his collaborators and the Comittee and I think I speak also on behalf of others who speak english in this respect.

*Allocution d'ouverture de la Conférence par* PIERRE–P. GRASSÉ

Mesdames, Messieurs,

C'est un grand honneur que de vous souhaiter la bienvenue. Pour venir à nous vous avez affronté les fatigues d'un long voyage et avez consenti à nous consacrer quelques jours d'un temps dont nous connaissons tout le prix. Au plaisir de vous accueillir s'ajoute un sentiment de sincère gratitude.

Nous mesurons aussi l'honneur que vous faites en acceptant de participer à cette conférence qui est vraiment vôtre. Plusieurs d'entre vous portent les noms les plus glorieux de la science contemporaine et rendent, par leur collaboration, un hommage aux chercheurs français qui en sont à la fois heureux et fiers.

Le comité organisateur de cette conférence tient à ajouter à ses remerciements l'assurance que tout sera mis en oeuvre pour que votre séjour à Versailles soit agréable et fructueux.

La science d'aujourd'hui perd de plus en plus le caractère strictement individuel qu'elle a présenté pendant des siècles. Ce n'est point tant au travail en équipe que nous faisons allusion qu'à l'impossibilité pour l'homme d'absorber, de s'assimiler une quantité illimitée de connaissances. Manque de temps, insuffisance de la mémoire, pour tout dire impossibilité physiologique de faire enregistrer par le cerveau la masse de faits, lois et expériences qui constituent une science. Le temps où un Buffon possédait la quasi-totalité du savoir scientifique de son temps, des mathématiques à l'histoire naturelle, est bien révolu. L'homme de science, si érudit soit-il, devient, souvent malgré lui, un spécialiste.

Pour que l'indispensable synthèse ne soit pas étouffée sous le poids des connaissances dont l'accroissement va s'accélérant du fait de l'augmentation continue du nombre des chercheurs, il faut, de toute nécessité, recourir aux moyens propres à la maintenir possible. La multiplication des colloques, la rédaction de livres collectifs tendent à pallier les inconvénients de l'inéluctable spécialisation.

On ne peut douter que la rencontre d'aujourd'hui ne tende vers ce but, mais selon moi, elle a une fin plus importante et plus originale.

La biologie, science encore au début de son développement, considère les objets les plus complexes existant dans notre système solaire. Ses moyens ne sont pas proportionnés à la grandeur et à la difficulté de sa tâche; elle travaille presque toujours à tâtons car elle ne connaît que rarement la totalité des facteurs et des causes qui interviennent dans le déterminisme et le déroulement des phénomènes qu'elle étudie.

Les progrès des sciences de la vie furent très lents et c'est seulement depuis quelques années qu'ils se sont précipités. N'oublions pas que pendant des siècles, les phénomènes vitaux furent tenus comme irréductibles aux phénomènes physiques ou chimiques: la vie étant la propriété de corps, ou organismes essentiellement différente des propriétés que possèdent les corps inertes. Aujourd'hui, nous savons que les phénomènes vitaux s'inscrivent sans aucun doute parmi les manifestations physico-chimiques mais que les réactifs, les appareils dont disposent les êtres vivants diffèrent de ceux qu'utilise ordinairement le physicien ou le chimiste dans son laboratoire. Cette originalité qui, sans doute, a sa valeur, n'autorise nullement à isoler les phénomènes vitaux dans une catégorie particulière.

Ainsi, d'emblée, l'union du physicien, du chimiste et du biologiste s'impose pour toute étude exhaustive et synthétique des phénomènes vitaux. Tandis que le biochimiste découvrait la vraie nature des constituants des êtres vivants et poussait son investigation jusqu'au niveau de la molécule, les biologistes, grâce au microscope électronique observaient, enfin, ces mêmes constituants sous leur forme réelle et, dans bien des cas, en voyaient les macromolécules constitutives.

Ainsi, peu à peu, l'intérêt s'est porté sur les phénomènes biologiques considérés à l'échelle moléculaire d'où la naissance d'une nouvelle discipline: *la biologie moléculaire*. La connaissance de la molécule de diverses protéines, d'hormones et des acides nucléiques a donné un grand élan à cette jeune science, qui, tout en suivant le déroulement des phénomènes au sein de la cellule, en définit les phases successives en termes physiques et chimiques.

Mais, dès maintenant, bien des chercheurs estiment que les constituants des êtres vivants doivent être étudiés au niveau des corpuscules. De ce fait, ce n'est plus à la physique classique que l'on doit recourir mais à la physique quantique. Nombreux sont les physiciens qui considèrent celle-ci comme étant la plus apte à expliquer les mécanismes biologiques.

Les phénomènes tels que les attractions qui s'exercent entre les chromosomes et entre les gènes homologues, l'influence des centrosomes, la production des mutations ressortissent, semble-t-il, à la fois de la biologie moléculaire et de la physique quantique. On pressent qu'il en va de même pour les phénomènes de la vision et de la mémoire. Les premières applications de cette physique à l'étude des acides nucléiques sont fort encourageantes et on a, dès maintenant, l'impression qu'elles ouvrent de nouvelles voies à la biologie. L'ère de l'application de la physique théorique à la biologie est donc ouverte.

Les nouvelles techniques, les nouveaux appareils ont permis à la biologie de progresser à pas de géant, mais ils ne doivent plus être les seuls à l'aider. Quelques hommes se sont demandés si une forme particulière de biologie qui ferait pendant à la physique théorique ne pourrait pas être édifiée et projeter une vive lumière sur des parties encore obscures du domaine de la vie. Bien que les premiers essais n'aient pas été entièrement concluants, ils ont été suffisamment positifs pour que de nouvelles tentatives soient entreprises avec des chances de succès

La physique théorique peut apporter un concours extrêmement précieux à la jeune biologie qu'elle soit théorique ou non; son champ d'action déborde largement les propriétés de la matière inerte et tend à atteindre les manifestations de la vie. Pour ces raisons, notre rencontre prend à mes yeux une valeur particulière, en quelque sorte symbolique. Elle annonce que physique théorique et biologie contractent une union étroite et définitive. Elle marquera une date dans l'histoire des sciences.

Je vous remercie, Messieurs, de l'attention que vous m'avez prêtée. Je déclare la conférence ouverte et je transmets la présidence au Professeur Rosenfeld.

## *Allocution d'ouverture de la première séance par* L. ROSENFELD

We must be extremely grateful to the Institut de la Vie for bringing us here together. This is not a conference of the usual kind. It is actually a kind of experiment in establishing contact between two branches of science which so far have developed rather independently of each other, with different methods, but which are now converging towards a common ground. It is with great expectation that I personally am looking forward to the way in which this dialogue is going to take place. It is up to us to make a success of it, as I very much hope we shall. It may be the beginning of a co-operation of which one may hope that it will accelerate the spectacular progress that has been made in recent years.

As a physicist I may add that so far the biologists have managed quite well without the help of the physicists. I have the greatest admiration for the way in which they have disentangled the very deep-lying problems of molecular biology, and especially the logical problems connected with the reading of the genetic code. I think this is a tremendous achievement. We physicists, if I may also speak for my colleagues, are approaching this conference in a very humble spirit. We shall see how things will develop, according to the program in which the physicists have been given the word right at the start.

# CONCEPTS THEORIQUES

*Ière séance*

PRÉSIDENT L. ROSENFELD

H. FRÖHLICH

Quantum Mechanical Concepts in Biology

I. PRIGOGINE

Structure, Dissipation and Life

Discussions

*Theoretical Physics and Biology,* © 1969, *North-Holland Publ. Co., Amsterdam*

## QUANTUM MECHANICAL CONCEPTS IN BIOLOGY

H. FRÖHLICH

*Chadwick Laboratory, University of Liverpool, Liverpool, England*

## 1. Introduction

In many macroscopic properties of materials quantum mechanics shows its influence only in an indirect way namely through the properties of the atoms of which the substance consists. A notable exception is formed by the specific heat at very low temperatures which for the first time demonstrated quantization on a macroscopic scale. In recent years discussion of certain other low temperature phenomena has shown them to exhibit effects which might best be described in terms of macroscopic quantization of a more subtle nature giving rise to an order of a new kind based on the concept of phase correlations. I believe that this concept may have a much wider range of application in particular in systems which are relatively stable but not near thermal equilibrium, and which show an organised collective behaviour which cannot be described in terms of an obvious (static) spacial order. In the following I shall first describe these concepts in a general way, and then present a speculative model which might be applied to biological systems.

## 2. Long range phase correlations

All materials are composed of atoms. Complete solution of the equations describing their dynamic behaviour would, however, yield such an immensity of irrelevant information that selecting from it the features of interest would be prohibitive. For the number of states of a system increases exponentially with the number of particles it contains; adding two particles would at normal temperatures increase the number of its states about 100-fold!

Clearly establishment of a useful connection between micro and macro physics requires introduction of relevant macroscopic concepts, and their expression in terms of microscopic (atomic) features. In terms of the latter, macro concepts are always "collective" properties. This programme is well advanced when dealing with behaviour near thermal equilibrium. Relevant concepts are the thermodynamic ones (free energy, entropy, etc.), hydrodynamical fields, properties of spacial order (crystal structure, etc.) etc. The study of the peculiar phenomena of superconductivity and superfluidity have revealed, however, existence of a macroscopic spacial order which is not of such an obvious nature and which is best described in terms of certain quantum mechanical concepts on a macroscopic scale.

Quantum mechanics treats the dynamic behaviour of any system in terms of a state vector (or wave function) $\Psi$. For a single particle this is essentially a de Broglie wave, but it is much more complicated for a system of many particles. It is essential that this $\Psi$ is a complex quantity, and that its absolute square describes a certain probability and is the quantity normally used in comparison with experiments. Multiplication of this $\Psi$ by a phase factor say $e^{i\alpha}$ does thus not alter the relevant probability. A second essential feature of quantum mechanics is, however, that the state vectors of two (or more) states can be superimposed linearly to form a combined state; the absolute square (probability) of this combined state thus depends on the difference of the phases of its components. This expresses a typical wave interference, characteristic for quantum mechanics.

In very large systems, owing to thermal disorder such phase differences frequently average out and lead to no observable result. In fact it can be shown that establishment of a "pure" state, i.e., one exhibiting a definite phase, in a macroscopic body would require times which are long even compared with cosmological ones. Nevertheless significant exceptions to the absence of a definite phase in a macrosystem exist. They are, however, always connected with a very strong excitation of one or a few modes of motion; and their phases then show correlations over macroscopic regions. It is not the state function $\Psi$, therefore, that we are concerned with but rather a much simpler quantity, a macroscopic wave function, say $\Phi(x)$ which persists after the appropriate thermal averaging has been performed. While $\Psi$ depends in a complicated way on the coordinates of all the particles composing the substance, the macrowave function $\Phi(x)$ simply depends on space, time and a few parameters.

A formal procedure exists of how one would derive existence of macro wave functions from the most general microscopic description of the system (use of Von Neumann's density matrix to form reduced density matrices [1]). This can be applied to specific systems but a general treatment has so far not been found. Clearly establishment of one or a few strongly excited modes of motion requires their stabilisation. In the mentioned low temperature phenomena this is achieved for energetic reasons. At normal temperatures, however, such excited states can be only metastable. Thus e.g. if a macroscopic sound wave is imposed on a material then clearly macroscopic classical phase relations exist for the density at various space points. This means that a phase correlated motion is superimposed on the irregular thermal motion of the atoms, and a macroscopic wave function then exists — formally derived by a procedure similar to the one used for superconductors.

A more subtle example is provided by laser states [4]. A strongly excited (transverse) electromagnetic wave imposes here the relevant long range phase correlations on the thermal motion. A material thus excited will ultimately emit as light the energy stored in the relevant modes. Another possibility exists, however, namely strong excitation of longitudinal electric waves. Light waves are transverse electromagnetic waves. Longitudinal waves cannot exist in vacuum and hence cannot be emitted. They exist, however, in materials at selected frequencies. These frequencies are in some way correlated to the resonance frequencies for transverse waves, but they are higher by a definite amount. If the corresponding transverse frequency is very low (or zero) then the correlation between longitudinal and transverse modes becomes rather artificial. Such longitudinal modes once strongly excited have a relatively high stability because they cannot lose energy by emitting radiation. If their frequency is relatively low, however, then energy may be lost by ordinary friction. In this case, however, another stabilising mechanism arises, in certain circumstances, in terms of internal deformations.

### 3.   Application to biological systems

Biological systems exhibit relative stability in a way in which some modes of behaviour remain very far from thermal equilibrium although from an atomic point of view the majority of the degrees of freedom behave like being close to thermal equilibrium. It is tempting, therefore, to postulate the existence of long range quantum mechanical phase correlations in biological systems. (Per-Olov Löwdin has informed me that he has made such a proposal three years ago). It must be realised, however, as pointed out in the previous section that the phase of the quantum mechanical state vector is averaged out by the thermal motion of the many atoms. Nevertheless we have seen that phase correlations may persist even after the averaging over thermal motion, provided a few modes of motion are very strongly excited. The strong polar character of biological objects suggests longitudinal electric oscillations as stabilising modes.

One can imagine two types of sources for these oscillations. One would be connected with certain molecular processes involving ions. The second would be characteristic of the cell as a whole. The two types might be in resonance.

The frequencies involved in molecular processes cover a very wide range. An estimate of the frequencies of the second type of oscillations might be given because some of these at least should involve the cell membrane which has a strong dipolar layer of thickness of order $10^{-6}$ cm. A localised vibration would correspond to a frequency $\nu$ of order $10^{11}$ sec$^{-1}$ (assuming elastic constants corresponding to a sound velocity of order $10^5$ cm/sec). If some molecules in the cell would contain effectively free electrons then resonance would require a number density $n$ of order $10^{14}\varepsilon$ per c.c. ($\varepsilon =$ effective dielectric constant; $\nu^2 = e^2 n/\pi m \varepsilon$; $m =$ electron mass). Some molecular processes would, however, lead to much higher frequencies.

Establishment of strong excitation of some modes would require that certain biological processes feed energy into these modes, and that frictional losses are relatively small so that a stabilisation can take place. This could be achieved through elastic deformations which change the dielectric properties in such a way that the strongly excited longitudinal electric vibrations would become "frozen in", i.e., a metastable state of equilibrium would be achieved. For a system of cells this is feasable on condition that (i) the vibration is sufficiently strongly excited, and (ii) the cells are not too densely packed.

What conclusions would one draw from establishment of these vibrations?

(i)   On a molecular basis they would impose the "correct" frequency though what this actually implies would require detailed investigation.

(ii)   On an individual cell, the deformation required for stabilisation would impose a stress at the surface which would increase with increasing size. It could thus act as a stimulus for cell division.

(iii)   When the cells are densely packed then the stabilising mechanism through deformation no longer works, and the electric field must then collapse. With it would then vanish the stimulus for cell division.

(iv)   Possible absorption in the medium will define a range for the correlations and might hence influence macroscopic size effects.

Clearly it would be desirable to attempt measuring the conjectured longitudinal electric vibrations. Their effectiveness is based on the long range coherence which implies that the energy content of these vibrations is concentrated into very narrow frequency ranges. This makes measurement, of course, difficult because although the energy per unit frequency would be high, the total energy contained in such a mode would be very low, compared with say the total thermal energy.

In conclusion I wish to remark that the above suggestions are meant to be highly speculative. They should demonstrate, however, that application of quantum mechanical concepts can lead to new points of view which might be used as guiding points in the search for undiscovered regularities.

**Appendix I:**   *Shape dependence of the free energy in the presence of longitudinal electric oscillations*

Consider a model in which a number of units (cells) are distributed in a continuous medium. If these units have a longitudinal resonance frequency then the total polarisation can be considered as superposition of contributions from the background, $P_0$, and from the resonance, $P$, such that the free energy becomes also a sum of two such contributions. A particularly simple example of such a superposition is treated in the theory of polarons [2]. In the region of linear response the contribution of the resonance oscillations to the free energy is

$$U_P = \sum cf(a) \, a^3 \overline{P^2}. \tag{1}$$

Here $a^3$ is of order of the volume of a given unit, $f(a)$ describes their size distribution such that

$$n = \sum f(a)$$

is their total number, and $c$ is a number; $\overline{P^2}$ is the mean of the oscillating $P^2$.

In the case of plasma oscillations in cells one would expect a non-linear response at fairly low polarisation. The direction of the influence of deformation can, however, be demonstrated even for still weaker polarisation for which (1) holds. The reason is that the constant $c$ depends on the shape of a unit provided the units are sufficiently separated. Thus if we assume the units in the absence of a polarisation to be spherical, the deformation into ellipsoidal shape will change the energy expression into

$$U_P = c \sum f(a) \, a^3 \overline{P^2}(1 - \alpha \eta_a), \qquad (2)$$

where $\eta_a$ is the eccentricity, and $\alpha$ is of order one. Note that $\eta_a$ may be positive or negative. The relevant free energy will thus be smaller than in the undeformed case if $\eta_a > 0$. This deformation is opposed by the elastic surface energy which in the absence of polarisation must have been a minimum with respect to deformation, and hence must be proportional to $a^2 \eta_a^2$. The total free energy thus becomes

$$U = \sum \{ ca^3 \overline{P^2}(1 - \alpha \eta_a) + \tfrac{1}{2} S a^2 \eta_a^2 \} \, f(a), \qquad (3)$$

where $S$ is related to the surface elastic constants.

We should now minimise $U$ with regard to the displacements $\eta_a$. This results in

$$\eta_a = \alpha c \overline{P^2} a / S, \qquad (4)$$

so that

$$\frac{1}{n} U = c \overline{P^2} a^3 - \tfrac{1}{2} \frac{\alpha^2}{S} (c \overline{P^2})^2 \, \overline{a^4}, \qquad (5)$$

where

$$\overline{a^n} \, n = \sum a^n f(a). \qquad (6)$$

Thus, owing to the deformation, expression (1) for the undeformed energy is reduced by an amount proportional to $(\overline{P^2})^2$. We also note that the stress parameter $\eta_a$ of an individual unit increases with its size $\propto a$.

One could now, of course, find the polarisation $P_m$ at which $U$ reaches a maximum and beyond which $U$ thus decreases with increasing $P^2$. This would yield

$$c \overline{P_m^2} = \frac{S}{\alpha^2} \frac{\overline{a^3}}{\overline{a^4}} \qquad (7)$$

and hence

$$\alpha \eta_{m,a} = \frac{a \overline{a^3}}{\overline{a^4}}, \qquad (8)$$

so that the first term in (3) would vanish at this point.

Such considerations, however, would be unrealistic. For not only would the linear and the quadratic in $\eta_a$ approximation used in (3) have become invalid, but also the assumption of linear response may no longer hold. Under such circumstances the neglected terms should influence the position of the maximum, and would also lead to a subsequent minimum unless the units were destroyed. At this minimum the assembly would then be in a metastable state with frozen-in polarisation waves.

The above considerations also break down if the units are closely packed. For if as a limiting case we would assume the units to be cubic then close packing would provide a homogeneous material and the question of shape dependence could not arise. In actual calculations the influence of the packing density would arise from considerations referring to the *local* field.

Finally it should be mentioned that on the basis of the polaron model mentioned above the constant $c$ would be proportional to $\varepsilon\varepsilon_0/(\varepsilon - \varepsilon_0)$ if $\varepsilon_0$ is the static dielectric constant of the background, and $\varepsilon$ that of the whole substance. The smaller $(\varepsilon - \varepsilon_0)$, therefore, the more sensitive would the energy be to changes in parameters.

## Appendix II

According to C. N. Yang [1] and to Penrose and Onsager [2] superfluids show a long range phase correlation which can be defined in terms of a reduced density matrix

$$\Omega_{\text{red}}(x', x'') = \text{Sp} \, \psi^*(x'') \, \psi(x') \, \Omega, \tag{1}$$

where $\Omega$ is von Neumann's density matrix, $\psi^*(x)$, $\psi(x)$ the wave operators of the particle field constituting the superfluid; $x'$, $x''$ are space points and Sp represents the trace. For superfluids then

$$\Omega_{\text{red}}(x', x'') = \Phi^*(x'') \, \Phi(x') + \chi(x', x''). \tag{2}$$

$\Phi(x)$ represents a macro wavefunction, and $\Phi^*(x'') \, \Phi(x')$ is large compared with $\chi(x'', x')$ if $|x'' - x'|$ is large compared with atomic dimensions.

A corresponding relation holds for superconductors but requires a second reduced density matrix (2nd order in $\psi$ and $\psi^*$).

We shall show that quite a similar relation exists for a material in which a macroscopic sound wave is excited. First note that for a material with sound velocity $s$ consisting of $N$ atoms of mass $M$ the operator

$$b = \left(\frac{Ms}{2\hbar k N}\right)^{\frac{1}{2}} \frac{1}{2} \sum_j \left\{ e^{-ikx_j}\left(1 + \frac{uv_j}{s}\right) + \left(1 + \frac{uv_j}{s}\right)e^{-ikx_j} \right\} \tag{3}$$

represents a destruction operator of phonons (quantised sound waves) of wave number $k$. Here $x_j$ is the coordinate of atom $j$, and

$$u = \frac{k}{k}, \qquad v_j = -\frac{i\hbar}{M}\frac{\partial}{\partial x}. \tag{4}$$

We note that the usual commutator

$$(b, b^*) = 1 \tag{5}$$

holds.

Also if $\Omega_0$ is the density matrix when no sound wave is excited macroscopically then

$$\mathrm{Sp}\, b^* b \Omega_0 = n_k \tag{6}$$

represents the mean number of thermal phonons of wave number $k$, i.e. a number of order 1, corresponding to an energy $\hbar k s n_k$.

Also we expect

$$\langle b \rangle = \mathrm{Sp}\, b\Omega_2 = 0, \quad \langle b^* \rangle = \mathrm{Sp}\, b^*\Omega_0 = 0, \tag{7}$$

or more general

$$\mathrm{Sp}\, b^{*n} b^m \Omega_0 = \delta_{n,m}\, n! \tag{8}$$

to hold in good approximation, as for harmonic oscillators.

If $V$ is the total volume then

$$b(x) = b\frac{e^{ikx}}{\sqrt{V}}, \qquad b^*(x) = b^*\frac{e^{-ikx}}{\sqrt{V}} \tag{9}$$

define wave operators at $x$. Their expectation values vanish in a state $\Omega_0$ in view of (7).

If a macroscopic sound wave is excited then the amplitudes $\langle b \rangle$, $\langle b^* \rangle$ should not vanish but be periodic in time, and the energy should be proportional to the volume $V$. Such a state can be described by a density matrix

$$\Omega = T \Omega_0 T^*, \tag{10}$$

where the operators $T$ and $T^*$ are given by

$$T = C e^{-\frac{1}{2}\beta\beta^*}\, e^{\beta b^*}, \quad T^* = C^* e^{-\frac{1}{2}\beta\beta^*}\, e^{\beta^* b}. \tag{11}$$

Here $\beta$ and $\beta^*$ are time dependent $c$-numbers,

$$\beta = \beta_0\, e^{ikst}, \quad \beta^* = \beta_0^*\, e^{-ikst} \tag{12}$$

and the numbers $C$ and $C^*$ are to be chosen such that

$$\mathrm{Sp}\, \Omega = 1 \quad \text{when } \mathrm{Sp}\, \Omega_0 = 1. \tag{13}$$

Particularly simple is the case in which $n_k = 0$, for then $C = 1$, $C^* = 1$. We note the relations

$$(b, T) = \beta T, \quad (T^*, b^*) = \beta^* T^* \tag{14}$$

and using (8) and (5)

$$\mathrm{Sp}\, T^* Tb\Omega_0 = \mathrm{Sp}\, Tb\Omega_0 T^* = 0. \tag{15}$$

We thus find from (13)–(15)

$$\langle b \rangle = \mathrm{Sp}\, b\Omega = \mathrm{Sp}\, bT\Omega_0 T^*$$
$$= \mathrm{Sp}\, (b, T)\, \Omega_0 T^* + \mathrm{Sp}\, Tb\Omega_0 T^* = \beta. \tag{16}$$

Hence also

$$\langle b^* \rangle = \beta^* \tag{17}$$

and

$$\mathrm{Sp}\, b^* b\Omega = \mathrm{Sp}\, b\Omega b^* = \mathrm{Sp}\, ((b, T) + Tb)\Omega_0((T^*, b^*) + b^* T^*) =$$
$$= \mathrm{Sp}\, (\beta + Tb)\, \Omega_0(\beta^* + b^* T^*) = \beta^* \beta + n_k. \tag{18}$$

An energy proportional to the volume thus requires $\beta$ proportional to $\sqrt{V}$ i.e.,

$$\beta = B\sqrt{V}, \tag{19}$$

where $B$ is volume independent. From (9) we then have

$$B(x) = \mathrm{Sp}\, b(x)\, \Omega = B\, e^{ikx}. \tag{20}$$

A reduced density matrix $\Omega_s(x', x'')$ can thus be defined as

$$\Omega_s(x', x'') = \mathrm{Sp}\, b^*(x'')\, b(x')\, \Omega \tag{21}$$

in terms of $\Omega$ and individual particle operators $x_j$, $\partial/\partial x_j$; $\Omega_s$ can also be written in terms of the particle field operators $\psi$, $\psi^*$; It then represents a second reduced density matrix. We note from (18)–(20) that

$$\Omega_s(x', x'') = B^*(x'')\, B(x') + n_k\, \frac{e^{ik(x'' - x')}}{V} \tag{22}$$

has the form required by (2).

## SUMMARY

Quantum mechanics has led to an understanding of the properties of atoms and of their interactions. It is often argued that the properties of all materials could, therefore, be derived automatically if only one could master the mathematical techniques required to handle systems composed of a very great number of individual atoms. This conclusion is at fault for complete solution of such complex problems would provide us with an enormous number of informational facts, nearly all of them irrelevant for most purposes. The task of selecting from them the features of interest to a physicist and to a biologist would be of the same order of complication, roughly speaking, as finding complete solutions in the first place. It has been realized, therefore,—by some physicists at least—that the first task in attempting treatment of a many body problem must lie in the formulation of appropriate questions.

Amongst the biological features particularly striking to physicists is the relative stability of systems in which some modes of behaviour remain very far from thermal equilibrium; another is the apparently strongly correlated behaviour of large systems consisting of a number of individual units — as demonstrated, for instance, in the growth of an organ (consisting of many units, the cells) which is discontinued after a certain roughly defined size and shape has been reached.

Physicists have, of course, for many years dealt with the collectively organised behaviour of assembleys of many units; but until quite recently such treatments were always connected with some long range spacial order of a rather obvious kind. More recently, however, new features have arisen which refer to fluid systems possessing no *obvious* long range spacial order. These features have been formulated in context with superfluidity (a property of liquid helium) and with superconductivity (a property of the electron "fluid" in metals). In the empirical formulation of C. N. Yang [1] they can be expressed as a long-range phase correlation in terms of the so called density matrix, the quantum mechanical quantity from which all features of a large system can be derived in principle. Existence of these phase correlations is, roughly speaking, equivalent with the existence of macro wave functions. They imply establishment of rather subtle, long-range, correlations in a phase which has no immediate physical significance, but from which superfluidity and superconductivity follow. These latter properties exist at very low temperatures only. I believe, however, that existence of these long range phase correlations (coherence) have a very much wider range of application, in particular in systems in which some modes of behaviour are very strongly excited and stabilized far from equilibrium. An example are laser states [4].

I shall try to apply to biological systems the general ideas expressed above namely phase correlations on a macroscopic scale connected with some strongly excited (far off thermal equilibrium) modes of motion. Naturally — as is the habit in theoretical physics — a very much simplified model must be used to represent the common features of biological systems. A common property of cells which attracts the physicists attention is the electric double layer at the cell membrane which supports electric fields of the order $10^5$ volts/cm. Oscillations of this membrane should, one can estimate, have frequencies of order $10^{11}$ sec$^{-1}$, or similar, and some of these oscillations must then have attached to them dipolar electrical oscillations; the frequencies correspond to electric waves in the mm or far infrared region. It might occur that some of the large molecules within the cell can resonate with these membrane oscillations and thus lead to electric dipole oscillations of the cell as a whole. The frequency or frequencies of these oscillations might be a characteristic property of a specific type of cell. An assembly of such cells suspended in a certain medium will no doubt exhibit the possibility of electrical oscillations. A long range phase correlation between cells would exist if one or a few of these modes were very strongly excited, far beyond the strength of thermal excitations, and if these excitations were relatively stable.

To investigate this possibility consider as a model a system of deformable spheres suspended in a medium of static dielectric constant $\varepsilon_0$. The spheres are considered capable of electric dipolar oscillations as described above, and they will hence increase the dielectric constant of the system to a value $\varepsilon > \varepsilon_0$. The system as a whole will exhibit a number of modes of oscillations, amongst them certain longitudinal modes based on the electric dipolar oscillations of the respective units. Now it can be shown that in view of the deformability of the single units, the possibility exists that a single mode of longitudinal electric oscillation gets stabilized provided: (i) it is sufficiently strongly excited, and (ii) the units are *not* too densely packed. This means that with regard to the particular mode of motion the system is very far from thermal equilibrium, i.e. that this mode contains relatively much energy. Stabilisation implies that this energy can only be lost by first supplying still more energy to this mode—much like a billiard ball can only move from the billiard table to the (lower energy) floor by first being lifted over the wall of the table to a higher energy. The stabilization is achieved by a certain deformation of the individual spherical unit under the influence of the high frequency electric polarisation $P$ of the excited mode. This deformation is proportional to the mean square $\overline{P^2}$ of the polarisation as it is too inert to follow the high frequency oscillations.

The above model consideration thus shows that a suspension of our units (cells) in an appropriate medium is capable of certain characteristic longitudinal electric vibrations and that strong excitation of one (or a few) of these modes may be relatively stable provided the density of our units is not too high.

Suppose we assume that existence of one or a few strongly excited relatively stable longitudinal electric modes in growing organs is established. What conclusions would we draw? Firstly "relatively" stable should mean that some supply of energy is still required to maintain and in particular to establish the vibration. I am told that a cell may be compared with a factory automatised for certain tasks expressed in terms of molecular processes.

(i)   Some of these processes might be connected with the pumping of energy into the particular normal vibration of the organ as a whole. In return, once excited, the vibration of the organ would enforce vibration of the individual cell of the "correct" frequency.

(ii)   Establishment of the strongly excited vibration would lead to a stress on the membrane surface and to the above mentioned deformation. The stress would increase with increasing size of the cell. It could thus act as a stimulus for cell division provided a particular cell is large enough.

(iii)   With continuing growth the cells get packed more densely; the quasi stability of the strongly excited mode then gets gradually lost and finally a degree of density will be reached when the vibration can no longer be maintained. If the excited mode provides the stimulus for cell division (as supposed in (ii)) then at sufficiently dense packing this stimulus disappears and growth will stop.

(iv)   Absorption in the medium will define a range for the phase correlations and might hence influence macroscopic size effects.

The above considerations are meant to be highly speculative. They should demonstrate, however, that application of quantum mechanical concepts to complex systems can lead to new points of view provided quantum mechanics is not used as the servant who tries to derive—with doubtful success—the value of some known parameters but is used as a guide capable of finding regularities hidden in a labyrinth.


### References

[1]   C. N. Yang, *Rev. Mod. Phys.* **34** (1962) 694.
[2]   H. Fröhlich, *Phil. Mag. Supplement* **3** (1954) 325.
[3]   O. Penrose and L. Onsager, *Phys. Rev.* **104** (1956) 576.
[4]   Cummings and Johnston, *Phys. Rev.* **151** (1966) 105.

*Note added in proof*: I have now been able to show that non-linear interaction will lead to the channeling of supplied energy into coherent modes (*Intern. J. Quantum Chem.*, in press). For possible application in photosynthesis cf. *Nature* **219** (1968) 743.

*Theoretical Physics and Biology*, © 1969, North-Holland Publ. Co., Amsterdam

# STRUCTURE, DISSIPATION AND LIFE

I. PRIGOGINE

*Faculté des Sciences, Université Libre de Bruxelles, Belgium.* *

## 1. Introduction

It is only too clear that there still exists a huge gap between biology and theoretical physics. This is not too astonishing: according to the opinion of some of the leading biologists [e.g. 1, 2], order both in its space and its functional aspect is the basic property of life. This is clearly a "many-body" problem as the formation and the maintainance of order involves the cooperation of a large number of molecules. However statistical physics which deals with such cooperative phenomena is still in its infancy. We only begin to understand even the simplest phase changes and the simplest hydrodynamic instabilities.

It is true that great advances have been made specially for phenomena near zero temperature (as superfluidity and superconductivity). However here we deal with essentially non dissipative "reversible" processes while in living systems metabolism and dissipation of energy is likely to play an essential role. For this reason my lecture will be based essentially on the use of phenomenological or thermodynamic methods. I would like to discuss in terms of such methods the source of biological order. I would also like to show that non-linear thermodynamics of irreversible processes as developed recently may narrow the gap between biology and physics.

As has been often noticed in macroscopic physics we may distinguish between two types of structure:

a) equilibrium structures

b) dissipative structures.

Equilibrium structures may be maintained *without any exchange of energy or matter*. A crystal is the prototype of an equilibrium structure.

Classical thermodynamics has solved the problem of the competition between randomness and structure (or "organization") for equilibrium situations: when we lower the temperature, the contribution of the energy to the Helmholtz free-energy

$$F = E - TS \tag{1.1}$$

becomes dominant. More and more complex structures corresponding to smaller

---

* Also Center for Statistical Mechanics and Thermodynamics, University of Texas, Austin, Texas, U.S.A.

entropy may then appear. To a phase transition such as crystallisation corresponds the loss of entropy

$$\Delta S = - \frac{\mathscr{L}_f}{T_m},$$ (1.2)

where $\mathscr{L}_f$ is the heat of melting and $T_m$ the melting temperature.

On the contrary, "dissipative structures" are maintained only through exchange of energy (and in some cases also of matter) with the outside world.

A very simple example is a thermodiffusion cell in which a gradient of concentration is maintained by a flow of energy. Here we have already a dissipative process (exchange of heat) which leads to an increase of organization [3]. However, in this example, this increase occurs gradually with the increase of the gradient of temperature.

But discontinuous changes in structure due to dissipative processes are also possible. A simple example is the so-called "Bénard problem" in classical hydrodynamics [4]. We heat a horizontal fluid layer from below. We create in this way a so-called "adverse gradient" of temperature. For small values of this gradient the fluid remains at rest. But for a critical value of this gradient, there is an abrupt onset of thermal convection.

The role which the temperature played in the liquid → solid transition is now played by the increase of the "constraint" which is here the adverse temperature gradient. The increase of this constraint leads from a situation in which the whole of the energy is in the thermal motion to a much more organized state in which part of it is in the form of macroscopic motion which is of course, a highly cooperative phenomenon from the molecular point of view.

The state of the fluid in the region of thermal convection may therefore be considered as an example of "dissipative structure". Its entropy is lower than that of a system in which the whole of the energy would be in the thermal motion. The corresponding loss of entropy is of the order

$$\Delta S = - \frac{E_{kin}}{\bar{T}},$$ (1.3)

where $\bar{T}$ is the average temperature characterizing the systems. The dissipative structure is achieved and maintained through the effect of the energy flow associated with the adverse gradient.

Are dissipative structures possible outside the range of hydrodynamics? What determines their occurrence? Before we try to give a more general answer to such problems let us consider a "chemical" example. We consider a sequence of reactions such as

$$A \underset{}{\overset{1}{\rightleftharpoons}} X \underset{}{\overset{2}{\rightleftharpoons}} B$$
$$\overset{}{\underset{3}{\Updownarrow}}$$
$$M$$ (1.4)

The concentrations of the initial product $A$ and of the final product $B$ are fixed; $X$ and $M$ are intermediate components. We expect that for given values of $A$, $B$ the concentrations of $X$, $M$ will take well defined values in the steady state. For example, if we assume the simple kinetics laws (we put equal to one all equilibrium and rate constants)

$$V_1 = A - X, \quad V_2 = X - B, \quad V_3 = X - M, \tag{1.5}$$

we obtain easily at the steady state

$$M = X = \tfrac{1}{2}(A + B). \tag{1.6}$$

As a special case, (1.6) includes the case in which the fixed ratio $A/B$ corresponds to thermodynamic equilibrium. Then $B/A = 1$ and (1.6) reduces to

$$M = X = A = B. \tag{1.7}$$

The equilibrium result (1.7) could of course have been directly derived using the law of mass action. We may say that in this simple case the "thermodynamic solution" (1.7) may be "extended" to the whole possible range of the constraint $A/B$ to obtain (1.6). However in general if we use a more complicated non-linear scheme of reactions there may be different time-independent solutions of the kinetic equations.

$$\frac{dX}{dt} = 0, \quad \frac{dM}{dt} = 0, \tag{1.8}$$

all satisfying obvious physical conditions such as the concentrations to be real, positive quantities. Let us call these solutions $X_1, ..., X_2$; $M_1, ..., M_2$. One of them, say $X_1$, contains as a special case the equilibrium solution corresponding to a minimum of free-energy. But, will it be the "correct" solution for all values of the relevant constraints, such as the ratio $A/B$?

What we have to require from the correct solution is that it is stable with respect to fluctuations. If we start with the initial concentrations

$$X_i + x, \quad \text{for} \quad t = 0, \tag{1.9}$$

when $x$ is considered as small, we have to show that

$$x \rightarrow 0, \quad \text{for} \quad t \rightarrow \infty. \tag{1.10}$$

In equilibrium thermodynamics or in the linear range of non-equilibrium thermodynamics such calculations are generally not necessary as it is sufficient to show that the relevant thermodynamic potential (e.g. the free-energy (1.1) or the entropy production) is minimum. However usually there exists no potential far from equilibrium and a direct proof of (1.10) based on stability theory has to be given. Such calculations are standard in hydrodynamics [4], and we shall see some examples for chemical systems later.

The occurrence of dissipative structures is associated with the existence of instabilities at which one branch of the kinetic equations (such as $X_1$, $M_1$ in the example above) becomes unstable and is replaced by a new branch. It is therefore immediately clear that dissipative structures may only exist in non-linear systems for which more than one solution of the kinetic equations exist.

Moreover dissipative structures will only occur at "finite-distance" from thermodynamic equilibrium as the stability of the "thermodynamic" solution must extend over at least some non-equilibrium region. In fact we shall show in § 4 that instabilities can only occur outside the range of linear thermodynamics of irreversible processes.

The appearance of a dissipative structure may be visualized in phase space in a way very similar to that used to visualize the appearance of an equilibrium structure.

In figs. 1a and 1b, we represent in a schematic way the phase space associated with the liquid and the solid phase (if we consider isothermal systems one has to weight each volume element by the corresponding Boltzmann factor $\exp(-E/kT)$). For temperatures $T$ above the melting temperature $T_m$ the phase space associated with the liquid is larger than that associated with the solid. The inverse is true for $T < T_m$.

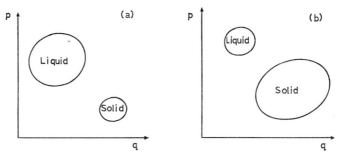

Fig. 1.   (a): $T > T_m$; (b): $T < T_m$.

Similarly let us suppose that there are two meaningful branches of the kinetic equations (that is such that the corresponding steady state concentrations are real, positive numbers). One of them is what we called the thermodynamic solution stable at equilibrium. As this solution corresponds to a maximum of entropy (or to a minimum of free-energy) it corresponds in the neighborhood of equilibrium to the biggest phase volume (see fig. 2a). If there is an instability it may still have the biggest phase volume but the stable solution is now on the second branch (see fig. 2b). The corresponding loss of phase volume then expresses the appearance of a dissipative structure.

Our main interest will be centered in this report on the specific class of instabilities which are symmetry breaking in the sense that they lead from a homo-

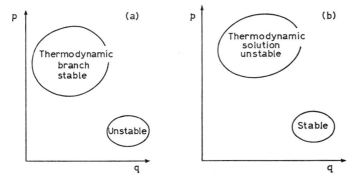

Fig. 2. Phase volumes corresponding to two branches of kinetic equations; (a): below the instability, (b): above the instability.

geneous system to an inhomogeneous one involving both structural and functional order.

We shall also devote some attention to the occurrence of time-order (biological clocks) in dissipative systems. As we shall see there is remarkable parallelism between the problems of structural dissipative order and time-order. Both will appear as possible consequences of large deviations from thermodynamic equilibrium in chemical systems undergoing chemical transformations of essentially the same type.

## 2. Entropy production and stability of equilibrium

As an introduction to the problem of stability of non-equilibrium steady states let us briefly consider the stability of equilibrium states with respect to small perturbations. We shall use a method based on the behavior of entropy production (for more detail see [5]). As is well known, the change of the entropy can be split into two parts. Denoting by $d_e S$ the flow of entropy due to interactions with the exterior and by $d_i S$ the entropy production inside the system we have

$$dS = d_e S + d_i S, \quad d_i S \geqslant 0,$$

or

$$\frac{dS}{dt} = \frac{d_e S}{dt} + \frac{d_i S}{dt}. \tag{2.1}$$

In the macroscopic range (validity of the Gibbs formula), we have

$$\frac{d_i S}{dt} = \int dV \sigma = \sum_i \int dV J_i X_i \geqslant 0. \tag{2.2}$$

Here $\sigma$ is the entropy production per unit time and unit volume. Moreover $J_i$ and $X_i$ are the flows (or rates) of the irreversible processes and the corresponding generalized forces (for more detail see [3, 6]). In the case of chemical reactions we have

$$J_i = V_i, \tag{2.3a}$$

$$X_i = \frac{A_i}{T} = -\sum_\gamma \frac{\nu_{\gamma i}\mu_\gamma}{T}, \tag{2.3b}$$

where $V_i$ is the reaction rate and $A_i$ the corresponding affinity related to the chemical potentials $\mu_\gamma$ through the second equality (2.3b). (Also $\nu_{\gamma i}$ is the stoechiometric coefficient of component $\gamma$ in reaction $i$.)

Similarily in the case of thermal conduction we have

$$\mathbf{J} = \mathbf{W},$$
$$\mathbf{X} = \text{grad } (1/T), \tag{2.4}$$

where $\mathbf{W}$ is the heat flow.

For chemical reactions we may also write

$$d_i S = \frac{1}{T} \sum_{i=1}^{r} A_i\, d\xi_i \geqslant 0, \tag{2.5}$$

where $d\xi_i$ is the change in the extent of reaction as has been defined by De Donder. We shall be mainly interested in the case of chemical reaction. It is well known that a formula such as (2.5) is only valid when the reactions are slow enough so that the collisions restore near equilibrium energy distribution functions [3, 6]. This condition is well met in dense media and we shall therefore assume its validity.

Let us use (2.5) to study chemical equilibrium and stability. At equilibrium we require $\delta_i S = 0$ for all independent variations of $\delta\xi_i$.

$$\delta_i S = \frac{1}{T} \sum_{i=1}^{r} A_i\, \delta\xi_i = 0. \tag{2.6}$$

This leads to

$$A_i = -\sum_\gamma \nu_{\gamma i}\mu_\gamma = 0, \quad i = 1, ..., r. \tag{2.7}$$

These are the well-known Gibbs conditions for chemical equilibrium. Moreover this equilibrium can only be stable if a spontaneous change (that is satisfying inequality 2.5) is impossible starting from equilibrium. In other words, for each possible change $\delta\xi_i$ starting from equilibrium we have to require

$$\delta_i S \leqslant 0, \tag{2.8}$$

or expanding $A_i$ around the equilibrium state and using (2.7)

$$\sum_i A_i\, \delta\xi_i = \sum_i \delta A_i\, \delta\xi_i \leqslant 0. \tag{2.9}$$

If this inequality is satisfied the changes $\delta\xi_i$ will correspond to *fluctuations* which give rise to a negative entropy production and the equilibrium state will be stable. These conditions are well known. At constant pressure and temperature they imply that the quadratic form [5].

$$\sum_{ij} \left(\frac{\partial A_i}{\partial \xi_j}\right)_{\mu T} \delta\xi_i\, \delta\xi_j < 0 \tag{2.10}$$

is negative definite. Now let us consider the inequality

$$\sum_{\gamma'} \left(\frac{\partial \mu_\gamma}{\partial n_{\gamma'}}\right)_{\mu T} \delta n_\gamma\, \delta n_{\gamma'} \geqslant 0. \tag{2.11}$$

This condition on chemical potentials is the so-called *stability condition with respect to diffusion* first established by Gibbs [5]. It expresses the statement that any fluctuation of composition which may appear in an initially homogeneous system will be followed by a response which will tend to restore the homogeneity. It is easy to verify that (2.11) implies (2.10). Therefore *stability with respect to diffusion implies stability of chemical equilibrium*. This is a theorem due to Duhem and Jouguet (see ref. [5]).

We shall see that even for steady states the Duhem–Jouguet theorem holds when the steady states are near equilibrium but that it fails for steady states far from equilibrium. This as we shall see is an essential point for the very existence of chemical instabilities.

Besides the diffusion stability condition there exist also stability conditions with respect to thermal and mechanical disturbances which are also used in the theorems we shall discuss in § 3.

Let us now show that these thermodynamic definitions of stability coincide with the kinetic stability conditions introduced in (1.9)–(1.10). Near equilibrium we may write the linear equations

$$\frac{d\,\delta\xi_i}{dt} = \sum_{i'=1}^{r} \alpha_{ii'}\, \delta\xi_{i'}. \tag{2.12}$$

Stability means that the eigenvalues $\omega$ of this system corresponding to the normal modes have a negative real part. Therefore if we write

$$\delta\xi_i = (\delta\xi_i)_0\, e^{\omega t} \qquad (\omega = \omega_r + i\omega_i) \tag{2.13}$$

the determinant

$$|\omega\delta_{ii'} - \alpha_{ii'}| = 0 \tag{2.14}$$

has to lead to roots such that for all of them

$$\omega_r < 0. \tag{2.15}$$

The equation (2.14) is called the dispersion equation. Let us first consider the case of real roots. Then

$$\omega = \omega_r, \qquad \omega_i = 0. \tag{2.16}$$

The inequality (2.2) together with (2.13) implies

$$\frac{d_i S}{dt} = \sum_i A_i \frac{d\xi_i}{dt} = \omega_r \sum_i A_i \, \delta\xi_i \geqslant 0. \tag{2.17}$$

Therefore the thermodynamic stability condition (2.9) indeed leads to the kinetic conditions (2.15) and vice-versa. Let us extend our discussion to the case of complex roots. We then have (the star means complex conjugate)

$$
\begin{aligned}
\frac{d_i S}{dt} &= \sum_i \left( A_i \frac{d}{dt}(\delta\xi_i)^* + A_i^* \frac{d}{dt}(\delta\xi_i) \right) \\
&= \omega_r \sum_i (\delta A_i \, \delta\xi_i^* + \delta A_i^* \, \delta\xi_i) + i\omega_i \sum_i (\delta A_i^* \, \delta\xi_i - \delta A_i \, \delta\xi_i^*) > 0.
\end{aligned}
\tag{2.18}
$$

The entropy production is in this way split into two parts, one is associated with the "radial" motion towards the unperturbed state and the other with the rotation around the unperturbed state. In the case studied here where the unperturbed state is an equilibrium state the second term vanishes as a consequence of the symmetry relation

$$\frac{\partial A_i}{\partial \xi_\varrho} = \frac{\partial A_\varrho}{\partial \xi_i}, \tag{2.19}$$

which follows from the relations between affinities and chemical potentials (2.3b). There could therefore be only "reversible", non entropy producing rotations around equilibrium. But even this is excluded, as Balescu and I proved [3] already some years ago that rotations are impossible in the linear range of thermodynamic of irreversible processes as a consequence of the Onsager's reciprocity relations (see 4.2). Rotational motions become only possible in the non-linear range of irreversible processes. We shall come back to this later.

## 3.   Evolution criterion and stability of steady states [9]

Let us study in more detail the time variation of the entropy production. Using the notation (2.2) (when not essential, we suppress the integration symbol over the volume) we obtain

$$P = \frac{d_i S}{dt} = \sum_i J_i X_i \geqslant 0. \tag{3.1}$$

We now decompose the time change $dP$ of $P$ into two parts: One is related to the change of the forces and the other to that of the flows. We therefore have by definition

$$dP = d_X P + d_J P = \sum_i J_i \, dX_i + \sum_i X_i \, dJ_i. \tag{3.2}$$

Glansdorff and I proved [3, 7, 8] that for time-independent boundary conditions

$$d_X P = \sum_i J_i \, dX_i \leqslant 0. \tag{3.3}$$

The proof is based on the stability conditions we have enumerated in § 2. As a special case we have for chemical reactions

$$T d_X P = \sum_i V_i \, dA_i \leqslant 0. \tag{3.4}$$

An important feature of (3.3) is the possibility of extending it to include flow processes such as convection [8, 9]. Therefore there exists a quantity $d\Phi$ of the form

$$d\Phi = \int dV \sum_i J_i' \, dX_i' \leqslant 0, \tag{3.5}$$

where the forces $X_i'$ and the flows $J_i'$ now include mechanical processes such as convection terms (see the example § 5). As we then in general deal with in-homogeneous systems we have written (3.5) as a volume integral.

The inequality (3.5) is so general that it can be really called a "universal" evolution criterion of macroscopic physics. However $d\Phi$ is *not* in general a total differential [7]. Exactly as equilibrium thermodynamics may be based on the study of entropy production and specially on (2.5), the study of steady states may be based on the evolution criterion (3.5). As an example let us show that we can use the evolution criterion (3.3) for chemical non-equilibrium situations exactly as we used $d_i S$ as given by (2.5) for chemical equilibrium. At the steady state we must have

$$T \delta_X P = \sum_i V_i \, \delta A_i = 0, \tag{3.6}$$

for *all independent* variations of the affinities. Let us consider as an example the sequence of reactions (1.4) where again the concentrations of $A$ and $B$ are fixed. There are two independent affinities because of the condition

$$A_1 + A_2 = \log \frac{A}{B} = \text{given}, \tag{3.7}$$

or

$$\delta A_1 + \delta A_2 = 0. \tag{3.8}$$

Therefore (3.6) leads to

$$\left. \begin{aligned} V_1 &= V_2, \\ V_3 &= 0, \end{aligned} \right\} \tag{3.9}$$

which are of course the usual steady state conditions and include as a special case the equilibrium conditions

$$V_1 = V_2 = V_3 = 0. \tag{3.10}$$

Again in complete parallelism with the study of equilibrium situations the steady state will be stable if [10]

$$\delta_X P \geqslant 0, \quad \text{or} \quad \delta\Phi \geqslant 0, \tag{3.11}$$

for all perturbations *compatible with the kinetic equations of the system*. In the chemical case this leads using (3.4) and (3.6) to the stability condition

$$T\delta_X P = \sum_i \delta V_i \delta A_i > 0. \tag{3.12}$$

The stability theory of steady states may be conveniently subdivided into two parts:

a)  the identification of phenomena which may give rise to *negative* contributions to $\delta_X P$ or more generally to $\delta\Phi$ (when hydrodynamic flow phenomena are included),

b)  the discussion of the numerical value of such negative terms and specially the discussion of the possibility that the sum involved in (3.11) or (3.12) may become negative *without violating* the kinetic equations.

The first part can at present be treated in a quite general way. We shall see in §§ 5, 6 that mechanical convection as well as certain auto- and crosscatalytic chemical reactions indeed give negative contributions. They are therefore the phenomena which are "dangerous" for the stability and of special interest for us here. The second part involves the solution of the dispersion equation for small perturbations or some approximate method (for example the use of the local potential method, [9]).

In the evolution criterion (3.4) or (3.5) the generalized rates and forces play a quite different role. No inequality for $\delta_J P$ outside the linear range (see 4.4) is known. It is therefore quite remarkable that in the stability condition (3.11) or (3.12) the rates and forces play again a symmetric role. For example in the chemical case the entropy production is

$$T\frac{d_i S}{dt} = \sum_i A_i V_i \geqslant 0, \tag{3.13}$$

which has to be compared to the quantity (3.12) which determines the stability of the system.

We may therefore also call $\delta_X P$ or $\delta\Phi$ the "excess entropy production" (more briefly E.E.P.) near the steady state we consider.

We see therefore that a sufficient condition of stability is that the E.E.P. is positive. In this *very specific sense* we may say that the theorem of minimum entropy production is valid for an arbitrary stable steady state.

Let us now consider a certain number of examples.

## 4. The linear range of thermodynamics of irreversible processes

Let us introduce the following assumptions:

a) Linear phenomenological laws

$$J_i = \sum_j L_{ij} X_j. \tag{4.1}$$

b) Validity of Onsager's reciprocity relations

$$L_{ij} = L_{ji}. \tag{4.2}$$

c) Phenomenological coefficients $L_{ij}$ may be treated as constants.
d) No mechanical flow processes.

Inequality (3.1) then becomes

$$P = \sum_{ij} L_{ij} X_i X_j \geqslant 0. \tag{4.3}$$

Moreover it can be easily shown that in this range [3]

$$d_X P = d_J P = \tfrac{1}{2} dP \leqslant 0. \tag{4.4}$$

Therefore in the linear domain the entropy production can only decrease and steady states are characterized by a minimum of entropy production [3].

Moreover it is clear that such steady states are automatically stable. Indeed in the neighborhood of the steady state we have (see 3.12, 4.1 and 4.3)

$$\delta_X P = \sum_i \delta J_i \, \delta X_i = \sum_{ij} L_{ij} \, \delta X_i \, \delta X_j \geqslant 0. \tag{4.5}$$

Note that we have here for the perturbation of the entropy production near the steady state valid till second order,

$$\delta P = \sum_i J_i^0 \, \delta X_i + \sum_i X_i^0 \, \delta J_i + \sum_i \delta J_i \, \delta X_i. \tag{4.6}$$

Here $J_i^0$ and $X_i^0$ are the flows and forces at the steady state. We have also used (3.6) (see also (4.4)). Therefore we have near the steady state including terms of second order

$$\delta_X P = \tfrac{1}{2} \delta P \geqslant 0. \tag{4.7}$$

As we see the positive sign of $\delta_X P$ is related to the theorem of minimum of entropy production. This sign shows that the stability criterion (3.11) is automatically satisfied. No instabilities may arise in the linear range.

To understand this result we have to remember that inequalities (3.3) and (3.5) are based on the validity of the stability conditions for equilibrium. Our result (4.5) simply indicates that once the *equilibrium* is stable, *steady states* in the neighborhood o˜ equilibrium are automatically stable as well. Using the terminology of § 1 we could say that the "thermodynamic solution" of the kinetic equations may be continued at least in some neighborhood of equilibrium.

Specifically for chemical reactions we may state that the Jouguet–Duhem theorem mentioned in § 2 includes not only chemical equilibria but also the whole range of steady states which belong to the linear range of non-equilibrium thermodynamics.

This shows that the problem of occurrences of new dissipative structures is outside the range of the theorem of minimum entropy production. This theorem only describes steady states which may be obtained through a continuous "deformation" of the equilibrium situation.


## 5.  Hydrodynamic instabilities

The case of hydrodynamic instabilities has been treated in the frame of this method by Glansdorff [8, 9]. We include here a brief summary in order to be able to compare hydrodynamic and chemical instabilities.

To be specific let us consider a two-dimensional laminar flow at constant temperature between two parallel walls. We have here (*see* 3.5)

$$\left. \begin{array}{l} \mathbf{J'} = -(\mathbf{P}+\varrho \mathbf{v}v), \\[2mm] \mathbf{X'} = \dfrac{1}{T}\dfrac{\delta \mathbf{v}}{\delta \mathbf{x}}, \end{array} \right\} \tag{5.1}$$

where $\mathbf{P}$ is the pressure tensor, $\mathbf{v}$ the convection velocity. The force $\mathbf{X'}$ is purely dissipative, but the corresponding flow has now besides the usual dissipative contribution $\mathbf{P}$ related to the viscosity, also a contribution due to convection.

After a few simple manipulations which are described elsewhere one finds [see ref. 9, page 50] in the neighborhood of the laminar flow

$$T\frac{d\Phi}{dt} = \eta N - \varrho M, \tag{5.2}$$

with

$$\eta N = \eta \iint y^2 \, dx \, dy, \quad y = \frac{\delta v}{\delta x} - \frac{\delta u}{\delta y}, \quad (u = \delta v_x, \ v = \delta v_y) \tag{5.3}$$

$$\varrho M = -\varrho \iint uv \frac{\delta v_x^0}{\delta y} \, dx \, dy = -\int \varrho \langle uv \rangle \frac{\delta v_x^0}{\delta y} \, dy. \tag{5.4}$$

Therefore $\eta N$ represents the viscous dissipation and $\varrho M$ the conversion of energy from the basic into the disturbance by a Reynolds shear.

The contribution of the viscosity to (5.2) is always positive. If we could neglect the inertial contribution $\varrho M$ we would always have a stable laminar regime.

On the contrary, the inertial term $\varrho M$ has no definite sign. Instability may occur if $\varrho M$ is positive and can compensate the dissipative term $\eta N$.

The relative magnitude of these two terms may be written in the form

$$\frac{\varrho M}{\eta N} = R \frac{M'}{N'}, \tag{5.5}$$

where $R$ denotes the Reynolds number and $M'$, $N'$ are dimensionless forms of $M$ and $N$.

Instability of the laminar flow will occur for $M > 0$ and for a Reynolds number larger than the critical value

$$R_C = \frac{N'}{M'}. \tag{5.6}$$

Then the transfer of kinetic energy into the fluctuations will exceed the decrease of kinetic energy due to viscosity and turbulence will occur.

These considerations are well known (for the references to the papers by Orr, Dryden, Murnaghan and Bateman, see [9]). We only wanted to show that we have here a special case of our general evolution criterion (3.5). It should be noticed that in the transition between laminar and turbulent flow the entropy production $d_i S/dt$ (due here to viscosity) *increases*. This is also true for all other hydrodynamic instabilities such as the Bénard and Taylor instabilities.

As a result (see 2.1) in the steady state when $d_i S/dt = 0$, (at least in an average sense if there are time-oscillations) the entropy flow $d_e S/dt$ increases in absolute value.

The dissipative structure involved in hydrodynamic instabilities has therefore a simple thermodynamic meaning: it appears as a response of the system to non-equilibrium conditions and leads to a more efficient mechanism for restoring equilibrium in the system as a whole.

## 6.  Chemical instabilities

The basic quantity we have to consider is according to (3.12)

$$T\delta_X P = \sum_i \delta V_i \, \delta A_i. \tag{6.1}$$

Instabilities may only occur if at least some terms in this sum are negative. Such terms would then play the role of the convection terms such as $\varrho M$ we have considered in hydrodynamics.

Let us consider the chemical reaction

$$X + Y \rightarrow C + D. \tag{6.2}$$

As we are now mainly interested in far from equilibrium situations we neglect the reverse reaction and write for the reaction rate (we again take equal to one all kinetic and equilibrium constants as well as $RT$)

$$V = XY. \tag{6.3}$$

The affinity is

$$A = \log \frac{XY}{CD}. \tag{6.4}$$

A fluctuation in the concentration $X$ around some steady state value gives rise to an excess entropy production

$$\delta V \delta A = \frac{Y}{X} (\delta X)^2 > 0. \tag{6.5}$$

Such a fluctuation would therefore not bring into danger the stability condition (3.12).

Let us now consider instead of (6.4) the autocatalytic reaction

$$X + Y \rightarrow 2X. \tag{6.6}$$

The reaction rate is still assumed to be given by (6.3), but the affinity is now

$$A = \log \frac{XY}{X^2} = \log \frac{Y}{X}. \tag{6.7}$$

We have now a "dangerous" contribution to the excess entropy production. Detailed calculations given in § 7 will show that indeed under well defined conditions such autocatalytic reactions give rise to instabilities and dissipative structures.

Of course (6.6) is meant here only as an example. As we shall see later there may be more subtle ways involving fluctuations of more than one component to produce negative contributions to the E.E.P.

Before we go to the study of specific examples we want to present the following observation. There are at least two types of chemical instabilities:

a)  instabilities in respect to homogeneous perturbations;

b)  instabilities in respect to space-dependent inhomogeneous perturbations;

In case a) we expect the system to go from a homogeneous steady state to another homogeneous state (which may be steady or not). In case b) diffusion plays an essential role. Diffusion appears in this theory in a twofold way:

1)  it gives a *positive* contribution to the E.E.P. similar to that of viscous dissipation in hydrodynamics (5.2); this effect can only *stabilize* the steady state;

2)  it increases the manifold of perturbations compatible with the macroscopic

equations of change (the kinetic equations). We have to test the stability of the system with respect to a wider class of situations.

If the second effect is dominant we may expect instabilities with respect to in-homogeneous perturbations and the appearance of new states which are no longer homogeneous in space.

We have then "symmetry breaking" instabilities [12] which, we believe are especially important from the point of view of the interpretation of biological order.

Let us now consider some explicit examples. A very simple scheme of reactions which may give rise to instabilities is the following

$$A \rightarrow X \qquad (1)$$
$$2X + Y \rightarrow 3X \qquad (2) \qquad \text{scheme I}$$
$$B + X \rightarrow Y + D \qquad (3)$$
$$X \rightarrow E \qquad (4) \qquad (6.8)$$

The overall reaction is
$$A + B \rightarrow E + D. \qquad (6.9)$$

This reaction scheme is physically unrealistic because it involves the trimolecular step (2). We shall indicate below other schemes which do not involve such a step. But the algebra involved in the discussion of (6.8) is much simpler as it involves only two intermediate components $X$, $Y$ and for this reason we shall discuss it in more detail and simply quote the results obtained for the other schemes (for more detail see [13]). The kinetic equations are

$$\frac{\partial X}{\partial t} = A + X^2 Y - BX - X + D_X \frac{\partial^2 X}{\partial r^2}, \qquad (6.10)$$

$$\frac{\partial Y}{\partial t} = BX - X^2 Y + D_Y \frac{\partial^2 Y}{\partial r^2}. \qquad (6.11)$$

As usual we maintain constant the concentrations of the initial and the final components $A$, $B$, $D$, $E$. The reverse reactions are neglected. The value of the non-vanishing kinetic constants is taken equal to one. We take into account diffusion. To simplify we assume a one-dimensional medium. There exists always a time-independent homogeneous solution

$$X_0 = A, \qquad Y_0 = \frac{B}{A}. \qquad (6.12)$$

This is the continuation of the thermodynamic solution (see § 1). To investigate its stability we first investigate the dispersion equation (see 2.14). We consider perturbations of the form

$$\left. \begin{array}{l} X = X_0 + x\, e^{(\omega t + ir/\lambda)}, \\ Y = Y_0 + y\, e^{(\omega t + ir/\lambda)}, \end{array} \right\} \qquad (6.13)$$

with

$$\left|\frac{x}{X_0}\right| \ll 1, \quad \left|\frac{y}{Y_0}\right| \ll 1.$$

Inserting (6.13) into (6.10) and (6.11) we obtain

$$\left.\begin{array}{l} (\omega - B + 1 + a)\, x - A^2 y = 0, \\ Bx + (\omega + A^2 + b)\, y = 0, \end{array}\right\} \tag{6.14}$$

with

$$a \equiv \frac{D_X}{\lambda^2}, \quad b \equiv \frac{D_Y}{\lambda^2}. \tag{6.15}$$

The corresponding dispersion equation is

$$\omega^2 + (A^2 + 1 - B + a + b)\, \omega + A^2(1 + a) + (1 - B)\, b + ab = 0. \tag{6.16}$$

There exists a critical value

$$B_c(\lambda) = \frac{1}{b}(1 + a)\, (A^2 + b), \tag{6.17}$$

which separates a root $\omega < 0$ from a root $\omega > 0$. We now have to look for the critical value of the wave length $\lambda_c$ at which the instability begins. To do this we have to calculate the wave length which gives to (6.17) its minimum value. This leads immediately to

$$(\lambda_c)^2 = \frac{1}{A}(D_X D_Y)^{\frac{1}{2}} \tag{6.18}$$

and substitution in (6.17) gives

$$B_c = \left[A\left(\frac{D_X}{D_Y}\right)^{\frac{1}{2}} + 1\right]^2. \tag{6.19}$$

To discuss these expressions it is useful to repeat the calculations without putting the kinetic coefficients equal to one. If we call $k_i$ the kinetic coefficient corresponding to step $i$ in (6.9) we obtain

$$\lambda_c^2 = \left(\frac{k_4}{k_1^2 k_2}\right)^{\frac{1}{2}} \frac{(D_X D_Y)^{\frac{1}{2}}}{A} \tag{6.18'}$$

and

$$B_c = \left[\frac{k_1}{k_4}\left(\frac{k_2}{k_3}\frac{D_X}{D_Y}\right)^{\frac{1}{2}} A + \left(\frac{k_4}{k_3}\right)^{\frac{1}{2}}\right]^2. \tag{6.19'}$$

We see that $\lambda_c$ and $B_c$ depend in an intrinsic way on both the various reaction rates and on the diffusion constants. If diffusion as compared to reaction rates becomes small, instability occurs for short wave length perturbations. In the inverse case the instability occurs for long wave lengths. This is quite reasonable and we shall come back to this point below.

We see that we have here a symmetry breaking instability as the system beyond the instability can no longer be expected to be homogeneous. We shall study this question in greater detail in § 7.

Let us indicate three other reaction schemes which also may lead to chemical symmetry-breaking instabilities:

$$
\begin{array}{ll}
A + X \rightarrow 2X & (1) \\
X + Y \rightarrow 2Y & (2) \\
Y + V \rightarrow V' & (3) \\
V' \rightarrow E + V & (4)
\end{array}
\qquad
\begin{array}{l}
\text{scheme II} \\
(6.20)
\end{array}
$$

The overall reaction is

$$
A \rightarrow E. \qquad (6.21)
$$

There are only one- and two-molecular steps. In this scheme both $X$ and $Y$ catalyze their own formation. It is a modification of a scheme suggested by Lotka [14] to investigate chemical oscillations (we discuss Lotka's scheme in § 8). We have simply added the "appendix" (3)–(4) involving the intermediate compound $V'$. Such an appendix also plays a role in the next schemes, such as

$$
\begin{array}{ll}
A \rightarrow X & (1) \\
X + Y \rightarrow C & (2) \\
C \rightarrow D & (3) \\
B + C \rightarrow Y + C & (4) \\
Y \rightarrow E & (5) \\
Y + V \rightarrow V' & (6) \\
V' \rightarrow E + V & (7)
\end{array}
\qquad
\begin{array}{l}
\text{scheme III} \\
\\
(6.22)
\end{array}
$$

This is a simplified form of Turing's original scheme [11] (see also [12]). We have the two overall reactions

$$
\left\{
\begin{array}{ll}
A + B \rightarrow D & (1) \\
B \rightarrow E & (2)
\end{array}
\right.
\qquad (6.23)
$$

Let us finally indicate Turing's scheme

$$
\begin{array}{ll}
A \rightarrow X & (1) \\
X + Y \rightleftarrows C & (2) \\
C \rightarrow D & (3) \\
B + C \rightarrow W & (4) \\
W \rightarrow Y + C & (5) \\
Y \rightarrow E & (6) \\
Y + V \rightarrow V' & (7) \\
V' \rightarrow E + V & (8)
\end{array}
\qquad
\begin{array}{l}
\text{scheme IV (Turing's)} \\
\\
(6.24)
\end{array}
$$

The Turing scheme IV differs from III only by the presence of the intermediate component $W$. Also the step (2) is supposed to be reversible. In an earlier paper [12] this scheme has been studied in great detail and it has been shown that in agreement with our general conclusion of § 4 no instability can occur when the affinities of the overall reactions (6.23) are smaller than some critical values.

One may broadly describe schemes III and IV by saying that the initial products $A$, $B$ are transformed into the final (or "waste") products $D$ and $E$ through the intermediate products $X$, $Y$ and by the action of the catalysts $C$, $V$, $V'$, $W$ following the general scheme

$$A \longrightarrow X \longrightarrow C \longrightarrow D$$

$$B \xrightarrow{(C,W)} Y \xrightarrow{(V,V')} E \tag{6.25}$$

Let us now consider the thermodynamic aspects of the chemical instability in terms of the basic quantity (6.1). We again go into more detail for the reaction scheme I. Using the kinetic equation (see (6.10, 11)) we find for (6.1), including the effect of diffusion

$$\delta_X P = (1-B)\frac{x^2}{A} + \frac{A^2}{B}y^2 + \frac{a}{A}x^2 + \frac{bA}{B}y^2. \tag{6.26}$$

Two interesting remarks have to be made:

1) in agreement with our general discussion there appears in the E.E.P. (6.26) the negative term $-Bx^2/A$ due to the autocatalytic action of $x$; this is the "dangerous" contribution we have mentioned earlier;

2) the explicit contribution of diffusion to the E.E.P. is positive and proportional to $D/\lambda^2$ (the last two terms in 6.26). Therefore if there is an instability, increasing values of $D$ must give rise to increasing values of the critical wave length. If not, the contribution of diffusion to (6.26) would become dominant and $\delta_X P$ would be always positive. This is in agreement with the formula for $\lambda_c$ (6.18) derived from the dispersion equation. But the diffusion has as we already mentioned a second role: the manifold of perturbations which we may introduce into (6.26) is now increased by the consideration of inhomogeneous systems. Let us verify that indeed the perturbations $(x, y)$ which satisfy (6.14) at the marginal state $\omega = 0$ lead to the vanishing of the E.E.P. Indeed we first obtain using (6.14) and (6.17)

$$y = -\frac{(1+a)}{b}x. \tag{6.27}$$

If we now substitute in (6.26) and again use (6.19) we obtain indeed

$$(\delta_X P)_{B=B_c} = 0. \tag{6.28}$$

There is therefore complete agreement between the *kinetic theory of chemical instabilities based on the dispersion equation and the thermodynamic theory developed in* § 3.

The situation for the reaction scheme II described in (6.25) is quite similar. The autocatalytic character of $X$, $Y$ again introduces negative terms into the E.E.P. It is interesting to comment briefly on the role of the appendix (steps 3 and 4 in 6.25). It appears that its role is quite similar to that of diffusion (for more details see [13]): it gives a positive contribution to the E.E.P. However at the same time it increases the range of permissible perturbations. Instability appears if the steady state concentration of $V'$ is larger than some critical value.

The schemes III and IV again lead to an E.E.P. which contains dangerous contributions. However here the negative terms are of the form ($\alpha$ is a positive constant)

$$- \alpha y c \qquad (6.29)$$

and *not*

$$-\frac{B}{A} x^2, \qquad (6.30)$$

as in (6.26). This is quite natural as we have in these schemes a *crosscatalytic* effect: $Y$ catalyzes the formation of $C$ and inversely $C$ the formation of $Y$.

We believe that the situation for chemical instabilities as compared to that described in our earlier paper [12] has been greatly clarified. We see indeed that:

1)  there are many reaction schemes which may lead to chemical instabilities; in fact it would be easy to imagine others than those we have listed;

2)  they are all characterized by negative contributions to the E.E.P. due to autocatalytic or crosscatalytic effects;

3)  the essential and unexpected role of diffusion is clearly understood.

Let us now consider the physical situation beyond the chemical instability point.

## 7.  Chemical dissipative structure

The study of the time evolution of the steady states beyond an instability is always rather lengthy and difficult [11]. Again scheme I studied in § 6 provides us with a relatively simple example [13].

However instead of considering disturbances of arbitrary wave length $\lambda$ we shall now consider a system of two boxes. On the contrary the concentrations of $X$ and $Y$ may be different. Instead of (6.10, 11) we now have the four equations

$$\frac{dX_1}{dt} = A + X_1^2 Y_1 - BX_1 - X_1 + D_X(X_2 - X_1),$$

$$\frac{dX_2}{dt} = A + X_2^2 Y_2 - BX_2 - X_2 + D_X(X_1 - X_2),$$

$$\frac{dY_1}{dt} = BX_1 - X_1^2 Y_1 + D_Y(Y_2 - Y_1),$$  \quad (7.1)

$$\frac{dY_2}{dt} = BX_2 - X_2^2 Y_2 + D_Y(Y_1 - Y_2).$$

As in (6.12) we have a single time-independent homogeneous solution

$$X_i = A, \qquad Y_i = \frac{B}{A} \, (i = 1, 2). \tag{7.2}$$

We now want to study the stability of this solution in respect to solutions in which the two boxes have different compositions. Let us write

$$\left.\begin{array}{l} X_2 - X_1 = x \, e^{\omega t}, \\ Y_2 - Y_1 = y \, e^{\omega t}. \end{array}\right\} \tag{7.3}$$

We again obtain a second order dispersion equation (see 6.16)

$$\omega^2 + (1 + A^2 + 2D_X + 2D_Y - B)\,\omega + A^2(1 + 2D_X) + 2D_Y + 4D_X\,D_Y - 2BD_Y = 0. \tag{7.4}$$

This leads to the critical value

$$B_c = \frac{1}{2D_Y}(A^2 + 2D_X\,A^2 + 2D_Y + 4D_X\,D_Y). \tag{7.5}$$

The basic difference from (6.19) is that here instability arises only for a finite range of values of the diffusion coefficients. Both for

$$D_X, D_Y \to 0 \tag{7.6}$$

and for

$$D_X, D_Y \to \infty, \tag{7.7}$$

we have

$$B_c \to \infty. \tag{7.8}$$

This is clearly related to the fact that the wavelength is here artificially imposed by the size of the system.

To continue the calculations we make the following choice of numerical values

$$A = 2, \qquad D_X = D_Y = 1, \tag{7.9}$$

therefore

$$B_c = 9. \tag{7.10}$$

After a few simple manipulations the time-independent solutions of (7.1) are shown to satisfy the equations

$$-3X_2^5+30X_2^4-(2B+102)\,X_2^3+(12B+132)\,X_2^2$$
$$-(16B+96)\,X_2+96=0 \qquad (1)$$

$$X_1=4-X_2 \qquad (2)$$

$$Y_2=\frac{1}{X_2}\left(B+3-\frac{6}{X_2}\right) \qquad (3)$$

$$Y_1=Y_2+X_2^2\,Y_2-BX_2 \qquad (4)$$

$$(7.11)$$

We have therefore to solve the fifth order equation (1) and to replace $X_2$ by its value in the other equations. This can of course be done easily using standard methods of numerical calculation.

The results are the following:

a) for $B \leqslant B_c$ equation (1) has only one real root

$$X_2=2. \qquad (7.12)$$

This is the homogeneous solution (see 7.2). All other roots are complex and therefore devoid of physical meaning;

b) for $B > B_c$ equation (1) has three real roots:

1) the homogeneous solution (7.12);
2) an inhomogeneous solution;
3) an inhomogeneous solution obtained by permutation of $X_2$ with $X_1$.

On fig. 3 we have represented the homogeneous and inhomogeneous roots above the critical point.

We have here a striking case of a dissipative structure as defined in the introduction.

In one compartment say 2 the concentration of $Y$ is larger than in the other, at the same time $X$ is smaller. If we refer to the reaction scheme (6.8) we see that the consumption of $B$ will essentially proceed in compartment 1. The system begins to look like a kind of "factory" with characteristic structural and functional order.

The stability of the dissipative structure may in turn be analyzed using equations (7.1). The study of the corresponding dispersion equations indicates that it is stable for

$$9.8 \leqslant B \leqslant 11.2. \qquad (7.13)$$

Outside this range at least one root of the dispersion equation has a positive real part.

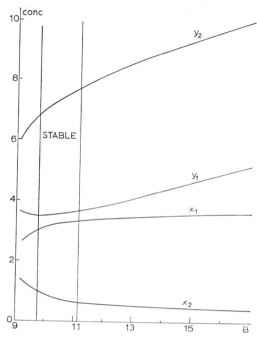

Fig. 3.   Steady States for systems (7.1) beyond the critical point (for the numerical values adopted see text).

We don't know yet what happens there: it seems likely to us that this instability arises from the artificial division of the systems into two homogeneous boxes. The real steady state may correspond to a much more complicated distribution of matter.

Certainly the behavior of matter beyond the instability deserves a much more careful investigation (specially as we are not able to reproduce some of the results given in Turing's paper [11]). However the existence of dissipative structures arising beyond symmetry breaking chemical instabilities seems to us established with a reasonable degree of certitude.

## 8.   Chemical oscillations

The last paragraphs were mainly devoted to the study of structural order. Let us now consider the problem of order in time and specially the problem of oscillations around steady states. We go back to our general evolution criterion (3.4) which we write in the neighborhood of the steady state (see also 2.18)

$$T\frac{d_X P}{dt} = \tfrac{1}{2}\sum_i \left( \delta V_i^* \frac{d\,\delta A_i}{dt} + \delta V_i \frac{d\delta A_i^{\,*}}{dt} \right) \leqslant 0. \tag{8.1}$$

Indeed we now have to consider complex normal modes. We obtain exactly as in (2.18)

$$T\frac{d_X P}{dt} = \omega_r \tfrac{1}{2} \sum_i (\delta V_i{}^* \delta A_i + \delta V_i \delta A_i{}^*)$$
$$+ i\omega_i \tfrac{1}{2} \sum_i (dV_i{}^* \delta A_i - \delta V_i \delta A_i{}^*) \leqslant 0. \tag{8.2}$$

The first term in (8.2) may be associated with the approach to the steady state, the second with the rotation around the steady state. Now it is easy to see that equilibrium stability conditions imply that each of these terms is *separately* negative (see also [27]). Indeed (see 2.3b)

$$\delta A_i = - \sum_{\gamma\gamma'} \nu_{\gamma i} \frac{\partial \mu_\gamma}{\partial n_{\gamma'}} \delta n_{\gamma'},$$

$$\frac{dn_\gamma}{dt} = \sum_i \nu_{\gamma i} V_i, \quad \frac{d\delta n_\gamma}{dt} = \sum_i \nu_{\gamma i} \delta V_i. \tag{8.3}$$

Therefore we obtain

$$\omega_r \sum_i (\delta V_i{}^* \delta A_i + \delta V_i \delta A_i{}^*) = - \omega_r \sum_{\gamma\gamma'} \frac{\partial \mu_\gamma}{\partial n_{\gamma'}} \left( \frac{d\delta n_\gamma^*}{dt} \delta n_{\gamma'} + \frac{d\delta n_\gamma}{dt} \delta n_{\gamma'}^* \right)$$

$$= - \omega_r^2 \sum_{\gamma\gamma'} \frac{\partial \mu_\gamma}{\partial n_{\gamma'}} (\delta n_\gamma^* \delta n_{\gamma'} + \delta n_\gamma \delta n_{\gamma'}^*) \leqslant 0, \tag{8.4}$$

where we used the stability condition (2.11). In the same way it is easy to prove that the second term in (8.2) is separately negative.

We have therefore two inequalities, the first is the same as we discussed in connexion with the stability of steady states, the second gives us the *direction of rotation* around the steady state (see also [3]).

Very little is known about the general conditions for the appearance of rotations. We have already quoted in § 2 our theorem [3] that rotations can appear only far from equilibrium. Also Bak [15] and others have proved that rotations cannot appear for reaction schemes involving only monomolecular reactions.

Our evolution criterion (8.2) shows that rotations are only possible if

$$\sum_i (\delta V_i{}^* \delta A_i - \delta V_i \delta A_i{}^*) \neq 0. \tag{8.5}$$

A necessary and sufficient condition is that the matrix which relates, near the steady state, the rates $\delta v_i$ to the affinities $\delta A_j$, is not symmetric.
Indeed let us write

$$\delta V_i = \sum_j L_{ij} \delta A_j = \tfrac{1}{2} \sum_j L_{(ij)} \delta A_j + \tfrac{1}{2} \sum_j L_{[ij]} \delta A_j, \tag{8.6}$$

with

$$L_{(ij)} = L_{ij} + L_{ji}, \quad L_{[ij]} = L_{ij} - L_{ji}. \tag{8.7}$$

Indeed (8.5) may be written (the $L_{ij}$ are real)

$$\sum_{ij} L_{[ij]} \delta A_j^* \, \delta A_i \tag{8.8}$$

and vanishes when $L_{ij}$ is a symmetric matrix. However it is not easy to use this criterion to discuss the mechanisms which give rise to rotations.

When the system presents pure, undamped rotations the first term in (8.2) has to vanish

$$\sum_i (\delta V_i^* \, \delta A_i + \delta V_i \delta A_i^*) = 0. \tag{8.9}$$

This condition is very similar to that used in the stability theory (see §§ 4, 6): it can only be satisfied if there appear negative contributions to the excess entropy production. Therefore most of our discussions of § 6 apply as well to the problem of rotations. It is necessary to have steps involving autocatalytic or crosscatalytic reactions. This is in complete agreement with the conclusions reached by Chance and his coworkers [16, 17].

In fact all the schemes we discussed in § 6 lead for some range of the constraints involved to rotations. However the inverse is not true: even undamped rotations may occur on the thermodynamic branch and are not necessarily related to instabilities.

To illustrate these conclusions, let us consider two simple examples [18]. We first consider the autocatalytic Lotka-process [14]

$$\left. \begin{array}{c} A+X \underset{k}{\overset{1}{\rightleftharpoons}} 2X \quad (1) \\[2ex] X+Y \underset{k}{\overset{1}{\rightleftharpoons}} 2Y \quad (2) \\[2ex] Y \underset{k}{\overset{1}{\rightleftharpoons}} E \quad (3) \end{array} \right\} \tag{8.10}$$

The difference from (6.20) is that here we have suppressed the "appendix" involving $V'$ which is necessary to obtain a chemical instability. As before we put all *direct* kinetic constants equal to one. However we do not neglect the inverse kinetic constants which we put equal to $k$. In this way we may discuss the behavior of (8.10) for all values of the overall affinity. The concentrations of the initial and final products $A$, $E$ are maintained constant. The overall affinity is

$$\mathscr{A} = \mathscr{A}_1 + \mathscr{A}_2 + \mathscr{A}_3 = RT \log \frac{A}{k^3 E}. \tag{8.11}$$

At thermodynamic equilibrium (total affinity is vanishing)

$$\left(\frac{A}{E}\right)_{eq} = k^3. \tag{8.12}$$

It is easy to show that the time-independent solutions of the kinetic equations satisfy the equations

$$X_0 = 1 + kY_0 - \frac{kE}{Y_0},$$

$$k^3 Y_0^4 + (1 - kA + 2k^2) Y_0^3 + (k - A - kE - 2k^3 E) Y_0^2 + (kEA - 2k^2 E) Y_0 + k^3 E^2 = 0. \quad (8.13)$$

The dispersion equation can also be easily obtained. It is

$$\omega^2 + (Y_0 - X_0 + 2kX_0 + 2kY_0)\,\omega + X_0 + 2kX_0$$
$$- 1 - 2kX_0^2 - 2kY_0 + Y_0 + 4k^2 X_0 Y_0 = 0. \quad (8.14)$$

Numerical calculations have been performed for

$$k = 10^{-2}, \qquad A = 1. \quad (8.15)$$

It has been shown that the thermodynamic solution remains stable *for the whole range of values of the total affinity*. In addition there exists a critical value of the total affinity corresponding to the appearance of two *complex* frequencies in the dispersion equation (8.14)

$$\omega = \omega_r \pm i\omega_i, \quad \text{with} \quad \omega_r < 0. \quad (8.16)$$

The critical value of the affinity is

$$\frac{\mathscr{A}_c}{RT} = 9.2. \quad (8.17)$$

In addition if one now takes $k = 0$ (no inverse reactions i.e. $\mathscr{A} \to \infty$) one finds $\omega_r = 0$; one has then pure, undamped rotations.

Similar conclusions may be reached with the glycolytic scheme proposed by Chance and Higgins [17, 18]. We have investigated the following slightly simplified version of this mechanism

$$
\begin{aligned}
A &\rightleftharpoons C_1 & (1) \\
C_1 + D_1 &\rightleftharpoons D_2 & (2) \\
D_2 &\rightleftharpoons C_2 + D_1 & (3) \\
C_2 + D_3 &\rightleftharpoons D_1 & (4) \\
C_2 &\rightleftharpoons E & (5)
\end{aligned}
\qquad (8.18)
$$

Here $C_1$, $C_2$ are intermediate compounds and $D_1$, $D_2$, $D_3$ some enzymes. The critical step is the reaction (4) between enzyme $D_3$ and the substrate $C_2$ to form the active enzyme $D_1$ which interacts with $C_1$. We have shown that the "thermodynamic solution" is stable over the whole range of the overall affinity $\mathscr{A}$ and moreover there exists a critical value of $\mathscr{A}$ above which rotations begin to occur.

These examples which are in agreement with our thermodynamic theory confirm that time-organization is exactly as is dissipative structure a response to far from equilibrium conditions.

## 9. Fluctuations and instabilities

Let us start with the classical Einstein formula [19] based on the Boltzmann principle: the probability of a fluctuation around an equilibrium state is related to the entropy change $\Delta_i S$ due to the fluctuation by (apart from a normalization constant)

$$P \sim \exp \frac{\Delta_i S}{k}. \tag{9.1}$$

For an isolated system there is no exchange of entropy with the outside world. Therefore

$$\Delta_i S = \Delta S. \tag{9.2}$$

Moreover the entropy is then a maximum and $\Delta S$ begins with a second order term as

$$\Delta S = \tfrac{1}{2}(\delta^2 S) + \dots \tag{9.3}$$

When higher order terms may be neglected (9.1) reduces to the Gaussian form

$$P \sim \exp \frac{1}{2k}(\delta^2 S). \tag{9.4}$$

The entropy change $\delta^2 S$ can easily be calculated (see e.g. [20]). For example in the case of temperature fluctuations one finds

$$\delta^2 S = -\int dV \varrho \frac{C_v}{T^2}(\delta T)^2 < 0. \tag{9.5}$$

Because of the thermal stability condition $C_v > 0$ (see § 1) this expression is definite negative.

As has been indicated elsewhere [20] formula (9.4) may also be used to calculate fluctuations around non-equilibrium steady states. In the case of temperature fluctuations the only difference would be that $T$ in the denominator of (9.5) would refer to the local temperature $T(x)$ which is now space-dependent.

We could say that (9.4) provides us with a *thermodynamic definition of fluctuations*. To be consistent we have of course to show that a fluctuation would regress on the average with an *increase* $-\delta^2 S$ of entropy.

Indeed one may show (see ref. [10]) that

$$\frac{d_i \delta^2 S}{dt} = -\int dV \, \delta J \, \delta X. \tag{9.6}$$

The analogy with (2.2) is striking. The excess entropy production we used to discuss stability theory is precisely the time change of the excess entropy $\delta^2 S$. Therefore in the whole range of stability of the thermodynamic solution of the kinetic equations the fluctuations will regress (for time-independent boundary conditions).

The situation changes drastically beyond an instability. The thermodynamic definition (9.4) of a fluctuation is no more valid around the unstable state. The evolution of the system goes spontaneously to situations which are *less probable* from the point of view of the Einstein formula. This is a completely general conclusion valid for all instabilities whatever their origin.

We may say that after the instability the role of the basic structure and of fluctuations are exchanged: what was a fluctuation before now becomes a stable structure maintained through the flow of energy and matter.

## 10.  Chemical instabilities and biochemical mechanisms

We believe that the discussion presented in the preceding paragraphs establishes firmly the existence of chemical instabilities and consequently of chemical dissipative structures. It may therefore be stated that a theoretical basis exists for the understanding of structural and functional order in chemical systems.

Are these considerations relevant for biological systems? Clearly the answer can only come from biologists. However I would like to present two general arguments in favor of an affirmative answer:

1)  the general picture of a dissipative structure as presented for example in § 7 has a striking similarity with the description of biological order as has emerged from the progress in molecular biology [1];

2)  the chemical mechanisms considered in molecular biology are often precisely of the type considered in §§ 6–7 and may indeed lead to chemical instabilities.

As an example let us discuss in more detail the Chance–Higgins mechanism for oscillatory reactions [13, 16, 17]. This mechanism may be written.

$$
\begin{aligned}
A &\to C_1 & (1) \\
C_1 + D_1 &\to D_2 & (2) \\
D_2 &\to C_2 + D_1 & (3) \\
D_3 + C_2 &\rightleftharpoons D_1 & (4) \\
C_2 + V &\to V' & (5) \\
V' &\to E + V & (6)
\end{aligned}
\qquad (10.1)
$$

As in (8.18) $C_1$, $C_2$, are intermediate compounds, $D_1$, $D_2$, $D_3$ enzymes, $A$ the initial compound and $E$ the final compound. The concentration of $D_3$ is maintained constant. Here we have two supplementary intermediate compounds $V$, $V'$. In the specific case studied by Chance and al. [16, 17] $A$ is glucose and $E$ glyceraldehyde phosphate.

As Lefever has shown (the calculation will be given in ref. [13]), this scheme does not lead to chemical instabilities. This can be understood as follows. We have two resultant reactions

$$\begin{cases} A \rightarrow E & (1) \\ D_3 + C_2 \rightleftharpoons D_1 & (2) \end{cases} \qquad (10.2)$$

While the reaction (1) is taking place in non-equilibrium conditions (its affinity may be taken as infinite), the reaction (2) may be considered at equilibrium as the steady state conditions lead to

$$V_4 = 0$$

for the rate of step (4) of the scheme (10.1). Now in a similar situation Nicolis and I have shown [12] that this prevents the occurrence of chemical instability (which as repeatedly emphasized is a *far* from equilibrium phenomenon).

If we now add to (10.1) the step

$$D_1 \rightarrow D_4,$$

where $D_4$ is some other form of the enzyme $D_1$, reaction (2) goes out of equilibrium and chemical *instability may occur*. However we have not yet investigated what the state of the system beyond the chemical instability is.

Anyway we see that the idea of chemical instabilities is certainly compatible with the types of chemical reactions presently investigated in biochemical processes.

## 11.  General discussion

It is certainly tempting to define living systems as open systems presenting a dissipative structure due to chemical instabilities. This definition is clearly incomplete as much more precise conditions on the type of chemical reactions involved would be necessary to explain essential features of life such as replication phenomena.

Incomplete as it is, this definitions has two advantages:

1)  it may lead to experimental investigations. Indeed no chemical instability has to our knowledge been observed *experimentally* till now;

2)  it puts together various ideas and point of views which may be found in the literature and which at first seem contradictory. Life no longer appears as an island of resistance against the second law of thermodynamics; as the work of some Maxwell demons, but on the contrary as following the laws of physics appropriate to specific kinetic laws and to far from equilibrium conditions. These specific kinetic laws permit the flow of energy and matter to build and maintain functional and structural order.

Some authors have insisted that life follows the second law of thermodynamics [21], others have insisted that the laws of biology have a somewhat special status. Elsasser has coined the term biotonic for a causal relationship which involves an increase of "information".

Both views find a justification in our approach: indeed the entropy production remains positive both before and after the instability (9.4) but the relation between kinetics and thermodynamics changes radically as discussed in § 9.

In this paper we used a physico-chemical language. We spoke about catalytic reactions. Others may prefer to speak about negative feedback, auto-regulation and so on [23]. It will therefore be feasible to closely link our approach to information theory.

Again some authors insist on the chemical aspects of life [24, 25]. Clearly our conclusion is not in contradiction with this as precise chemical conditions are to be met to produce the chemical instabilities. Also active transport appears here as a consequence of dissipative structure. All symmetry requirements valid in the linear range of irreversible thermodynamics fail here. The maintainance of inhomogeneities ipso facto means active transport through chemical reaction rates.

But the most exciting perspective is the possibility of combining the unity of matter with a clear distinction of what is life and what not as these two states of matter would be separated by an instability corresponding to a critical affinity.

Therefore there appears at least some hope of reconciling the basic duality of our experience with the unity of the laws of nature.

I am deeply indebted to Prof. P. Glansdorff. It is in collaboration with him that the thermodynamic concepts used in this paper have been developed. I am also much indebted to Dr. G. Nicolis and M. R. Lefever for continuous help and discussion.

This work was sponsored in part by the "Fonds National de la Recherche Collective" (Belgium), by the Air Force Office of Scientific Research under Grant AF EOAR 64–52 through the European Office of Aerospace Research (OAR). United States Air Force and by Dupont de Nemours International (Geneva).

# References

[1]  A. Lwoff, *Biological Order*, (M.I.T. press, 1965).
[2]  A. Szent-Györgyi, Introduction to *The Structure and Properties of Biomolecules and Biological Systems*, Ed. J. Duchesne, (Interscience, 1964).
[3]  *see* e.g. I. Prigogine, *Introduction to Thermodynamics of Irreversible Processes*, 2d ed., (Interscience, 1961).
[4]  S. Chandrasekhar, *Hydrodynamic and Hydromagnetic Stability*, (Oxford Un. Press, 1961).
[5]  I. Prigogine and R. Defay, *Chemical Thermodynamics*, English translation by D. H. Everett, (Longmans Green, 1954): chp XV.
[6]  S. de Groot and P. Mazur, *Non-equilibrium Thermodynamics*, (North-Holland, 1962).
[7]  P. Glansdorff and I. Prigogine, *Physica* **20** (1954) 773.
[8]  P. Glansdorff and I. Prigogine, *Physica* **30** (1964) 351.
[9]  *Non-equilibrium Thermodynamics, Variational Techniques and Stability*, Eds. R. J. Donnely, R. Herman and I. Prigogine, (Un. of Chicago Press, 1965).
[10]  I. Prigogine and P. Glansdorff, *Physica*, to appear.
[11]  A. M. Turing, *Phil. Trans. Roy. Soc.* (London) **B 237** (1952) 37.

[12]  I. Prigogine and G. Nicolis, *J. Chem. Phys.* **46** (1967) 3542.
[13]  I. Prigogine and R. Lefever, *J. Chem. Phys.* **48** (1968) 1695.
[14]  A. J. Lotka, *J. Am. Chem. Soc.* **42** (1920) 1595.
[15]  T. Bak, *Contribution to the Theory of Chemical Kinetics*, (Benjamin, 1963), cf. 3.
[16]  B. Chance, A. Gosh, J. Higgins and P. K. Maitra, *Ann. N.Y. Acad. Sc.* **115** (1964) 1010.
[17]  J. Higgins, *Proc. Nat. Acad. Sc.* **51** (1964) 989.
[18]  R. Lefever, G. Nicolis and I. Prigogine, *J. Chem. Phys.* **47** (1967) 1045.
[19]  *see* for example L. Landau and E. Lifshitz, *Statistical Physics*, (Addison Wesley, 1958).
[20]  I. Prigogine and P. Glansdorff, *Physica* **31** (1965) 1242.
[21]  J. A. V. Butler, *Nature* **158** (1946) 153.
[22]  W. M. Elsasser, *The Physical Foundation of Biology*, (Pergamon, 1958).
[23]  B. C. Goodwin, *Temporal Organization in Cells*, (Academic Press, 1963).
[24]  M. Calvin, *Chemical Evolution*, reprinted in Interstellar Communication, Ed. A. G. W. Cameroun, (Benjamin, 1963).
[25]  A. I. Oparin: *The Origin of Life on the Earth*, third ed., (Oliver & Boyd, 1957).
[26]  W. Strieder, *Ac. Roy. Belg., Bull. Cl. Sci.* **50** (1965) 318.

**Additional references**

More details about the scheme I in § 6 may be found in Lefever, R., *Acad. Roy. Belg., Bull. Cl. Sci.* **54** (1968) 712 and *J. Chem. Phys.*, to appear.

The sufficient conditions for chemical instabilities are discussed in papers by Perdang, *Acad. Roy. Belg., Bull. Cl. Sci.*, (to appear) and B. Edelstein (submitted to *J. Theor. Biol.*).

The theory of fluctuations in non-equilibrium steady states has been developped by G. Nicolis and A. Babloyantz (submitted to *J. Chem. Phys.*).

For interesting biological applications of the ideas, see B. Lavenda and G. Nicolis (submitted to *J.A.C.S.*) and B. Lavenda, *Acad. Roy. Belg., Bull. Cl. Sci.*, (to appear).

*Theoretical Physics and Biology,* © 1969, *North-Holland Publ. Co., Amsterdam*

# DISCUSSIONS

E. G. D. COHEN: Dr. Fröhlich, in the preprint of your paper as well as in your lecture quantum mechanical concepts as long range order in super-conductors and superfluids seem to play an essential role.

However, as far as I can make out, the example you gave does not directly involve any quantum mechanical concepts at all and is entirely classical. This raises the question, whether classical rather than quantum mechanical analogies may not be useful for understanding some aspects of f.i. cooperative phenomena of cells.

M. KATCHALSKY: The case discussed by Dr. Fröhlich has a classical counterpart in the organization of small molecules in the formation of micelles, well known from colloid chemical studies. Recent investigations mainly by Luzatti have shown that lipid molecules in aqueous solution form two types of micelles (a) bilayer sheets or (b) infinite cylindrical structures. My colleague, Dr. Adrian Parsegian has shown that the stability of these micelles could be fully explained by the electrostatic double layer interaction (of the type evaluated by us for polyelectrolytes) and by short range forces determining the conventional surface tension. This field of colloid chemistry seems therefore to be adequately covered by a classical treatment.

I would like to ask Dr. Fröhlich whether he sees in the quantum mechanical wave treatment the possibility of predicting new phenomena which are not understood by the classical approach. Since the biologists are very interested in long range interactions, which govern the behaviour of lipoprotein membranes, such a prediction would be very valuable.

H. FRÖHLICH: Mon impression est que c'est là un cas classique, dont j'ai parlé moi-même. Bien entendu, les forces que vous pouvez avoir dans telle ou telle succession peuvent être traitées de la manière que je vous ai dite. Plus loin, vous avez de toute façon la mécanique quantique, dans ce domaine.

Je dois dire que j'adhère à tout ce dont vous avez parlé. Cependant, il y a certaines limites statistiques. J'ai moi-même puisé dans un des principes du M. Prigogine, à savoir qu'à travers le système (nous pourrons en reparler), il se pourrait qu'il y ait, en effet, . . .

I. SEGAL: What is the biological counterpart to the quanta which should be associated with this model in accordance with the general principles of quantum mechanics? Perhaps it is due to my ignorance of biology, but I don't understand what the biological counterpart of the photon is proposed to be.

H. FRÖHLICH: I am not introducing single photons in connection with the example I gave. What I proposed here is, instead of taking phonons, to take what sometimes is called "plasmons". But the individual quantum would be negligible if there is a similarity with lasers or with coherent sound waves; the longitudinal plasma would oscillate essentially as classically.

E. G. D. COHEN: Do you believe that it is *always* useful to consider biological systems from a quantum mechanical point of view? Or would you be prepared to say that in certain cases — where f.i. a large number of entities is involved — it might be more simple and more adequate to forget the quantum mechanical basis and use a classical model or analogy for the understanding of biological phenomena?

H. FRÖHLICH: La situation est compliquée. Depuis des années, les gens qui s'occupent des "lasers" ont affaire à la mécanique quantique, que d'autres ont pu laisser de côté. Aussi, vous pouvez oublier ce côté et vous pouvez fort bien étudier les choses de manière classique. Mais on ne peut pas toujours le faire!
De toute façon, tout cela n'intéresse pas les biologistes.

L. ROSENFELD: There is a simple criterium to distinguish classical from quantal effects: if in the expression obtained for an observable quantity Planck's constant appears explicitly, one may say that it represents a quantal effect. If, however, the observable quantity does not depend on Planck's constant, it means that one is in the classical domain.

H. FRÖHLICH: I am afraid we are getting off the biologists. But there is a very interesting example I shall give you privately where the $h$ does not appear.

H. HAKEN: Let us consider an arrangement of atoms which are excited and are starting to emit spontaneously light. In general these spontaneously emitted light waves are not in phase (e.g. they are incoherent). This emission process is a purely quantum mechanical effect. It may happen that a simple photon hits another excited atom and causes the atom to recombine emitting a second photon by the process of stimulated emission. This process is still a quantum mechanical effect. It may be continued several times so that a macroscopic number of photons is generated. Now the question is: is this phenomenon still a quantum mechanical one or does it already belong to classical physics. In the mathematical treatment one starts with the description of the light by a photon creation operator which contains all the quantum mechanical fluctuations. But if the photon number increases one indeed ends up under certain conditions with a classical lightfield amplitude. Thus one starts with the random process of spontaneous emission and ends up with a new structure, a new organization on a microscopic scale.

H. L. LONGUET-HIGGINS: I should like to raise a point which Dr. Fröhlich and I discussed during the coffee break. In the course of his talk Dr. Fröhlich suggested that electrical resonance between cells might trigger cell division. But a cell can divide when there are no other cells around, and this process is a very complicated business in which the individual parts form a highly organized whole, rather like a cocktail party at which the guests divide themselves into two groups. The relation between this process and the double layer oscillations proposed by Dr. Fröhlich is quite unclear to me. Perhaps Dr. Fröhlich would care to comment on this point?

H. FRÖHLICH: First point. I did say it stimulates, not causes. But it is not a stimulation through an oscillation which would be much too fast for something mechanical to follow. It is stressed through deformation.

Biological systems are highly organised. There should be resonances with certain molecular oscillations in some way or other.

H. C. LONGUET-HIGGINS: The other point which I wanted to discuss is the question of the resonance of the individual molecules inside a cell. There are various degrees of freedom that one might consider — translational, rotational, torsional and electronic. In metals there are electronic oscillations which are quite distinct from individual particle motions, but it does seem that most molecules of biological importance, particularly the colourless ones which make up the bulk of most cells, are electronically quite normal. It does seem rather far-fetched to suggest that such molecules have plasma oscillations in the microwave region, or that such oscillations could arise spontaneously, in view of the viscous damping due to the neighbouring molecules. Such questions have been the subject of heated discussion among physical chemists, in particular whether proteins and nucleic acids have electronic excitations which are important biologically. My own view is that the case is a very dubious one at the moment.

G. CARERI: I believe one can extend the Fröhlich model to the intracellular long range interactions, which perhaps are responsible for the unity of the cell as a whole. As a matter of fact inside the cell one can identify in the proteins the polarizable entities which can interact in the collective longitudinal mode according to the Fröhlich model. Most likely the oscillators can be found in the hydrogen bonded NH...O chains which should exhibit some longitudinal modes in the very far infrared (say 10 to 50 $cm^{-1}$). A detailed theoretical calculation is lacking, but one has good reason to expect these modes in that range, following the same arguments which usually explain hypocromism in a chain of coupled oscillators, because the stretching frequency of the single hydrogen bond is at about 200 $cm^{-1}$. So far the polarizability of the biological matter has not been studied in this region of the electromagnetic spectrum, because of the obvious difficulties; experiments of this kind are now in progress in our laboratory. Therefore as yet there are not experimental data available to prove this possibility for intracellular interactions, but let me emphasize that it is precisely in this range of e.m. radiation that one must look for collective effects. This is because at lower frequencies (namely in the microwaves) the continuous Debye relaxation of free water molecules displays its dissipative absorption, while at higher frequencies there are many absorption peaks due to the molecular modes which would trap the collective modes. It is gratifying that the Fröhlich estimated membrane frequency also falls in this region.

The need for collective intracellular effects has been already suggested on a different molecular basis by Elsasser (J. Theor. Biology 1962) to create in the organism a high degree of physical impredictibility. The way they are introduced here has the advantage of a clear similarity in the pattern or the collective behaviour both inside the living cell and among the cells in the organism, displaying just that hyerarchy character between the ordered units and subunits as expected in biology (see for instance, P. Weiss, Rev. Mod. Phys. 1959). However, it must be remembered that up to today it has not been proved that there are such long range effects between the cell subunits.

L. ROSENFELD: I notice that we are deviating from specific questions to questions of principle. So I should like to suggest that, unless somebody is really very impatient to ask Dr. Fröhlich a specific question, we now postpone the discussion about the Fröhlich paper and pass to the preliminary discussion of Dr. Prigogine's paper.

K. MENDELSSOHN: Dr. Fröhlich has mentioned the superfluids in connection with more general questions of the solid state. Having regard to the high degree of orderliness which is a basic feature of living matter, the superfluids present an order pattern which is rather unusual. These phenomena, superfluidity in liquid helium and superconductivity, occur at very low absolute temperatures. This in itself indicates that the entropy is low, i.e. the degree of order high. Moreover, measurement of the entropy reveals that at the onset of these states a further sharp decrease of entropy takes place.

The usual feature which we associate with the cooling of matter is the phenomenon of condensation in which the drop in entropy is expressed by the fact that the atoms take up an orderly pattern in a crystal lattice. This is an order in respect of *position* and, indeed, ordering in position space is the only form of an orderly arrangement which we can easily comprehend.

However, no such thing happens at the onset of superfluidity; helium remains a liquid and the electrons in a superconductive metal remain free. Orderliness, which we know exists, must be manifest in another way than by a pattern in position. The obvious place to look for it, is not the space of positions but that of momenta. Indeed, the unusual properties which we observe in the superfluids are all connected with the phenomena of motion.

We know not nearly enough about the superfluids to be able to treat the onset in the same way as the formation of a crystal. However, most people will agree that in the superfluids we witness a new type of aggregation of matter in which some sort of condensation according to the momentum co-ordinates seems to take place. As I just said, the relevant phenomena are given by a frictionless form of motion and this velocity structure is evidently spread over the whole volume of the superfluid.

We thus know structures which are ordered in position, the crystals, and structures which are ordered in momenta, the superfluids. Inevitably the question arises whether we can conceive of structures which show *both* types of order simultaneously. From our knowledge of the solid state and of the superfluids we can hazard some conclusions as to the properties of such structures. First of all their statistical order must be very high, that is they must be statistically extremely improbable. Secondly they must have form and shape like a solid. Thirdly, they must have the property of superfluids that is, what happens to one part of the structure will affect the whole. To give an example: if you have a metal ring with a supercurrent flowing through it, local quenching of super-conductivity at any part of the ring will destroy the current everywhere in the ring.

It now appears, when we summarize all these features, that our hypothetical structures which are ordered in position *and* momentum possess features which we associate with the structures of living matter.

H. FRÖHLICH: I agree. What there is to be said I think is equivalent to Dr. Haken's remarks. The remark I wanted to make on Prigogine's paper concerns the use of the entropy concept. Now what do we know about life? One thing that it is statistically improbable, but the structures of life are certainly an improbable structure. But this does not mean that the entropy concept helps us entirely. The entropy concept may not reflect the correct improbability which we need.

Let's see. Consider a mouse or a flea or a protein molecule. Then it is made of a certain number of atoms. Now you can actually arrange the same number of atoms of the mouse and an equally improbable structure and an infinite number of equally improbable structures only the mouse will not live. So you see somehow the entropy concept is insufficient. The concept which I needed I think is one of a selective organization referring to few degrees of freedom only, and hence not relevant for entropy.

I. PRIGOGINE: Je crois que vous me demandez trop. La notion de structure dissipa-tive n'est sûrement pas suffisante à elle seule pour expliquer la vie. D'autres éléments fondamentaux doivent intervenir pour expliquer par exemple la duplica-tion. Le point important que je voulais souligner c'est que la physico-chimie permet actuellement de comprendre l'apparition de situations hautement improbables du point de vue entropie.

Quant à la superfluidité qui a été évoquée à plusieurs reprises, je crois que l'analogie avec les phénomènes vitaux est fort trompeuse. La caractéristique des phénomènes à basse température telle que superfluidité ou superconductivité est précisément l'absence ou la faible importance de phénomènes de dissipation.

Au contraire les structures de la biologie moléculaire exigent un flux d'énergie et de matières continuelles. Ce sont des structures de non-équilibre et je ne crois dès lors pas que l'analogie avec les phénomènes à basse température nous apprendra grand chose.


H. FRÖHLICH: I will answer to Dr. Mendelssohn that this is the crucial point. The point which I made is that the deviation from thermal equilibrium refers to a very few degrees of freedom only. But it's the essential point, nevertheless, at least in my thinking.

Secondly, there is no suggestion of saying that you have in our system the same type of stabilisation as you have in superconductors and in superfluids. Quite on the contrary, as I mentioned explicitly, the stabilisation there is either energetical or quantum mechanical. Here something else is required which requires a supply of energy and secondly it requires non-linear effects.

Nevertheless the deeper structure of this correlation has very great similarity in both cases. Dr. Haken has also mentioned in detail the case of lasers which also show a similar type of correlation.


J. POLONSKY: J'aimerais apporter ici quelques remarques d'un ingénieur en électronique sur le "many body problem" évoqué par M. Fröhlich, tel qu'il se pose pour des systèmes hautement organisés, artificiels ou biologiques.

Il est connu que tout système organisé doit être alimenté en régime dynamique par une source d'énergie externe capable de compenser l'accroissement d'entropie totale produit dans le système au cours de son fonctionnement. La thermodynamique des systèmes non en équilibre, dont nous a entretenu M. Prigogine, permet d'étudier les échanges d'énergie ainsi que les contraintes thermodynamiques extensives et intensives qui interviennent au cours de ces échanges.

Il est connu, par ailleurs, que dans un système hautement organisé, le réseau des contraintes à caractère informationnel prend le pas sur le réseau d'énergie et sur les contraintes purement thermodynamiques. Le rôle prioritaire joué par les contraintes informationnelles provient du fait que celles-ci, avec une dépense d'énergie minime, exercent des asservissements hiérarchisés dans le système et déterminent le mode fonctionnel du système organisé. La structure du réseau informationnel dans un dispositif cybernétique (comme par exemple le pilotage automatique d'un avion ou d'un satellite) ne peut pas être étudiée par la thermodynamique. Il est un fait qu'il nous manque une véritable science des systèmes organisés. En attendant que celle-ci voie le jour, il me semble utile, pour l'objet de notre discussion, d'exposer brièvement la méthode d'approche utilisée par l'ingénieur dans ce domaine, d'examiner dans quelle mesure celle-ci peut être étendue à la biologie et quel pourrait être le rôle des concepts quantiques dans cette approche.

L'ingénieur procède en deux étapes:

1.  Au cours du projet d'un dispositif cybernétique artificiel, il opère dans l'ordre suivant:

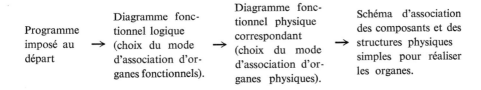

Programme imposé au départ → Diagramme fonctionnel logique (choix du mode d'association d'organes fonctionnels). → Diagramme fonctionnel physique correspondant (choix du mode d'association d'organes physiques). → Schéma d'association des composants et des structures physiques simples pour réaliser les organes.

Dans cette étude, les deux premières étapes sont d'ordre logique, tandis que les deux dernières se situent sur un plan physique.

2.  Pendant la phase de réalisation de l'appareil, l'ingénieur opère en sens inverse.

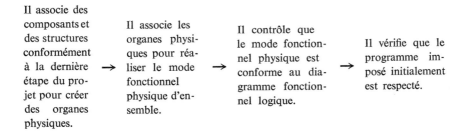

Il associe des composants et des structures conformément à la dernière étape du projet pour créer des organes physiques. → Il associe les organes physiques pour réaliser le mode fonctionnel physique d'ensemble. → Il contrôle que le mode fonctionnel physique est conforme au diagramme fonctionnel logique. → Il vérifie que le programme imposé initialement est respecté.

Ce dernier diagramme pourrait convenir à l'étude de la formation d'un système biologique en substituant, dans les associations, la finalité par des lois physiques. Néanmoins, en électronique, le modèle fonctionnel précède le choix de la structure physique correspondante tandis qu'en biologie, la structure physique précède et crée ipso facto la fonction biologique. La philosophie qui se dégage de cette méthode d'approche pourrait se résumer comme suit:

Toute étape de formation ou de complexification d'un système organisé est caractérisée, sur le plan analytique, par l'élimination sélective des degrés de liberté ou l'introduction dans le système global d'un nombre croissant de corrélations sélectives sous forme de contraintes des probabilités régies par des lois stochastiques. La distribution spécifique des contraintes (et par voie de conséquence le mode fonctionnel du système) est réalisée grâce au schéma particulier d'association des composants et des sous-ensembles. Ceci reste vrai en électronique comme en biologie.

Un système organisé, placé dans une telle optique, permet, me semble-t-il, de mieux apprécier le rôle qui pourrait être dévolu aux concepts quantiques en biologie. En effet, dans la mesure où l'on doit tenir compte des lois quantiques dans un système, celles-ci introduisent, d'une façon naturelle, une distribution fine des contraintes des probabilités, chose inconnue en physique classique. La discontinuité d'états et des transitions, les règles de sélection et d'exclusion, imposées par les concepts quantiques, sont équivalentes, du point de vue de la théorie de Shannon, à des matrices des probabilités d'états et des transitions d'états dans un processus stochastique. Comme il est connu, par ailleurs, que les contraintes quantiques sont d'autant plus sévères que les dimensions du système sont plus réduites et que l'échelle d'organisation biologique est à base moléculaire, il parait logique que les contraintes quantiques jouent un rôle très important en cybernétique biologique. Le langage de l'information et du codage utilisé dans les communications entre molécules et macromolécules biologiques pourrait ainsi être exprimé en termes quantiques. De même l'ensemble des corrélations d'origine quantique qui régissent les interactions entre structures moléculaires jouent, sur le plan fonctionnel, le rôle de redondance prise au sens de la théorie de l'information, c'est-à-dire d'une information fixe à priori.

J'espère pouvoir, au cours de la discussion de cet après-midi, préciser davantage le rôle cybernétique des contraintes quantiques en biologie moléculaire en me référant, en particulier, au concept de cellules dans l'extension en phase pris au sens de la mécanique statistique.

H. C. LONGUET–HIGGINS: I think I now understand Dr. Fröhlich's hypothesis. First, if I am not mistaken, one can state it in classical terms. Secondly, he postulates the excitability of certain oscillations, and it is the oscillatory character that he is stressing. The situation seems mathematically similar to the blowing of a flute, where one has a non-linear coupling between the jet of air and what happens inside the flute; by blowing one excites the principal vibration of the air in the flute. The oscillations which Dr. Fröhlich is particularly thinking about are electrical oscillations associated with the double membrane layer. The hypothesis is really that the non-linear coupling between the metabolic dissipation and these particular modes is sufficient to keep them in a state quite far from equilibrium; this is the essence of the hypothesis. The quantum-mechanical aspect of the matter does not seriously disturb the description which I have given—or does it? I ask you this question to make sure whether I have fully understood you or not.

P. MAZUR: I assume that within the framework of your theory you obtain a formal expression for the length which characterizes the structure that arises as a consequence of the stability conditions within a chemical system of the type you discuss. Do you have any estimate of the order of magnitude of this length. I believe this would tell us something about the implications of such a theory for biological systems.

J. Duchesne: Si j'ai bien compris M. Prigogine, sa théorie suppose que le métabolisme est la vie elle-même. Or, on sait bien qu'à très basse température, au voisinage du zéro absolu, et dans un état de dessèchement profond, les cellules vivantes ne sont pas détruites, bien que leur métabolisme soit supprimé. Je crois, par conséquent, que, pour saisir complètement les phénomènes vitaux, il faudrait notamment pouvoir identifier les structures limites à partir desquelles la vie ne peut se rétablir. Ceci permettrait sans doute d'élargir notre conception du phénomène vital et en même temps d'englober cet aspect des choses dans les admirables inférences de Prigogine.

I. Prigogine: Pour répondre à la question de M. Duchesne, 'survivre n'est pas vivre'. Le problème que j'ai discuté est la création de structures, non le maintien de ces structures, ce qui est un tout autre problème.

W. M. Elsasser: I would like to make a comment on the memoir of Dr. Prigogine. He has two types of systems, stable and unstable. The latter have positive exponentials so that the instability is more or less violent. I would like to point out that there exists a third category of instability which is much more general and which for some reasons biochemists have not considered. I have searched for many years but I have not been able to handle the mathematical tools.

That is the case of systems which have stereo asymmetry without an inversion appearing. These systems have the ability of going off slowly from equilibrium. As I happened to be a geophysicist for many years, I am very familiar with these systems because all of the hydrodynamics of geophysics is dominated by asymmetrical forces (the Corioly forces). I believe it would be worthwhile to look into these systems first.

Pasteur has said something very profound when he said that the stereoasymmetry of molecules is a fundamental property of living matter. It leads immediately to problems of probability which are asymmetrical and this will lead to deviation from a stable state of equilibrium. I believe that you now have the tools to handle this. In this beautiful presentation you have given us the mathematical tools so that perhaps one can now begin to study this third intermediate category. I would not be surprised that it is perhaps as important for biology as the kind of instability which you have considered.

# CONCEPTS THEORIQUES

*2ème séance*

PRÉSIDENT S. L. SOBOLEV

J. D. COWAN

Some Remarks on Neurocybernetics

W. M. ELSASSER

What is Theoretical Biology?

General discussions

*Theoretical Physics and Biology*, © 1969, *North-Holland Publ. Co., Amsterdam*

# SOME REMARKS ON NEUROCYBERNETICS

JACK D. COWAN

*Committee on Mathematical Biology, The University of Chicago, U.S.A.*

The last quarter of a century has seen the emergence and development of cybernetics, the science of information and control. As a result, a new language has evolved out of the continued efforts of scientists and engineers to make precise such concepts as "communication channel", "automaton", and "control system". This language has been widely used in those sciences which have to deal with the interaction of animals and machines with their environments. One science that has drawn heavily on this language is neurology. Cybernetic concepts have been used in many different ways to investigate how the brains of men and animals might work.

Perhaps the earliest example of the use of this language in neurology is to be found in the 1943 publication by W. S. McCulloch and W. H. Pitts of "A logical calculus of the ideas immanent in nervous activity" [1]. In this paper, the concept "formal neuron" was introduced, essentially an abstraction from then current details of neuronal operation. These formal neurons operated upon and emitted, at specified times, binary "all-or-none" signals. Their junctions, called "synapses", were either excitatory or else totally inhibitory. They functioned by computing the algebraic sum of the values of their inputs, 1 and 0, subsequently emitting an output signal if, and only if, this sum exceeded a certain specified threshold. Formal neurons could be made to represent the elementary operations of two-valued logic, and *a fortiori*, formal neuronal networks to represent complicated logical formulas [2]. To the extend to which the concept "formal neuronal network" symbolizes the process whereby brains respond to and represent stimuli, the McGulloch–Pitts theory is similar to that of K. J. Craik's [3] regarding the nature and function of neuronal networks. A corollary of the McCulloch–Pitts theorem is also of interest: formal neuronal networks plus "receptors" and "effectors" are equivalent to Turing machines. A Turing [4] machine is itself a formalization of, and an abstraction from, the processes underlying computing. The equivalence of the two concepts, therefore, led immediately to the idea that in at least some aspects of signal processing, brains and computers are similar. N. Wiener, in his now classic book "Cybernetics" [5] developed this analogy, incorporating chapters on computing machines and the nervous system, on gestalts and universals and on cybernetics and psychopathology.

This research was sponsored in part by the Physics Branch, Office of Naval Research, Washington D.C., under Contract No. F61052 67C 0061.

Any direct analogy between formal neuronal networks and brains requires the existence in brains of specific circuits, so that there has to be a process of selection of nerve-cells and of the pattern of interconnection, the anatomy or the "wiring-diagram", and a rejection of all other possibilities. However, ablation studies on the brains of many animals [6] apparently indicated that such complex signal processing tasks as visual integration were independent of the specific details of the wiring diagrams of the visual cortex. Even extensive destruction of this tissue did not produce marked disintegration of function, although those activities that persisted were somewhat retrenched. This suggested that perhaps the circuits were redundant throughout areas of nervous tissue and that only gross parameters of the tissue such as the mean number of cells and their interconnections were reliable measures of the tissues' functioning. Wiener [5] suggested that nervous tissue might be similar to large telephone exchanges in containing redundant cells and interconnections, and that messages might be simultaneously transmitted from area to area along many distinct pathways, and might be repeated several times along each pathway, the final decision concerning the validity of signals reaching any area of nervous tissue being determined by some kind of voting procedure. J. von Neumann [7] in the paper, "Probabilistic logics and the synthesis of reliable organisms from unreliable components" gave the first proof of the existence of designs for the construction of neuronal networks which might survive extensive malfunctions of components, or failures of individual formal neurons and of errors in and damage to their connections. Von Neumann's designs utilized the replication of many individual circuits, the simultaneous transmission of messages by many circuits, and "majority-voting" circuits to ensure the overall reliability of signal processing and transmission. Von Neumann was not satisfied with the results. He considered his treatment of error to be rather *ad hoc*, and he suggested that error should be the subject of a thermodynamical theory, as C. E. Shannon had treated the concept of information [8]. He was also aware that formal neuronal networks were digital in their mode of operation, whereas those comprising brains were not, despite their use of pulses in some operating modes [9].

S. Winograd and the author applied Shannon's theorem concerning the reliable transmission of messages through a noisy communication channel to include computation, and so provided part of von Neumann's suggested thermodynamical theory of error for formal neuronal networks [10]. The resulting design for the construction of formal neuronal networks that function correctly in spite of malfunctions of individual neurons, or of errors in and of damage to wiring diagrams, differs in interesting ways from von Neumann's. The ability of formal neuronal networks to function correctly in spite of viccissitudes depends on their composition by anastomotically redundant circuits. In von Neumann's prospectus, this kind of redundancy was sought by incorporating many copies of the one and only one circuit necessary for the representation of the specified functions which are intended to determine the behaviour of the network. In the Winograd–Cowan theory, however, the redundancy is obtained, not by multiplication of identical circuits, but by replacing the specified functions by a greater variety of more complicated functions which require circuits containing many more and richer formal neurons and interconnections than the circuits required for the original specified functions. The key to theory is that the rules governing this replacement constitute error-correcting codes (*see* [11]). The network that results from such an encoding is immune to many kinds of error, i.e., it is largely "error-insensitive". The degree of error-insensitivity obtained depends upon the complexity of the requisite behaviour, on the frequency of errors, and on the level of redundancy introduced by whatever code is used. Because the code operates on functions and not on circuits, the redundancy obtained is functional, i.e., more functions are represented in the encoded network than in its precursor. Because of the nature of error-correcting codes, any one specified function appears in many of these encoded functions, and any one encoded function is essentially a different mixture of many of the specified functions. It is the multiple representation of a multiplicity of specified functions which leads to error-insensitive operation, rather than the mere replication of circuits. In short, it is the diversity of the encoded function that is computed by each individual formal neuron comprising the encoded network which leads to the efficiency of design. So the number of formal neurons required to realize any given mode of behaviour at some level of error-insensitivity, despite probability of malfunction of individual formal neurons and of error in the wiring, is ultimately as small as Shannon's theorem indicates is possible. Naturally, the formal neurons required to represent encoded functions are much more complicated than those which would be required to represent only the original specified functions. The application of Shannon's theorem thus requires that the more complicated formal neurons be no less reliable than the simpler ones, a requirement which is equivalent to assuming that the extra "hardware" required for coding is completely error-insensitive, as in Shannon's theorem. The Winograd–Cowan theorem is, in fact, like von Neumann's theorem, an existence theorem; but it differs from it in requiring a minimal number of very complicated circuits and formal neurons rather than a large number of simple circuits of threshold formal neurons. Our theory is the other extreme to von Neumann's theory wherein complication is minimized at the cost of increased replication, in that redundancy is minimized at the cost of increased complication, to attain some requisite level of error-insensitivity. Suitable combinations of the two techniques, functional coding followed by the replication of the resulting circuits, plus a randomization of the interconnections between them, lead to efficient and practical error-insensitive networks.

However, these networks can persist only for lifetimes that are limited by the error and failure rates of their components. There is an ageing effect, so that the reliability of function of these networks degrades under the cumulating effects of uncorrected failures. W. H. Pierce [12] has shown how "adaptive" networks may be designed whose lifetimes are substantially longer than those of the above networks. A basic defect of the techniques outlined, in which majority voting effectively controls errors, is that, since those inputs to a given formal neuron which issue from a failed one will be permanently in error, a consistently reliable minority may be outvoted by a consistently unreliable majority. Such a limitation may be overcome by using more complicated formal neurons whose inputs are weighted according to their reliabilities. This requires computation by the recipient formal neuron of these reliabilities. Pierce has proved that such computations can be carried out. If input errors are statistically independent, the input weights $\alpha_i$ can be selected so that the output of the formal neuron is the digit most likely to be correct, on the assumption that what is required of the formal neuron is the representation of the majority function [13]. One variant of this is particularly interesting. In this, if $p_i$ is the number of coincidences and $q_i$ the number of disagreements between the $i$th input to, and the output from a given formal neuron in a cycle of $m$ operations, then the selections $\alpha_i = \log (p_i/q_i)$ lead to the computation of the most likely estimate of the reliabilities of the inputs for the representation of the majority function.

There are evidently good grounds for asserting that reliable networks of competent formal neurons can be constructed, using the techniques of functional coding followed by replication of the resulting circuits, together with feedback controlled selection of vote-weights, which would be efficient and long-lasting compared with their components. However, the question remains of how such networks are to be constructed. The existence of specific circuits in a network requires the selection of components and wirings. For these redundant networks the problem is crucial; i.e., the amount of selective information required to specify them is very large. Not only do the wirings and thresholds have to be specified, but also particular patterns of synaptic interactions. Vote weights do not have to be selected because the process is automatic, but the necessary wirings of formal neurons need to be more complicated than those of non-adaptive networks. Thus, a very complicated programme is required for the construction of these networks, that contains all the requisite information. An important and interesting question concerns the possibility of leaving the bulk of the selection process to be performed during the lifetime of the network, i.e., by adaptation. A. M. Uttley [14] has outlined how certain replicated circuits might arise by chance in a randomly interconnected network, thus diminishing the amount of selective information what has to be supplied either "genetically" or "epigenetic-ally". What remains is the problem of specifying the complicated functions required for functional coding. For high levels of functional coding, each formal neuron need only represent a random selection of the specified functions of the network, provided it correctly decodes its inputs. It is interesting that recent work on the design of adaptive machines capable of learning to classify, represent, and recognize patterns of stimuli, has given rise to machines whose functional organization is apparently very similar to the redundant networks we have designed. [*See* 15, 16].

The combination of such theories of automata and information-processing with what might be called machine theories of adaptation, provides a not unreasonable model of what might be the organization of those parts of brains concerned with perception, learning, and perhaps with memory. We have cited Lashley's work on ablation effects. Lashley concluded that it was not possible to demonstrate the isolated localization of the memory trace or "engram" anywhere within the nervous system. He supposed that there was no special reservoir of cells which would serve as the seat of special memories, every instance of recall requiring the activity of millions of neurons. Moreover, the same neurons which retained the engram must also participate in countless other activities. [*See* also 17 and 18]. Sperry has suggested a number of principles that might give rise to this property: multiple interconnections between nerve-cells, the fidelity of the wiring being controlled by specific biochemical factors, much overlapping of interconnections; multiple reinforcement of any given function from numerous different sources any one of which may itself be capable of sustaining the activity; reciprocal and surround inhibition between and among neurons; the arrangement of cortical circuits in vertical, rather than horizontal, dimensions and the bilateral duplication of the cerebral hemispheres. J. C. Eccles [19] and A. E. Fessard [20] have inferred from their own experimental studies that presynaptic and lateral inhibition between and among neurons are important features contributing to error-insensitivity. In considering changes in the "evoked potentials" in various areas of the brain, associated with the engramming of conditioned responses, E. R. John [21] demonstrated an effect related to the replication of circuits and the delocalization of function which we have discussed, namely that evoked potentials following the engramming have similar shapes, even over many anatomically distinct regions of the brain.

In the light of these experiments we note certain logical requirements on the structure of the error-insensitive networks obtained by functional coding techniques. These are the existence of large numbers of both excitatory and inhibitory synapses at all units, the existence of many presynaptic "axo-axonal" interconnections, a great deal of multiple interconnection and overlap, and the existence of lateral inhibition between groups of neurons. Furthermore, as a consequence of functional coding, there is an extensive representational system in these networks, any unit of which can be activated by many different patterns. Finally, the reliable activation of any complete pattern of activity would require the synergic activity of many units. All these features of functionally coded networks appear to have experimental correlates, most of which are the result of experiments performed on the central nervous system proper. We should not expect to find high levels of functional coding in peripheral areas, but rather many replicated circuits more in line with the "telephone-exchange" analogy. At some intermediate stage where the degree of synaptic interaction among neurons is sufficiently high to sustain functional coding as well as circuit replication there will emerge assemblies of synergic neurons, which may be taken to be the "functional units" of the network. Perhaps the vertically organized groups of neurons found in sensory projection areas [22, 23] may be taken to be the experimental correlates of these functional units.

We consider it to be of some importance that deductive models of the organization of neuronal networks be forthcoming which lead to the specification of such entities as functional unit, engram, and so on. R. L. Gregory [24] has made the point that a knowledge of function is required to classify observed biological features into "essential" or "accidental" properties, and that it is only when the functional units of the system being studied can be identified, that deductive inference and not mere description becomes possible. Gregory made the further point that the neurologist is never able to identify functional units directly by observing neurons and their interconnections; in all cases, knowledge is needed of what neurons do, and of how they do it. This raises a peculiar and difficult problem common to almost all attempts to apply cybernetics to biology. For, in order to do this, one must know precisely what is the ensemble of possibilities upon which operates the selection process that alone gives meaning and utility to the ideas of message and information. For the neurological problem, this is equivalent to saying that one needs to know the code, or codes, of the nervous system. Thus, any real application of theories of formal neuronal networks can follow only from a knowledge of what neurons do and of how they do it. This takes us far from the ablation experiments of the physiological psychologist, to the electrophysiologists who measure not behaviour deficits and the like, but the firing patterns of neurons and the changing electrical potentials of nervous tissue, the ECoG and EEG.

Once we enter the domain of experimental neurophysiology, however, it becomes difficult to see how the automata models apply to data.

It is clear that the automata approach is seriously deficient in many respects. In the automata, we have considered changes in the firing patterns occur only at instants specified by an external clock, i.e., they are already "synchronized" in the time domain. There is no clear evidence that the CNS operates in such a fashion. In any case, the formal neuron is a rather crude abstraction of neuronal behaviour, and the theory covers special aspects such as functional stability and does not serve to help us understand the responses of nets comprising many thousands or many millions of cells. The theory, in fact, was designed for the analysis of small-scale local interactions between abstract functional units. Digital computer simulations of neuronal nets [26, 27] together with some combinatorial analysis [27, 28] have given us some indication of the type of activity to be found in homogeneous networks of randomly interconnected formal neurons. If there are only excitatory synapses in the net, the only stable states of activity are either a large proportion of all the units in the net are active, or else the net is quiescent. This is the "switching effect", discovered by many workers. It has been suggested that a neuronal network which acts itself as a switch might serve as a functional unit, and so correspond to one of McCulloch and Pitts' formal neurons. In case there are also inhibitory synapses, the behaviour is more complex and several intermediate stable levels of activity can persist. Farley has shown that even small networks comprising about one hundred cells can exhibit quite complicated behaviour, recruiting responses, augmenting responses, rhythms and so on, that are reminiscent of experimental phenomena. What is lacking in this approach, however, are the concepts and mathematical tools that would further the analysis of the responses of very large networks, nor is there any real attempt to face the problems of coding in these networks.

There has been one interesting attempt at a mathematical treatment of the responses of large nets [29]. In this, neurons are assumed to be randomly distributed in a mass with a given volume density. The neurons have thresholds, synaptic delays, EPSPs and IPSPs, and a summation time constant. Attention is directed to the proportion of cells becoming "sensitive" per unit time. Sensitive cells are those which are not refractory and can, therefore, be fired by a sufficiently potent stimulus. Although the mathematical treatment is not rigorous, the conclusions are essentially correct. The switching effect is discovered, and the conditions for the distortion-free propogation of plane waves of neuronal excitation are given. It is shown that the connection function which gives such propogation is equivalent to that one found empirically by D. A. Sholl [30]. Since only excitatory synapses are present, this wave propogation is unstable and depends critically on the stimulus, on the cell properties, and on the local density of interconnections. Beurle introduced two mechanisms of interest. Waves were stabilized by "servo" control from external nets which acted by firing off cells ahead of the wavefronts. And cell thresholds were assumed to be dependent upon past activity. These extensions of the random network assumptions are important. The servo idea is an attempt to treat network interactions, the control of one network by another. Threshold modification by past activity changes the responses of a network, according to past responses. So the network can be trained, given the proper feedbacks, to act as a permanent store of messages. Indeed, all the possibilities of behaviour, functional stability, adaptation, and so on, which we considered to be present in functionally stable automata, are to be found in these networks. However, the scale has been changed, so to speak, in that complete networks play the role of functional units, and dynamical variables play an important role in the encoding of stimuli.

There is, therefore, the intriguing possibility that some kind of answers to the problems of modelling some aspects of the activity of neuronal networks in the CNS might be forthcoming from a mathematical formulation closely related to

Beurle's. But there are a number of problems that must be solved before this approach can be made useful. The analysis has some defects. For example, the effects of refractoriness, of the finite size of the network, and of delays, are not properly formulated. To some extent, these defects are not too serious. Correct formulations of similar problems have been given by J. S. Griffith [31] and by M. ten Hoopen [32]. Their conclusions are similar to Beurle's concerning switching and stability, if somewhat less far-reaching in scope. However, what is really lacking from the whole approach is that it does not make contact with the experimental variables of the neurophysiologist, the histograms, correlograms, ECoGs and so forth, and so, in this sense, it is not a testable model for the responses of neuronal network in the central nervous system. Moreover, while the image of interacting waves of neuronal excitation suggests many possibilities for the coding of messages, it has not been used so far to produce precise, quantitative predictions of how and where permanent changes take place in the networks of the central nervous system, and what they represent.

What is required, then, is a novel mathematical formulation of the responses of neuronal networks that takes account of many of the organizational features of the central nervous system that we have listed, in which the variables relate to measureable quantities by current experimental and data processing techniques, and in which there is a precise hypothesis concerning the nature of neuronal coding. It is not too much to hope that this theory will be forthcoming during the next decade.

## Acknowledgement

The author would like to express his gratitude to Lord Jackson of Burnley, F.R.S., of the Electrical Engineering Department, Imperial College for the warm hospitality he has enjoyed there as an academic visitor while the above research was carried out; and to the Director of the National Physical Laboratory, Teddington, and the Superintendent of the Autonomics Division there, for making available to him valuable experimental and computing facilities, while he was there as a guest worker.

## References

[1]  W. S. McCulloch and W. H. Pitts, *Bull. Math. Biophys.* **5** (1943) 115–133.
[2]  S. C. Kleene, *Automata Studies*, Eds. C. E. Shannon and J. McCarthy, (Princeton Un. Press, 1956) 1–42.
[3]  K. J. W. Craik, *The Nature of Explanation*, (Cambridge Un. Press, 1943).
[4]  A. M. Turing, *Proc. Lond. Math. Soc.*, **42**, 230 (1937) 137.
[5]  N. Wiener, *Cybernetics*, (M.I.T. Press, 1948).
[6]  K. S. Lashley, *Soc. Exp. Biol. Symposium* **4**, (Cambridge Un. Press, 1950).

[7]   J. von Neumann, *Au'omata Studies*, loc. cit. [2] 43–89.

[8]   C. E. Shannon, 1948 *Bell System Technical J.* **27** (1948) 379–423, 623–656.

[9]   P. Elias, *Sensory Communication*, Ed. W. A. Rosenblith, (M.I.T. Press, 1961).

[10]  S. Winograd and J. D. Cowan, *Reliable Computation in the Presence of Noise*, (M.I.T. Press, 1963).

[11]  W. Peterson, *Error-Correcting Codes*, (M.I.T. Press, 1961).

[12]  W. H. Pierce, *Redundancy Techniques in Computing Systems*, Eds. R. H. Wilcox and W. C. Mann, (Spartan Books, 1963) 229–250.

[13]  E. F. Moore and C. E. Shannon (J. Franklin Institute, 1956).

[14]  A. M. Uttley, *Information and Control* **1** (1959) 1–24; *Brain Research* **2** (1966) 21–50.

[15]  W. K. Taylor, 1956 *Information Theory*, Ed. E. C. Cherry (Butterworth, 1956) 314–328.

[16]  F. Rosenblatt, (1958) Cornell Aeronautical Labs., Report No. VG-1196-6-1, *Principles of Neurodynamics*, (Spartan Books, 1962).

[17]  R. Sperry, *Principles of self Organization*, Eds. H. von Foerster and G. W. Zopf, Jr., (Pergamon, 1960).

[18]  J. Z. Young, *A Model of the Brain*, (Oxford Un. Press, 1964).

[19]  J. C. Eccles, *The Physiology of Synapses*, (Springer, 1964).

[20]  A. E. Fessard, *Information Processing in the Nervous System*, Eds. R. W. Gerard and J. W. Duyff, (Excerpta Medica, 1964), 415–418.

[21]  E. R. John, 1964, Ibid.

[22]  V. B. Mountcastle, *J. Neurophysiol.* **20** (1957) 408–434.

[23]  D. H. Hubel and T. N. Wiesel, *J. Physiol.* **160** (1962) 106–154; *J. Neurophysiol.* **28** (1965) 229–289.

[24]  R. L. Gregory, *Mechanization of Thought Processes*, (H.M.S.O., 1959) 669–650.

[25]  J. D. Cowan, *Encyclopaedia of Information, Linguistics and Control*, (Pergamon Press, 1967); *Automata Theory*, Ed. E. R. Caianiello, (Academic Press, 1964).

[26]  B. G. Farley, *Computers in Biomedical Research* 2, (Academic Press, 1965).

[27]  D. R. Smith and C. H. Davidson, *J. Assoc. Computing Machinery* **9** (1962) 268–279.

[28]  J. S. Griffith, *Biophys. J.* **3** (1963) 299–308.

[29]  R. L. Beurle, *Phil. Trans. Roy. Soc. London* **240** (1956) 669.

[30]  D. A. Sholl, *Organization of the Cerebral Cortex*, (Methuen, 1956).

[31]  J. S. Griffith, *Bull. Math. Biophys.* **25** (1963) 111–120; *Bull. Math. Biophys.* **27** (1965) 187–195.

[32]  M. ten Hoopen, *Cybernetics of Neural Processes*, Ed. E. R. Caianiello, (C.N.D.R., 1965) Rome.

*Theoretical Physics and Biology*, © 1969, *North-Holland Publ. Co., Amsterdam*

# WHAT IS THEORETICAL BIOLOGY?

WALTER M. ELSASSER

*Department of Geology and Biology, Princeton, U.S.A.*

One may trace the beginning of modern efforts in theoretical biology to a celebrated paper of Niels Bohr that appeared in 1933. I should say here at once that my personal interest in theoretical biology was largely aroused by this work. My basic ideas began to develop during my stay in Paris in 1934/35. I am therefore particularly grateful to the Organizing Committee of this Meeting for the opportunity to return to the places of my early stimulation and development.

Bohr's paper indicates how one can apply the ideas of quantum theory to biological problems. What quantum physics has taught us on a more philosophical level is that each progress in our understanding of the behavior of matter is achieved at the price of a corresponding loss: We are no longer able to trace in detail the geometrical arrangements and clearcut causal relationships that govern motions in classical physics. It is a familiar fact that the concept of well-defined orbits of particles ceases to exist in quantum theory. Instead, one has a statistical theory in which prediction is profoundly limited. In theoretical biology, analogous arguments hold but they apply now to a higher level of organization.

This raises at once the question whether changes are required in the mathematical apparatus of quantum mechanics. Our answer is that such changes are unnecessary. Organisms differ from other bodies by their extraordinary complexity. Their dynamics, however, takes place at the energetic levels of ordinary chemical reactions. There are no plausible grounds for changing the laws of physics solely because of an increase in complexity. What then is the new element which appears in the analysis of living systems? We claim that this novelty lies precisely in complexity itself.

On combining the uncertainty relations with this complexity, Bohr concluded that the perturbations which result from quantum measurements are of fundamental importance in biology. Such measuring processes interfere with the delicate operation of a system of as complex a design as an organism. In order to gain a thorough knowledge of the system, which is indispensable for precise prediction, the measurements must be so thorough as to disturb the organism seriously; eventually the animal will become sick and die. But on the other hand, in the absence of prediction science ceases to be analytical and remains on a purely descriptive level.

In my own writings I have emphasized that predictions may be derived from two different procedures. One of these is the determination of initial values, as already mentioned. The other may be called the method of the sampling of classes. One can for instance predict the behavior of some variety of molecules by measuring other molecules of the same kind, notwithstanding the fact that the measured molecules may be destroyed in the process. This is of course a common method in physics and chemistry. I have been able to show that the application of this method to the classes of biology, e.g., species, leads also to fundamental limitations of prediction. These limitations arise from the fact that organisms of the same class differ from each other in very numerous particulars by virtue of the pervasive inhomogeneity of all living matter.

The totality of these limitations imposed on prediction corresponds to what may properly be called the semi-autonomous character of living matter. If we had chosen to apply a similar terminology in quantum theory, we could have said that the stability of atoms and molecules is a semi-autonomous phenomenon, in the sense that an explanation of this stability in terms of classical physics is not possible. In a similar fashion the stability of organisms and even more the tremendous precision of developmental processes are properties which cannot be fully explained by means of ordinary physics. Undoubtedly, a great part of this stability can be explained in mechanistic terms, for instance by feedback devices. But owing to the intrinsic complexity of living matter one cannot make the purely physical mode of explanation an exhaustive one.

An important reservation is needed at once. This is that physics and chemistry are never false in the organism. They are incomplete. What is meant here is that the initial conditions can never be determined with the required accuracy; nor can one select a biological class whose members would be sufficiently similar to each other to allow adequate prediction based on the method of sampling. The primary distinguishing characteristic of living matter is, therefore, the essential complexity, inhomogeneity, and variability of this matter. We are thus able to define theoretical biology in the following manner: Theoretical biology is the science of radically inhomogeneous classes. Physics, on the other hand, is the science of homogeneous classes, if not a priori, then at least in fact and by usage.

In the course of my investigations I have become convinced that complexity, inhomogeneity, and variability are not only necessary conditions for the existence of life; they are also sufficient in a simple sense: Any proposition about life inasmuch as it deals with it as a semi-autonomous phenomenon must be based upon this inhomogeneity, beyond the purely physico-chemical properties of the system considered. Again, we may be confident that this is the only new principle required when we want to characterize life in a general and essentially abstract manner.

We can exhibit the same ideas in a different form, mostly to show their relationship with what is usually called molecular biology. We know that the organism possesses many well-defined constituent parts which one can synthesize in the laboratory; or else one will certainly be able to do so in the future. We shall designate these as homogeneous components. They are primarily macromolecules. In the living organism, these homogeneous components are immersed into a radically inhomogeneous environment. This inhomogeneity of the internal environment arises out of the almost endless variation of geometrical relationships among parts, and out of the almost limitless variability of the low-energy chemical reactions which this entails. One may express this by saying that in the organism the dynamics of the homogeneous components needs must be coupled into the inhomogeneity of the internal environment.

In order to gain a better understanding of the implications one ought to compare inhomogeneity with what the physicist calls noise. Although noise is a statistical phenomenon, one presumes always that one can form averages over any variable whatsoever. I propose then to define an inhomogeneous class in the following manner: It is a set composed of a finite number of objects such that one can never obtain enough samples to give an operational significance to all averages that it may be possible to define from a purely mathematical standpoint. From this there results an essential limitation of prediction as compared to the case where homogeneous classes would be available.

Bohr had long ago proposed that every fundamental progress in our knowledge of the properties of matter is tied to a loss of so called explanation. This idea found its original application in quantum mechanics. Uncertainties of a quite analogous kind turn out to be basic in theoretical biology. The abstract apparatus which expresses these new uncertainties is the theory of inhomogeneous classes. One fact stands out: When we go from abstract models in the more conventional sense to a theory of abstract *classes* which are not necessarily homogeneous, we introduce a freedom and a generality which are not to be found in traditional theoretical science. I have convinced myself in the course of many years that we have in hand here the chief tool that is required in theoretical biology.

## References

Walter M. Elsasser, *The Physical Foundation of Biology*, 219 pp. (Pergamon, 1958).
Walter M. Elsasser, *Atom and Organism*, A new Approach to Theoretical Biology, 143 pp. (Princeton Un. Press, 1966).
See also a forthcoming paper by the author in vol. 3 of the series *Towards a Theoretical Biology*, ed. by C. H. Waddington (Edinburgh Un. Press, 1969).

*Theoretical Physics and Biology*, © 1969, *North-Holland Publ., Amsterdam*

## GENERAL DISCUSSIONS

A. LICHNEROWICZ: M. Grassé nous disait qu'il fallait trouver un langage commun: un premier thème est revenu, entre M. Prigogine et M. Fröhlich: l'expression "non linéaire". Je voudrais dire à la fois mon accord et les limitations qui me semblent s'imposer.

Il faut d'abord faire très attention, en parlant linéaire et non linéaire. Il s'agit de choses dans lesquelles le linéaire peut sécréter des êtres non linéaires et des effets non linéaires. Et par conséquent il faut dire exactement ce qui est, dans chaque cas, effectivement non linéaire. Mais enfin nous serons d'accord, je crois … Et je pense que le mot "non linéaire", qui embrasse le linéaire, est une des clés de notre dialogue.

Secundo: je voudrais rendre hommage, en particulier, à l'effort d'arrachement à la thermo-quasi-statique, en faveur d'une vraie thermodynamique, fait par M. Prigogine; là est probablement l'un des modes d'approche de nos problèmes. C'est l'étude de régimes permanents stables. Et je voudrais peut-être poser, en même temps, à M. Prigogine, une question. Un certain nombre des choses qu'il a dites sont en exact parallèle avec certaines des méthodes contemporaines et anciennes de l'économie mathématique. Nous savons d'ailleurs qu'il y a des parallèles entre thermodynamique et economie. Et en economie les fonctions convexes jouent un rôle important pour les problèmes de stabilité (de stabilités un peu plus globales que les siennes). En fait, Monsieur Prigogine, votre condition du second ordre est une condition de convexité locale.

Je crois que l'introduction de symboles mathématiques du type fonctions convexes, pour certains problèmes, avec contraintes, doit jouer dans l'avenir un rôle important.

Tertio: je voudrais affirmer ma foi dans la biologie théorique, mais je ne crois pas qu'elle doive avoir, à ce stade, un instrument privilégié. Je pense que nous devons également avoir confiance en ce qui a été dit tout à l'heure, par exemple par M. Elsasser, qui est une approche très intéressante des classes hétérogènes aussi dans la *vraie* thermodynamique. Sous les réserves qu'il faut tout de même bien prévenir nos amis biologistes que le mot "entropie", qu'on met à toutes les sauces, n'est pas toujours parfaitement aussi clair et aussi mesurable que nous souhaiterions. Je pense que ces approches convergentes doivent être poursuivies distinctement, et que nous ne pouvons pas, en ce moment, espérer *une* seule approche.

Bien entendu, la mécanique quantique jouera son rôle. Pour le moment, elle est linéaire. Essentiellement en ce sens que ses équations fondamentales sont linéaires. Mais on sait que des essais sont faits pour sortir du linéaire. En gros, j'ai l'impression que c'est la description qui s'applique aussi à M. Prigogine. Nous avons des équations, disons en gros, rigoureuses, non linéaires, comme équations fondamentales, dans beaucoup de situations; et nous étudions la stabilité à partir des équations aux variations qui forment le système linéaire canoniquement associé. C'est cela que nous faisons constamment, sous des formes et avec des discours, physiques ou chimiques, de type varié.

W. M. ELSASSER: Je suis tout à fait d'accord avec M. Lichnerowicz: j'ajoute l'idée que la théorie des classes hétérogènes joue un rôle similaire à l'usage qu'on fait de la géométrie riemanienne dans la cosmologie. Sans quoi, on aura la contradiction que Kant, le philosophe, constatait il y a deux cents ans: la contradiction entre le problème d'un univers infini et la possibilité de construire un univers sensible. Alors, avec l'introduction de la théorie des classes hétérogènes, des contradictions semblables, entre la nécessité et l'indéterminisme (c'est-à-dire la liberté et le potentiel créateur), ... ces choses-là disparaissent. Mais naturellement, il faut aussi avoir des méthodes beaucoup plus concrètes, qu'on trouvera en appliquant la mécanique statistique à la biologie moléculaire.

A. LICHNEROWICZ: Je veux dire mon accord. Votre théorie fournit un "background" fondamental pour la suite. L'image est excellente! ... "Back-ground" dans lequel on mettra des choses assez variées.

I. PRIGOGINE: Au sujet de l'intervention de M. Lichnerowicz, je voudrais dire qu'il existe effectivement une certaine analogie formelle entre les problèmes que j'ai traités et des problèmes sociaux et économiques. Je pense même qu'il doit être possible d'aboutir à de tels éléments d'une théorie des structures sociales. Toutefois à un point de vue précis les problèmes dont j'ai parlé sont plus simples car je puis utiliser les conditions classiques de stabilité thermodynamique. Il faudrait voir si de telles conditions existent dans les problèmes sociaux. Il faut aussi, je crois, souligner combien les notions d'hétérogénéité et de complexité sont ambigues. Quant à moi je ne sais pas ce qui est plus complexe, une particule élémentaire ou un être vivant.

TH. VOGEL: Je voudrais toucher à deux questions. On a beaucoup parlé du caractère quantique des modèles qui ont été présentés. Je sais bien que "per fas et nefas perseverare diabolicum ..." Cependant je n'ai pas été convaincu.

Il est bien évident que la théorie quantique doit couvrir, dans un certain domaine, tout ce que couvre la théorie classique, et doit en plus donner des renseignements complémentaires. Cela est vrai également de la théorie de la relativité, dont personne n'a parlé. Mais dans un domaine où la mécanique classique permet de rendre compte de tous les phénomènes qui ont été mentionnés, il est inutile de faire intervenir les quanta.

On nous a présenté, le cas d'illustration d'une lumière cohérente. Illustration qui pourrait être reprise, mot pour mot, des travaux de Bernouilli, sur la vibration d'une file de points matériels. Par conséquent, cela n'a rien de très nouveau. Il ne faut donc pas faire intervenir les quanta, tant qu'on n'a pas montré ce qu'ils peuvent donner, que la théorie classique ne peut pas donner.

Ma deuxième observation a trait à ce que disait M. Elsasser, et qui m'a vivement intéressé.

Il y a longtemps que je pense que l'outil le plus communément utilisé en physique —c'est-à-dire les équations différentielles—n'est pas un outil extrêmement approprié. Je pense, en tout cas, qu'en biologie mathématique, en biologie théorique, comme dans la plupart des théories que l'on peut être amené à bâtir, les équations différentielles, généralement utilisées en physique mathématique, ne sont pas l'outil le mieux adapté pour cette discipline. En tout cas, je suis convaincu — et M. Elsasser a achevé de me convaincre — qu'il n'est pas le plus adapté à la biologie théorique.

Alors, je voudrais signaler la possibilité et l'intérêt d'études sur les systèmes évolutifs sans unicité. Voici quelques années que je m'occupe des équations au paratingent qui peuvent couvrir une assez large part de ces problèmes, mais ces équations ne sont elles-mêmes qu'un cas particulier: il faut considérer des inégalités fonctionnelles qui permettraient d'avoir toutes les solutions possibles, l'ensemble des solutions possibles d'un système incomplètement déterminé.

La solution qui consiste à faire intervenir la probabilité est une fausse solution, à mon sens, parce que l'introduction de la probabilité suppose une axiomatique qu'on laisse généralement implicite et qui est beaucoup plus difficile à défendre que tout le reste de la théorie. Autrement dit, on fait une théorie qui est claire, facile, mais elle n'est valable que moyennant des hypothèses que généralement on ne spécifie pas et qu'il serait extrêmement difficile de vérifier par expérience.

L. Tisza: The discussions concerning the relation of physics and biology seem to be dominated by the intricacies of biology as if the nature of physics were clear to everyone. In reality, I believe, that a satisfactory characterization of physics is a difficult task, and one not to be ignored if the interdisciplinary discussion is performed with a reasonable measure of precision. The task is difficult because of the wide ramification of the branches of physics, and it is aggravated by the fact that all this is in a state of rapid evolution. I will have to confine myself to the discussion of a few characteristic points.

Let me consider the case of quantum mechanics that in the course of a few decades has undergone already essential changes. I am tempted to compare the course on quantum mechanics I took almost forty years ago in Göttingen with the course now being offered to second year undergraduates at M.I.T. The old course, among the first to be offered anywhere on the new discipline, was nominal that cf Max Born, but actually it was given by his assistants, among them

Dr. Rosenfeld who is here with us today, and who may check up on my recollec-
tions. Well, the old course seemed paradoxical and outright mysterious, whereas
the new one is being accepted with striking ease; in discussing the resolution of
some of the "paradoxes" of quantum mechanics sometimes the hardest point is
to convince the students that there is anything paradoxical involved at all. Needless
to say, this comparison is not intended to cast any reflection on Professor Rosenfeld's teaching ability. It is, of course, to be expected that the novelty of a discipline
wears off, the rough edges polished, and there are plenty of reasons why its
teaching should become a great deal simpler. The point I wish to make is that the
metamorphosis of quantum mechanics is more profound than might be expected
on such grounds.

An elementary introduction to quantum mechanics necessarily starts with
classical physics and a great deal depends on the kind of bridge that we construct
between the two disciplines. The most striking difference between the above
mentioned two stages of quantum mechanics is that they are tied in with very
different parts of classical physics. In fact, I am convinced that the key to the
clarification of most of the paradoxes of quantum mechanics is that we have to
come to grips with some of the paradoxes and illusions of classical physics.
The first illusion to be given up is that there *is* such a thing as a coherent body
of classical physics. It is, indeed, well known that Newtonian physics had from
the outset two main departments: first, an inductive phenomenological one in
which the objects of everyday life and of the laboratory are taken for granted;
and, second the analytical mechanics of rigid bodies and mass points. To the
extent that this mechanics is applied only to macroscopic motion, these two
departments are compatible with each other. This is no longer the case for the
wider interpretation of Newtonian mechanics in which the validity of this discipline
is postulated even for the smallest, atomic constituents of matter, and where it is
assumed, or rather taken for granted that the entire phenomenological physics
is reducible to mechanics. For the sake of brevity I shall refer to the two departments of classical physics as PCP (phenomenological classical physics) and MCP
(mechanistic classical physics) respectively. The wide range of achievements of
classical physics belong primarily to PCP, but its philosophy is altogether
dominated by MCP. We know at present that PCP is not reducible to MCP.
However, the origin of this failure is not as well understood as it might be. I wish
to introduce a few concepts in the hope of shedding some light on the situation.

There are two types of regularities discernible in natural phenomena that are
in a way complementary to each other. The first is the well known determinism
of classical mechanics. I like to call it *temporal determinism* in order to emphasize
that we are dealing with temporal sequences in which the arbitrarily chosen
initial state determines the state at a later time. The initial state itself is arbitrary
in the sense that it is not restricted by a law of nature within this discipline. This
kind of ordering of natural phenomena was distilled from celestial mechanics.

The second type of order manifests itself most obviously in chemistry, say the periodical table of the elements, in which we find the systematization of certain configurations of nuclei and electrons forming the chemical elements. Such configurations are very much favored by stability among infinitely many other configurations that are not observable except in a most transient fashion. Some years ago I have suggested that this type of regularity be called *morphic*. I believe that the success of quantum mechanics is in a large measure due to the fact that this discipline provides a mathematical expression to morphic ideas. Thus, the quantum mechanical "pure state" can be conceived as the ultimate of chemical purification and the samples of a class of systems in the same pure quantum state are absolutely identical to each other. This property may be designated as "morphic invariance". It is also noteworthy that in quantum mechanics temporal and morphic considerations appear jointly in a consistent scheme. In strong contrast with this situation MCP has a purely temporal character, whereas PCP deals with the world as it is and implicitly contains morphic elements. Experimental spectroscopy is a good example for morphic aspects appearing in experiments set up entirely by classical means.

The granting to morphic invariance an independent conceptual status that cannot be reduced to temporal determinism enables us to envisage some of the problems raised from a fresh perspective.

The replacement of MCP by quantum mechanics no longer appears as the replacement of one mechanics with a somewhat different variant, but a fundamentally new conceptual element, namely the morphic point of view, is added to the theory. This explains why the traditional introduction of quantum mechanics as an analog of classical mechanics has a purely formal character giving no allowance to intuitive conceptual understanding. In contrast, the new quantum mechanics referred to above builds a bridge between quantum mechanics and PCP and the transition is a great deal smoother since both disciplines contain morphic elements.

I wish to comment now from the point of view just outlined on Dr. Elsasser's discussion of homogeneous classes. This concept has a clearly morphic character and I agree with Dr. Elsasser that it is of crucial importance for the discussion of the relation of biology and physics. However, I must take exception to the claim that physics deals *only* with homogeneous classes. First of all, we must not speak summarily of all of physics. Thus, MCP knows nothing of homogeneous classes. This concept emerged only within quantum mechanics as the set of systems in the same pure state. Second, quantum mechanics deals not exclusively with homogeneous classes, but contains actually rules for constructing more complicated situations. Let me mention only that we take the homogeneous classes of electrons, protons and neutrons and we build up the classes of nuclei, atoms and molecules. All of these systems can exist in homogeneous classes of pure quantum states, but in actual practice we are more likely to encounter them as various kinds of mixtures of homogeneous classes, or more or less random selections from such.

There can be no doubt about it that biology deals *par excellence* with morphological properties and that morphic quantum mechanics has a great deal better chance to serve it as a fundamental theory than MCP ever had; it certainly goes a long way in providing a conceptual basis for molecular biology. Nevertheless, Dr. Elsasser may have a point in emphasizing that biological systems present us with a degree of complexity much beyond anything we have learned to cope with on the basis of the present day quantum mechanics geared to deal with physical systems. Yet would it not be foolhardy to claim that quantum mechanics has reached its final form? While no one can predict with confidence the evolution of quantum mechanics in the future, the prediction that is most likely to come to grief is one that would rigidly limit its adaptability and eventual scope!

L. ROSENFELD: This was a friendly challenge to me to express an opinion on the present state of quantum theory, compared to what it was forty years ago. Now, we have learned very much in all these years, of course, but in a sense we may also say that quantum theory, in its conceptual frame-work, is much the same to-day as it was then. What we have achieved, essentially, is a better delimitation of its domain of validity. All our physical theories operate with strongly idealized concepts, and the problem is to determine the limits within which such idealizations are useful tools for the analysis of the phenomena. Within these limits, we may well say that a theory that has stood the test of experience has acquired a perennial validity. So it is with the theories of classical physics, so it is with quantum mechanics.

We know that the biological processes, in their molecular aspects, fall within the limits of validity of quantum mechanics — often even within the range of application of the classical approximations to the quantal laws — and this makes quantum mechanics directly relevant to molecular biology, as has been emphasized by my colleagues this morning.

There is, however, another, more indirect way in which quantum theory may be of some help to biology: it concerns an epistemological problem which has been the object of long and confused debate in the course of history and is still widely regarded as a bone of contention among biologists. Will a physico-chemical analysis of the molecular processes underlying the biological phenomena provide a complete, exhaustive description of these phenomena, or must one expect that specific, non-physical concepts will be needed for this purpose? In this connexion, the concept that immediately comes to mind is that of "function", the usefulness of which in biological investigation nobody will deny. And the idea of function, with its implication of finality, seems to be incompatible with the type of causality exhibited by a purely physical description.

Now, in trying to account for quantal phenomena physicists have been led to recognize the existence of a logical relationship between concepts, to which the name of complementarity has been given, which makes it possible to use without danger of contradiction concepts which are mutually exclusive, but nevertheless adequately describe certain aspects of experience. This possibility arises from the fact I stressed before, that concepts are idealizations of limited validity; contradictions only arise when they are stretched beyond such limits, and are avoided if the limits are properly recognized.

In this sense, there is certainly room in biology both for a full, unrestricted physico-chemical analysis of the molecular processes on the one hand, and a description of the functional aspects of the biological processes on the other. These two points of view should not be opposed to each other, but fruitful use could be made of them in a complementary fashion.

E. G. D. COHEN: I would like first to raise a general question. In many lectures and discussion remarks it has been tacitly assumed that the main connection of biology and theoretical physics should be through quantum mechanics. This may well be so but this morning already I raised the question on account of Dr. Fröhlich's lecture whether not in some cases analogies with *classical* statistical mechanics might be helpful for the understanding of some biological phenomena. On account of Dr. Cowan and Dr. Elsasser's lectures I would like to raise another question, namely one concerning the proper macroscopic description of a many particle system in physics. It seems to me that both Dr. Cowan's and Dr. Elsasser's lectures dealt with an analogous problem in biology. If in physics, the system is in thermal equilibrium the macroscopic variables to describe it are obvious. However, for a system not in thermal equilibrium the proper macroscopic description is in general far from obvious and usually unknown. The problem here is not only: what are the appropriate macroscopic variables for the system but also how are these variables related to the microscopic properties of the individual particles of which the system consists.

A second question relates more in particular to Dr. Elsasser's lecture. Could he perhaps say again in how far a biological system differs *essentially* from a many particle system in physics. His wording of the properties of a biological system could – as far as I can see – be taken directly from a textbook of statistical mechanics.

W. M. ELSASSER: Je ne crois pas que la différence entre la mécanique quantique et la mécanique classique soit considérable à cet égard. Il y a toujours deux niveaux dans l'exemple standard, nous avons le niveau microscopique. Ensuite, le niveau macroscopique, la pression, la température, etc. . . .

Je ne sais s'il y a ici des biologistes empiriques (ils sont d'habitude rares dans les réunions comme celles-ci): ils conviendront avec moi et vous diront qu'un organisme a une structure extrêmement compliquée à tous les niveaux: au niveau des angströms, au niveau des microns, au niveau des millimètres, etc. Là, il y a une structure dans une structure — dans une structure — dans une structure — : comme on l'a déjà dit, c'est la caractéristique essentielle d'un organisme. Donc, vous ne pouvez pas appliquer la mécanique statistique: le processus d'organisation est trop compliqué pour le faire. Naturellement on tentera cette application puisqu'il n'y a pas d'autre outil. Mais il faut y être extrêmement prudent.

Ma deuxième observation est un peu plus subtile: dans le passé il y avait une distinction très claire entre ces deux grandes catégories: le déterminisme et les statistiques. S'il y a des mathématiciens ici, j'espère qu'ils ne se vexeront pas du terme "statistiques" au lieu de "calcul des probabilités". Il y avait donc déterminisme et statistiques. L'hétérogénéité qui apparait ici est une catégorie essentiellement différente de la statistique. Car, en statistique, vous supposez, selon l'axiomatisation acceptée par les mathématiciens, que vous pouvez former n'importe quelle moyenne, quelle qu'elle soit. L'essence d'un système hétérogène est que vous ne pouvez point prendre toutes les moyennes. Vous pouvez en prendre certaines. Si je pouvais prendre toutes les moyennes, je serais reduit immédiatement à la biologie mécanistique. Quant à la mécanique quantique, je dirai, par exemple: j'ai une matrice statistique et je peux définir tous les opérateurs que je veux, et je peux déterminer, mathématiquement, la moyenne de ces opérateurs dans l'ensemble des échantillons. L'hypothèse admise est qu'on peut toujours le faire. Von Neumann a toujours fait cela, comme mathématiques. L'axiomatique de Von Neumann est fausse, à mon avis du point de vue physique: vous n'avez pas assez de spécimens, pas assez d'homogénéité pour pouvoir produire ces moyennes par des procédés opérationnels. Je sais que c'est extrêmement difficile à saisir. Mais c'est la différence essentielle entre l'hétérogénéité et les statistiques. A moins que vous n'ayez cette différence et que vous ne l'ayez saisi, on ne peut pas comprendre ce qui importe en matière de biologie. Mon temps n'était pas assez long pour développer cette pensée en détail.

H. C. Longuet–Higgins: I want to make one or two disconnected remarks first, and then to say something more connected. First of all, a comment on Dr. Elsasser's contribution. It seems to me to be too simple to say that physics deals with homogeneous classes. Solid state physicists deal with such things as imperfect crystals, and no two imperfect crystals are alike.

My second point concerns a comment made by Dr. Rosenfeld on Dr. Fröhlich's contribution which we have been discussing with much interest. Dr. Fröhlich seems to have contributed a suggestion about a possible new phenomenon to look for. It seems to me that the problem we really have in front of us is not to find new phenomena but to find concepts for the interpretation of existing phenomena. Plenty of strange things happen in Biology, without introducing any more. I would suggest that what we really need to do is to find some way of coping with the phenomena we do know about. Can I just make a very short list of a very few things which seem to me to be outstanding characteristics of biological systems that are not found in the inanimate world.

(1)   Evolution,
(2)   The processing of information by organisms — for example in the central nervous system of vertebrates,
(3)   Morphogenesis.

I think we would not feel the need of new fundamental concepts for discussing the theory of evolution; there are already very good existing concepts such as variation, mutation, selection and so forth which have proved themselves in this connection. But I would like to make one or two suggestions about the other two phenomena.

First of all, the question of information processing by the individual organism. In what way should we look at the relation between the stimuli and the responses? Here it seems to me that one could use the sort of ideas that Dr. Cowan has been describing to us. These come from control engineering, where one talks not so much about the physical nature of the processes but their mathematical description.

Now a word about morphogenesis. We are all of course amazed by what happens when we sow a seed in the garden and up comes a rose. It is a most fantastic phenomenon. Now what kind of ideas do we need in order to understand this? May I suggest that the kind of idea one wants is that of a computer program and its output. Perhaps this sounds rather low-brow, but the mathematicians are now busy developing the theory of computer programs and the theory is concerned with the classification of different types of program and their different modes of implementation. There is a quite remarkable formal resemblance between feeding a paper tape into a certain environment and getting out a complicated data structure, such as a set of mathematical tables, and on the other hand pushing into a cell a long double molecule with bases all along it and getting out a complicated physical structure which is an organism.

My feeling is, therefore, that the sort of concepts which we need for understanding morphogenesis — which is an outstandingly biological kind of thing — are perhaps the same sort of concepts that one needs for understanding the scope and the limitations of computer programs. And it seems that there is some hope of making progress in this direction, because, after all, computer programs are a human invention and people who work with computers may be expected to have a pretty clear idea of what they are up to. In fact, the problems are really not physical problems at all, but mathematical and logical ones.

J. POLONSKY: Ma première remarque concerne la "querelle" entre l'entropie thermodynamique et l'entropie de l'information. Cette "querelle", évoquée par M. Lichnérowicz, me fait penser à une petite histoire que je vais vous confier:

Un profane demande à un spécialiste en électronique de lui expliquer d'une façon simple la différence qui existe entre le télégraphe avec et sans fil. Imaginez, lui répond le spécialiste, un chien très long dont la tête serait à Paris et la queue à Marseille. Vous tirez la queue à Marseille et le chien aboie à Paris. Ceci est le télégraphe avec fil. Le télégraphe sans fil, c'est la même chose, mais sans chien.

L'entropie de l'information est souvent représentée dans la littérature comme équivalente à une entropie thermodynamique, mais sans support matériel ou énergétique. En effet, les deux notions d'entropie expriment le degré d'incertitude ou la part aléatoire d'une structure donnée; et il est exact que dans les problèmes traités par l'ingénieur, l'entropie thermodynamique est toujours liée à une certaine énergie, tandis que celle qui se rattache à l'information peut être souvent utilisée sans référence énergétique.

Néanmoins, peut-on affirmer que l'entropie de l'information peut se passer d'un support énergétique? Szilard, Brillouin et d'autres ont montré qu'il n'en est rien, et que la différence entre ces deux types d'entropies est essentiellement d'ordre quantitatif, aussi bien sur le plan du degré d'incertitude qui règne sur une structure donnée qu'au niveau de l'énergie qui s'y rattache. A la limite, un bit d'information peut être créé ou transmis avec une dépense d'énergie de l'ordre de $10^{-16}$ erg/°K, c'est-à-dire que le support énergétique d'un très grand nombre d'informations reste toujours négligeable par rapport aux énergies mises en jeu dans un système à l'échelle macroscopique. Il est, par contre, facile de montrer qu'à l'échelle microscopique, pour des raisons quantiques, il n'en est rien. A cette échelle d'organisation, l'énergie rattachée à l'entropie d'information * est de même ordre de grandeur que l'énergie mise en jeu dans un travail microscopique. Il suffit de rappeler que l'ATP, source universelle d'énergie en biologie, fournit des quanta d'énergie de 0,4 eV environ, aussi bien dans le cadre d'une catalyse informationnelle que dans une conversion d'énergie chimique en énergie mécanique ou électrique.

Ma deuxième observation se réfère à la question posée par M. Vogel: "La mécanique quantique peut-elle fournir des armes nouvelles au biologiste?".

Je suppose que la question posée par M. Vogel va au-delà des méthodes développées ces dernières années en biochimie quantique et auxquelles est consacrée la journée de demain.

Pour ma part, je voudrais évoquer ici une méthode d'approche que l'on pourrait qualifier de cybernétique quantique, où l'information et les contraintes dans les messages échangés entre molécules biologiques se trouvent directement rattachées à des concepts quantiques. Nous avons vu ce matin que sur les plans logique et physique, le mode fonctionnel collectif d'un système organisé se trouve défini par les corrélations entre les contraintes et qu'à leur tour, ces corrélations sont déterminées par le mode d'association des composants.

---

*   Dans un "dialogue" entre molécules biochimiques.

Une question centrale posée par la cybernétique quantique pourrait être formulée comme suit:

Quelle est l'origine physique des contraintes collectives qui apparaissent dans un système polyatomique? ou, en d'autres termes:

Les contraintes collectives dans un système étant équivalentes à une certaine entropie négative, la question se pose: Qui paie le prix de l'entropie négative qui apparait dans le système global?

Je vais essayer de résumer ici quelques éléments de réponse qui découlent de l'étude préliminaire sur un plan qualitatif (l'étude quantitative est en cours):

1. L'ordre, ou toute forme d'organisation dans un sytème polyatomique, *provient d'une délocalisation des contraintes des probabilités* dans les composants atomiques moléculaires ou supramoléculaires, par voie de couplage. On peut parler d'un véritable transfert d'ordre ou d'invariance de structure de l'échelle des composants vers l'échelle du système.

2. Grâce au couplage, la structure collective devient moins aléatoire, en revanche, la structuré individuelle de chaque composant devient plus aléatoire. Les contraintes électroniques hautement quantifiées dans chaque atome pris isolément, diminuent dans la structure polyatomique au bénéfice des corrélations créées entre les divers états électroniques et au bénéfice des contraintes vibration-nelles et rotationnelles.

3. La délocalisation des contraintes au cours du couplage découle directement de la délocalisation des cellules quantiques prises dans l'extension en phase à 6 dimensions, cellules occupées par les électrons les plus mobiles appartenant aux structures atomiques ou moléculaires. Les nombreuses corrélations qui apparaissent dans le volume commun à 6 dimensions (3 de position, 3 de moment) et l'énergie d'échange qui s'y rattache, sont à l'origine d'effets coopératifs spatio-temporels et forment ensemble la base de l'organisation structurale et fonctionnelle du système.

On pourrait évidemment se poser la question suivante: Quel intérêt pratique peut présenter une telle approche pour le biologiste?

Je vais tenter de répondre brièvement à cette question.

L'organisation en biologie est souvent représentée par une pyramide d'organisa-tions faiblement couplées entre elles. Le transfert d'ordre au cours d'associations successives devrait, de ce fait, y être beaucoup plus modéré que dans les structures inorganiques. On peut ainsi procéder pas à pas au cours de l'étude de la com-plexification des structures biologiques, en partant de molécules vers des macro-molécules, des macromolécules vers des assemblées de macromolécules, etc. et évaluer à chaque étape sous quelle forme apparait l'ordre collectif toujours égal ou inférieur à la perte de l'ordre inviduel dans les composants (sans tenir compte de l'entropie négative complémentaire apportée éventuellement par une source d'énergie externe).

A titre d'illustration, examinons quelques modèles simples. Précisons tout d'abord la nature physique des contraintes ou d'invariants dans une structure à électrons très localisés (atomes ou petites molécules). Il est connu que les cellules occupées par des électrons localisés sont caractérisées par un très petit volume de position $(\Delta x \cdot \Delta y \cdot \Delta z)$ et par un très grand volume de moment $(\Delta p_x \cdot \Delta p_y \cdot \Delta p_z)$. Dans le cas d'une telle structure, on peut dire que les contraintes des probabilités, ou l'information structurale, sont essentiellement de nature spatiale et sont très faiblement impulsionnelles. Inversement, si l'on examine une structure riche en électrons très délocalisés (par exemple des cristaux), les cellules de phase occupées par des électrons délocalisés sont caractérisées par de très petits volumes de moment et par de très grands volumes de position. Il en résulte que les contraintes d'information dans une structure à électrons fortement délocalisées sont essentiellement de nature impulsionnelle et se trouvent caractérisés par la structure de bandes. Dans le cas de systèmes biologiques, on a souvent affaire à des structures conjuguées où la délocalisation des électrons mobiles $\pi$ (et des cellules correspondantes) est différenciée et s'étend entre les deux limites examinées tout à l'heure. Dans de telles structures, les contraintes d'information sont mixtes: elles sont spatio-impulsionnelles. Plus la délocalisation d'une cellule augmente, plus sa structure et son information tendent à être impulsionnelles. Par contre, son information spatiale perd en importance et vice-versa. Aux limites, on peut dire qu'en chimie l'information est spatiale tandis qu'en électronique macroscopique, l'information est impulsionnelle et la seule variable significative devient le temps. En biologie, il faut tenir compte de l'espace et des impulsions conjointement.

L'approche par la méthode du transfert d'ordre permet d'obtenir d'autres résultats qualitatifs intéressants. En particulier, on sait qu'une cellule très localisée est protégée par des barrières quantiques élevées. Dans un atome, par exemple, l'intervalle entre l'état fondamental et le premier état excité d'un électron périphérique varie de 5 à 15 eV. Pour obtenir un bit d'information sur la structure atomique, il faut payer un prix énergétique élevé. Par contre, dans le cas du cristal, les cellules délocalisées sont protégées par des barrières quantiques très faibles de $10^{-20}$ eV environ. En l'absence de fluctuations thermiques, le prix du bit d'information dans un tel cas aurait été très bas. Néanmoins, à la température normale, il est nécessaire de tenir compte des fluctuations et toute cellule à caractère informationnel doit être protégée par une barrière supérieure à $kT$ ($kT = 0.025$ eV pour $T = 300°K$). De cet examen qualitatif, on peut tirer quelques conclusions intéressantes sur le plan de la cybernétique biologique.

1. A la température normale, des structures optimisées (capables de fournir des informations structurales suffisamment sûres avec le minimum de dépense d'énergie) devraient être constituées par des cellules de phase protégées par des

barrières quantiques de 4 à 10 *kT*, c'est-à-dire de 0.1 à 0.25 eV environ, selon la probabilité d'erreur admissible.

D'autre part, les quanta d'énergie de la source biologique devraient être légèrement supérieurs à 0.25 eV, compte tenu du rendement de conversion énergie/information.

2. Les conditions ci-dessus sont parfaitement réalisées dans le cas des systèmes biologiques.

En effet, dans les structures biochimiques, le bit d'information est souvent déterminé par la rupture d'un pont hydrogène (énergie de 0.1 à 0.25 eV) et la source universelle d'énergie, l'ATP, fournit des quanta de 0.4 eV environ.

Comme autre exemple d'application de cette méthode d'approche, examinons le modèle cybernétique de l'ADN. On peut distinguer, dans l'ADN, deux structures électroniques collectives: celle du *squelette* comportant tous les électrons localisés de la double hélice et des plateaux basiques et celle des *domaines informationnels* constituée par des pools d'électrons $\pi$. L'invariance du squelette est essentiellement de nature spatiale et provient d'une perte partielle d'invariance spatiale dans chaque atome faisant partie du squelette. Le gain en contraintes collectives dans le système est l'équivalent d'une très basse température statistique de l'ADN.

D'autre part, l'invariance de la structure collective des électrons $\pi$ est de nature spatio-impulsionnelle, et les nombreuses corrélations entre les cellules de phase des électrons $\pi$ sont à l'origine de la grande spécificité informationnelle de chaque plateau basique. Le couplage, enfin, entre les électrons $\pi$ des plateaux basiques voisins fournit une nouvelle invariance collective de nature informationnelle d'où peut dériver un mode multiplex dynamique très spécifique de l'ensemble vis-à-vis des molécules environnantes.

*Nota*: Aux effects coopératifs d'états électroniques il faut ici, bien entendu, ajouter les effets coopératifs des vibrations et rotations moléculaires.

*En résumé*:

1. Dans les molécules, le transfert d'invariance crée une spécificité chimique essentiellement spatiale tridimensionnelle qui confère à ces molécules une *fonction de nature informationnelle*. Dans les cristaux, le transfert crée une spécificité électronique essentiellement impulsionnelle caractérisée par la structure des bandes. Cette structure ordonnée confère aux cristaux leurs excellentes propriétés de *convertisseurs d'énergie* à l'échelle macroscopique. Dans les molécules et agrégats des molécules, le transfert développe une invariance mixte spatio-impulsionnelle où un squelette tridimensionnel abrite contre des fluctuations, une riche information hexadimensionnelle spécifique. Une telle structure collective forme une organisation et peut exercer sur son environnement moléculaire, des effets puissants

*d'asservissement* ou de catalyse, c'est-à-dire modifier profondément les probabilités des réactions chimiques et des conversions d'énergie à l'échelle microscopique.

2.    Au fur et à mesure de l'association des structures différenciées, les effets coopératifs des cellules de phase créent des nouvelles propriétés physiques (voire biologiques, physiologiques, etc.); celles-ci ne sont que des *effets intégratifs qui dérivent de l'ordre quantique transféré des composants vers le système collectif.*

Dans un système où l'ordre et l'information potentielle atteignent un degré tel que le système est capable d'asservir une énergie d'origine externe (capacité de transformer une énergie au bénéfice de sa propre organisation, comme la photosynthèse et la phosphorylation oxydative en biologie) nous nous trouvons en présence d'un système cybernétique naturel. Ce pas est décisif car il offre à une structure organisée des perspectives immenses d'asservir l'environnement, soit en se multipliant au détriment du milieu, soit en organisant ce milieu à son profit. Les limites de l'asservissement se trouvent essentiellement déterminées par les ressources du milieu et par la compétition avec des systèmes analogues. La "finalité" biologique et la sélection naturelle dérivent ainsi directement de la capacité d'asservissement des systèmes biologiques.

P. O. LÖWDIN: I would like to make a short comment on the lectures given by Dr. Fröhlich and Dr. Prigogine. In the discussion, there seems perhaps to be some contradiction between the points of view expressed in these two lectures. In my opinion, there is no contradiction, since in reality the two topics discussed refer to different parts of the theory of the treatment of biological systems.

In order to explain what I mean, I would like to refer to fig. 1, where I have drawn an axis representing the degree of "order" of a system. On the extreme left of this axis, one has a point describing a system in "complete order", for instance a crystal at absolute zero of temperature. On the other end of the axis, one may have a system characterized by complete "disorder" or random-phase distribution. It is well known that all the important concepts of thermodynamics are associated with systems characterized by a certain disorder, and some of the properties of these systems are described by the concepts of entropy, free energy, etc. Phenomena occurring in such a disordered system are often described as "incoherent". Dr. Prigogine has discussed some of the fundamental properties of such disordered

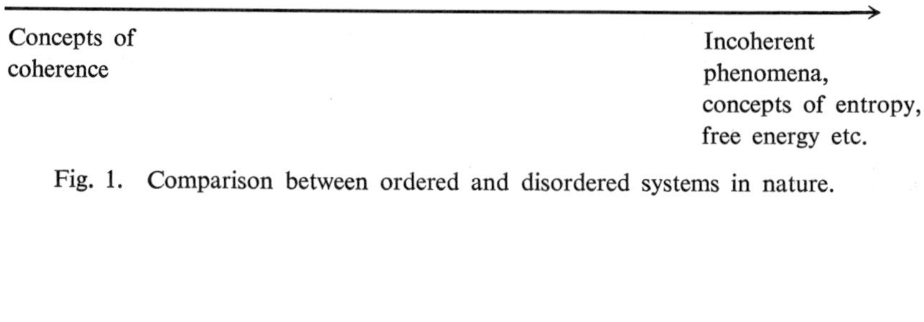

| Well-ordered systems | | Dis-ordered systems |
|---|---|---|
| Concepts of coherence | | Incoherent phenomena, concepts of entropy, free energy etc. |

Fig. 1.   Comparison between ordered and disordered systems in nature.

systems, whereas Dr. Fröhlich has in his talk described phenomena which may best be characterized as coherent or wellordered with respect to phases etc.

In quantum mechanics, the properties of random-phase systems were first discussed by Pauli and Dirac, and Oskar Klein showed that one could derive an irreversible law in quantum mechanics from the randomphase postulate. In biology, the degree of order may vary from one type of system to another. Certain types of substances, like blood, are to a large extent random mixtures and may necessarily be described as liquids, whereas certain cells are so well-ordered, that they are often described as "fluid crystals". Different types of theories are hence needed for treating different types of biological systems.

In treating biological bodies, one should certainly study the macroscopic behaviour by means of classical mechanics and, in the next approximation, one may need thermodynamics and classical statistical mechanics, to understand the general behaviour of the transport phenomena and similar processes inside the body. However, when going inside the cell, it is necessary to go into molecular and submolecular biology, and one is then going to look at rather well-ordered systems of fundamental particles which are subject to the laws of modern quantum theory.

In theoretical biology, the use of statistical mechanics is rather well established, whereas the importance of coherent phenomena is not so well understood. In his example, Dr. Fröhlich has chosen a rather "macroscopic" coherent phenomenon in analogy to superconductivity and superfluidity, and I would here like to mention another example of an even simpler character – the covalent chemical bond.

The coherent nature of the ordinary chemical bond, of fundamental importance in chemistry and biochemistry, becomes particularly clear in the molecular-orbital picture. Let us consider a diatomic molecule, where the valence electrons may occupy the atomic orbitals $a$ and $b$. By linear combination, one may here construct either a "bonding" molecular orbital of the type $\varphi = a + b$ or an "anti-bonding" molecular orbital of the type $\chi = a - b$; see fig. 2. From the hydrogen

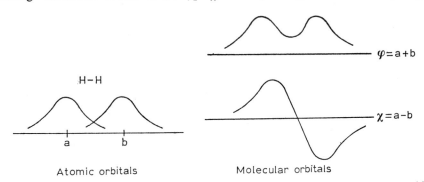

Fig. 2. Comparison between the bonding orbital $\varphi = a + b$ and the antibonding orbital $\chi = a - b$ in the molecular orbital model of the hydrogen molecule; note that the phase factor $\pm 1$ in front of $b$ leads to an energy difference of about 8eV.

molecule, one knows that the orbital energies may differ by as much as 8eV, which is an enormous quantity in comparison to the thermal energy fluctuations given by $kT$, and which depends essentially on a phase factor — the coefficient $\pm 1$ in front of $b$.

The quantum-mechanical coherence can hence give a certain definite order and chemical structure to a biological system, in spite of thermal fluctuations and statistical irregularities. This point was strongly emphasized by Schrödinger in his famous lecture series in 1943 about "What is Life?". He said that many biological processes are so precise that it seems as if they would occur at absolute zero temperature ($T=0°K$), in spite of the fact that they occur at $T=310°K$. This can only depend on the fact that, even at the latter temperature, the thermal energy $kT$ is very small in comparison to the quantum-mechanical energy differences.

From this point of view, quantum mechanics becomes of importance in biology when one starts studying the innermost structure of the cells in terms of elementary particles — electrons, protons, neutrons, photons, etc. — and in terms of atoms and molecules, linked together by covalent bonds, ionic bonds, and hydrogen bonds. One difficulty is connected with the fact that most of our familiar macroscopic concepts — entropy, free energy etc. — are well-defined for disordered systems, whereas the corresponding conceptual framework for ordered systems is still partly missing.

I believe that these questions will be of even greater importance in the future, and one may wonder what contributions quantum mechanics may have to offer to the development. It should be remembered that quantum mechanics is basically only a theoretical tool for handling experimental information. It is based on the Schrödinger equation:

$$\mathcal{H}\Psi = -\frac{h}{2\pi i}\frac{\partial \Psi}{\partial t},$$

and the problem is to determine the solution $\Psi$ corresponding to the "initial condition" $\Psi = \Psi_0$ for $t=t_0$. The solution may be written in the form

$$\Psi(t) = U(t; t_0)\,\Psi_0,$$

where $U = U(t; t_0)$ is the evolution operator connected with the system. Today one knows many methods to determine the evolution operator mathematically, and the big question is instead: "How does one determine the initial wave function $\Psi_0$ to absolute value and phase for a given physical situation at $t=t_0$"?

The stationary states of a system are determined by the eigenvalue problem

$$\mathcal{H}\Psi = E\Psi,$$

underlying the theory of atomic and molecular spectra, and, even if this specia. problem is of a more ab-initio-character, one should remember that most problems in molecular biology are of time-dependent nature and require initial conditions obtainable only from the experimental experience.

In this connection, one should also remember that quantum mechanics is a special case of quantum statistics, which is based on the use of density matrices $\Gamma$ satisfying the time-development equation

$$\Gamma(t) = U\Gamma_0 U^\dagger,$$

where $\Gamma_0$ is the density matrix for the "initial state" at $t = t_0$. This little analysis implies that, even if quantum mechanics can provide a certain conceptual framework for the discussion of the behaviour of elementary particles in molecular and submolecular biology, all the initial conditions have to come from experiments.

J. D. COWAN: I would like to comment on Dr. Cohen's and Dr. Longuet–Higgin's remarks concerning the microscopic and macroscopic aspects of biological modelling.

There are three cases where statistical mechanics has been applied to biological problems without any connection whatsoever with quantum mechanics. One is the generalization of Volterra's population mechanics carried out by E. H. Kerner. Here it is very natural to use classical statistical mechanics to describe the interaction between predatory species and their prey. It turns out that it is very useful to use statistical mechanics, because if one examines ecological data, for example, the incidence of foxes caught in Labrador over a long interval, unless one knows what to measure, what parameters of the ensemble to choose, it is very difficult to determine what is going on. Corbet, Fisher and Williams worked on population mechanics without really understanding the nature of the underlying ensemble, and very cleverly produced empirical distributions for the catches that fitted the data. It was only when Kerner introduced the Gibb's ensemble and showed that the concept of temperature was so useful that he was able to deduce the CFW theory and to infer just exactly what was going on in the population as a whole.

Recently I have done something very similar for neuron populations. Microelectrode recordings are currently being made of individual neuronal activity without sufficient knowledge of the relevant population activity. Here, again, one needs to know something about the ensemble before one can study interactions. B. C. Goodwin has recently worked out a Gibbsian statistical mechanics for Protein synthesizing systems, using the Jacob–Monod circuits. He has shown that there are oscillations in Protein concentrations that result from the feedback inhibition of the production of Messenger RNA by the by-products of metabolism. Again, there is a useful ensemble here, but it starts to become a little too difficult, because the interaction between units, i.e., between protein synthesizing circuits, starts to become a lot "stronger" than in the other cases.

But all this is in the equilibrium case. It is in the non-stationary case that things really become difficult, and I think it is here that the basic biological problems arise, and here that the concept of "information" enters. I use the term "information" not in the static sense of the communication engineer, but as originally introduced by Wiener. He defined information as the name for the process by which the organism adapts to its environment and makes the results of its adaptation effect the environment. Information, then, is a relation between organism and environment. But this means that associated with information processing, there must be a change of state of an organism. It must "switch" from one state to another. However, apart from the necessary minimum requirement of about one "$kT$" per "bit" to signal the switching, there is no other fundamental connection between information and energy, and I, therefore, view the equivalence of information with "Negentropy" postulated by Brillouin with reservation. The main issue is that of the stability of the states between which switching is to occur, and I was therefore very interested in Prigogine's remarks, in which he showed how both dissipative and non-dissipative phenomena are important for the stability of states of interacting substances. It is this kind of approach that will lead us to an understanding of the relation of microscopic to macroscopic biological phenomena, rather than any extension of quantum mechanics.

H. HAKEN: Dr. Fröhlich has stressed the importance of the concept of order in physics, particularly in superfluidity and also in lasers. We have further seen a diagram of order and disorder, which Dr. Löwdin suggests also for the formation of biological systems. On the left-hand side ("disorder") we have for instance the concept of entropy and random phases.

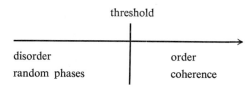

On the other side ("order") we have in this group the phenomenon of coherence. We had another discussion as to where quantum mechanics comes in and where classical physics begins. I want to present a model which we have treated exactly and which shows the complete transition from disorder to order. It means one can show how a completely random system organizes to become a completely organized (or coherent) system. If one makes such specific statements one has to have a very simple model which is fortunately provided in physics by the laser.

First I have to make a few comments how lasers work. I am apologetic to the experts and to the other participants because one can read it in every newspaper how lasers work. But nevertheless we have to make a few comments in order to show the essential points. A laser consists of two mirrors with an active material in between. The role of the active material is to emit light. If you excite the atoms they start to emit light in all directions. Thus the energy is distributed over an enormous number of degrees of freedom of the light-field.

In order to make the laser work we have to reduce the number of degrees of freedom. This is one of the points Dr. Fröhlich has made. If you have these two mirrors you can select those waves which go in the axial direction. If such a wave hits the mirror it is reflected back and forth. So this mode of excitation has a chance to stay a rather long time in the laser. This is the first means to reduce the number of degrees of freedom, because all other modes are now unimportant. In radio technique you would say you have a filter. The question to the biologist would be what is the analogue of a filter. We don't know.

Next we look only at light which is emitted in the axial direction and we consider the emission process in more detail. We have atoms which are excited at random and we know they emit light spontaneously. That means one atom emits a light wave at a certain moment while another atom emits another light wave, etc. So you have a random superposition of waves which corresponds to complete disorder.

Now let me turn to a mathematical description which shows quite clearly how the transition from disorder to order comes in. As you know we can describe a light field by electromagnetic waves with an electric vector $E$ and we can decompose the electric vector in an amplitude which is time dependent $b(t)e^{-i\omega t}$ ($\omega$: mode frequency) and a factor $g(x)$ which describes the spatial behaviour of the wave:

$$E = (b(t)e^{-i\omega t} + b^*(t)e^{i\omega t}) \cdot g(x).$$

I want to describe what happens to the amplitude $b(t)$ and how one may treat it in a quantum mechanical manner.

In quantum theory the light amplitude $b(t)$ is the creation operator of a light quantum. Then we can write the following equation for $b$ or $b^*$

$$\frac{d}{dt}b^* = -\varkappa b^* + F(t) + \text{const. } b^* \sigma. \tag{1}$$

Because the light may escape a little bit through the mirrors its amplitude decreases. This is described by $-\varkappa b(t)$ in eq. (1). Then we have the process of spontaneous emission, which we have taken care of by a fluctuating force $F(t)$ acting on $b^*(t)$. $F$ is a sum over the contributions of all atoms $\mu$ which emit light at random times with a certain "force" $F\mu$. I don't want to show how these forces describe spontaneous emission. I just mention that they are of both quantum mechanical and statistical nature.

The third term on the right hand side of eq. (1) represents the role of stimulated emission. An atom with two levels which is hit by a light wave absorbs light if it is in its ground state. We know that if the atom is in its upper state it performs the process of stimulated emission, i.e. the light amplitude is amplified.

According to eq. (1) the temporal change of $b(t)$ is proportional to the light amplitude and to the degree of inversion. $\sigma = N_2 - N_1$ when we have $N_2$ atoms in the upper state and $N_1$ in the lower state.

We have an absorption if $N_1$ is bigger than $N_2$ and vice versa, i.e. the sign of $\sigma$ is essential. If $\sigma$ is positive the light wave increases, if $\sigma$ is negative it decreases. Now we come to the second aspect of our problem which might be discussed to a large extent namely the concept of feedback, or with other words of self-control. We know that one can write a second equation for $\sigma$

$$\frac{d}{dt}\sigma = - \text{const}' \cdot b^*b + \frac{\sigma_0 - \sigma}{T}. \tag{2}$$

The time derivative of $\sigma$ decreases proportionally to the intensity of the light field. When thus the light intensity becomes too large $\sigma$ decreases. This is compensated by the external pumping. The last term in eq. (2) describes this pumping, i.e. $\sigma$ approaches a certain value $\sigma_0$ if there is no field $b$. But now as you may see we may integrate this equation in a simple way if the relaxation time $T$ is short so that $\sigma$ can follow adiabatically the light field:

$$\sigma = \sigma_0 - T \, \text{const}' \, b^*b. \tag{3}$$

When we insert (3) into (1) we obtain a non-linear quantum mechanical equation

$$\frac{db^*}{dt} = - \varkappa b^* + F(t) + \text{const} \cdot b^*(\sigma_0 - T \, \text{const}' \cdot b^*b), \tag{4}$$

which had been discussed in the *classical* domain by a number of physicists. What I want to show is that this equation with the noisy driving force $F(t)$ allows us to explain completely the transition of disorder to order.

a) *Disorder:* The force $F(t)$ is the sum of many contributions. When the light amplitude is very small we may neglect the non-linear terms. Then we have just a differential equation with a term which consists of many statistically independent contributions. As a consequence also $b$ is a superposition of random contributions. Thus we find complete disorder, which is usually observed. Thermal fluctuation or spontaneous emission produce complete disorder.

b) *Order:* We now change the parameters $\sigma_0$ which describes the degree of inversion. We then come to a state which is stable. In order to explain this let me rewrite eq. (4) a little bit differently and let me add a term $\varepsilon \ddot{b}$, where $\varepsilon$ is a small quantity

$$\varepsilon \frac{d^2}{dt^2} + \frac{db^*}{dt} + \varkappa b^* - \text{const} \cdot b^*(\sigma_0 - T \, \text{const}' \cdot b^*b) = F(t). \tag{5}$$

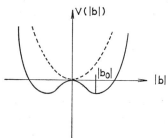

This is now the equation of a particle moving in the potential field

$$V = -\text{const}'' \cdot \tfrac{1}{2}|b|^2 + \text{const}''' \cdot \{\tfrac{1}{4}|b|^4\}; \text{const}'' = \sigma_0 \cdot \text{const} - \varkappa$$

and subject to an additional force $F(t)$.

If $\sigma_0$ is small the potential looks like the dotted line. That means when we excite the "particle" (the light field) it falls down again so that it performs a damped oscillation.

On the other hand if the parameter $\sigma_0$ becomes bigger and bigger we find the potential given by a solid line. That would mean that the "particle" (or the light field) is now stabilized at a certain point $b_0$. This is a case of more complete order. One can show exactly that a classical quantity $b_0$ arises out of the quantum mechanical quantity $b^*(t)$ so that the light field is a classical number. You see if we have thermal or spontaneous emission fluctuations the light field relaxes to a stable $c$-number amplitude. Thus we have only small quantum fluctuations around a stable macroscopic state. If we increase $\sigma_0$ the hills of the potential around $b_0$ become steeper and steeper and so the effective noise which is produced by this spontaneous emission becomes smaller and smaller.

I shall not speak more about these things. Perhaps I should conclude with the following remark. When one looks at the statistics in the region a) where we had many independent contributions, then one finds a Gaussian distribution of $b(t)$ which corresponds to Bose-Einstein statistics. On the other hand if the non-linearity becomes important you may neglect the fluctuations and then you get a Poisson distribution. That means if one measures the light coming out of a laser while one changes the pumping parameter one may find a transition from complete disorder to order. This prediction was indeed experimentally checked in the meantime and fully substantiated. I think we have a very simple but also realistic model of such a transition.

H. Fröhlich: I should like to make a very little remark. This is very instructive of course, mathematically a very simplified model and it tells you how such an ordered oscillation can build up order and disorder. I think a corresponding model could be made in connection with the longitudinal mode of which I have spoken of. Of course these models with reflecting mirror would be biological molecular processes which are stimulated by the radiations. There is no emission of radiation. Instead there would be frictionnal terms. I think such a model could be developed.

L. Rosenfeld: I think that it is misleading to describe this as a classical effect because the quantity $b_0$ contains the parameter $d$, which is a quantal parameter, depending essentially on Planck's constant.

M. Dalcq: Ce n'est pas sans appréhension qu'un biologiste préoccupé des problèmes du développement et de la vie cellulaire se risque à intervenir après les communications réellement transcendantes qui viennent d'être entendues, et qu'il n'a guère pu — il n'est probablement pas seul dans ce cas — saisir intégralement. Qu'il lui soit d'abord permis de remercier nos collègues spécialisés en physique théorique de l'effort qu'ils ont consenti pour permettre aux biologistes d'approcher la démarche de leur pensée.

J'imagine, à tort ou à raison, qu'il peut être utile de confronter avec leurs points de vue celui qui se dégage d'une fréquentation assidue de la vie cellulaire, notamment sous les aspects spécialement constructifs qu'elle prend dans le développement.

Accordons-nous d'abord sur cette notion très générale que l'être vivant, quel que soit le règne dont il relève, est, sauf si ses dimensions sont très réduites, d'une grande complexité, qu'il est caractérisé par une organisation comportant des niveaux emboités les uns dans les autres, que ses activités fondamentales sont extraordinairement diverses et variées. Permettez-moi maintenant d'envisager spécialement cette activité biologiquement universelle qu'est la division cellulaire. Cette activité si générale semblerait, à première vue, devoir être pratiquement univorme—Or, ce n'est pas le cas.

Ses prémisses, qu'il serait trop long de résumer, comportent déjà bien des variantes, tant au point de vue cytologique que biochimique, et ce qu'on en sait reste d'interprétation difficile. Elles aboutissent en général, à l'exception de types cellulaires inférieurs, à l'édification d'un appareil mitotique dont la forme, le plus souvent en fuseau, et la structure varient considérablement. Lorsqu'alors la division se réalise, on s'attendrait à observer uniformément un allongement du corps cellulaire dans le sens de l'axe du fuseau, puis un étranglement médian perpendiculairement à cet axe. Or, dans nombre de cas, il n'y a pas de déformation préalable, et c'est par un remaniement interne qu'une cloisonnement s'accomplit. On voit donc que la cellule recourt à des mécanismes divers pour assurer un même résultat.

Je voudrais maintenant faire une remarque de portée plus générale, à savoir que, dans l'existence de tout être vivant doué d'une organisation quelque peu complexe, il convient de distinguer deux phases, celle de devenir ou d'ontogenèse, celle d'état définitif, souvent dit adulte.

Dans ce dernier, l'organisation spécifique est acquise, il s'agit d'un système morphologiquement stabilisé, bien que tous ses matériaux constitutifs soient constamment renouvelés. Il n'en va pas seulement ainsi pour les populations de molécules disposées au sein des cellules, mais aussi à la surface de celles-ci, dans leur membrane limitante, dont la structure est si importante. Dans cette phase d'apparence statique, on peut, dans certains cas, comme M. Duchesne l'a déjà rappelé, suspendre momentanément le cours des processus vitaux soit par dessication, soit par refroidissement par un procédé adéquat. Il semble bien que la possibilité si remarquable de cet arrêt réversible soit liée à celle de respecter intégralement la microstructure, tout en suspendant les échanges. C'est dire que la notion de vie est, pour la période d'état, moins liée au métabolisme qu'à la forme.

C'est ce qu'affirmait implicitement ce matin M. Fröhlich, c'est ce dont Albert Brachet avait une vision très claire lorsque, il y a juste quarante ans, il intitulait un de ses ouvrages généraux: "la Vie créatrice des formes".

Cette allusion m'amène à considérer la période ontogénétique. Pour celle-ci, Courrier et Marois ont montré que l'hypothermie suspend l'ontogenèse chez le rat, ce qui n'étonne pas, car on connaît, dans la nature, certains exemples de période stationnaire dans le développement. En tous cas, la phase créatrice se caractérise par une évolution intense du métabolisme. Il ne s'agit guère de cycles, de processus se répétant périodiquement, comme c'est le cas dans l'organisme adulte, mais de progression, notamment quant à la synthèse rapide de macromolécules, particulièrement de protéines, ce qui implique un perfectionnement marqué de l'équipement enzymatique. Cependant, aux yeux des embryologistes, la complication progressive du chimisme, avec son perfectionnement graduel, n'est qu'un ordre de faits, sous-jacent à la morphogenèse en soi. Celle-ci suppose l'acquisition de structures localisées à certaines régions du germe et qui préparent l'apparition des organes primordiaux. Il est indispensable de rappeler ici le cas des Amphibiens, chez lesquels les premiers indices de structure morphogénétique surgissent par différenciation du *cortex* de l'oeuf fécondé. Ce qu'on nomme ainsi est plus que le plasmolemme; c'est à la fois celui-ci et une fine pellicule de cytoplasme périphérique. Avant la fécondation, on peut dire, en schématisant quelque peu, que sa structure est uniforme. A un moment précoce du développement, des particularités apparaissent dans une certaine région corticale en forme de croissant. Cette différenciation va régir la polarité céphalo-caudale et la symétrie bilatérale du nouvel organisme. Je n'en dirai pas plus à ce sujet, car je pense que M. Etienne Wolff le traitera plus amplement dans une séance ultérieure.

L'exposé si attachant que nous a fait M. Prigogine tend à montrer qu'une hypothèse relevant de la physique théorique pourrait aider à interpréter ces faits de structuration progressive de la cellule-oeuf. S'il en était ainsi, cette manière de voir éclairerait sans doute aussi l'extraordinaire capacité de régulation, c'est à dire de réaction finalisée à certaines agressions, que possède l'oeuf jeune. Bien que l'on puisse tenter d'interpréter une telle régulation en restant sur le plan embryologique, ce serait un progrès remarquable que d'intégrer cette conception dans une théorie beaucoup plus vaste.

J. POLONSKY: Je crains que l'expression utilisée par M. Haken ne puisse conduire à une confusion. M. Haken a parlé d'une transformation de désordre en ordre et l'on pouvait en déduire que dans un laser, la transformation s'opère de cette manière. Je ne pense pas qu'il en soit ainsi. Il ne faut pas négliger, dans le cas du laser, l'ordre apporté par la source de pompage. Le laser fonctionne, en réalité, comme un convertisseur d'ordre où l'entropie négative d'inversion des populations est transformée en entropie négative de cohérence du faisceau laser.

Le laser transforme une distribution improbable d'états énergétiques en distribution improbable de phase. Je pense que M. Haken est d'accord avec moi sur
se point.


H. HAKEN: I will contradict your opinion because the pumping occurs completely
incoherently that means that each individual atom is pumped completely differently
from each other and in a random fashion. Nevertheless, in spite of this completely
random process you find a completely coherent system.


S. L. SOBOLEV: Je voudrais faire une remarque visant les mathématiciens ici
présents. Il est curieux que nous ne soyons liés apparemment ni avec les biologistes, ni avec les chimistes, alors que toute la science nous montre — dans de
nombreux exemples — qu'avant toute découverte dans les sciences naturelles,
la physique, la chimie etc., une nouvelle théorie mathématique s'était édifiée,
provoquant une révolution générale dans toutes les disciplines. Mais il se passe
un demi-siècle avant d'appliquer cette théorie nouvelle au développement de la
science.

Les exemples? Je peux en fournir quelques-uns. Prenons l'histoire de la théorie
de la relativité. C'est une géométrie non euclidienne, éclose au XIXème siècle,
qui mena à cette théorie de la relativité, mais il se passa plus de cinquante ans
avant que celle-ci vit le jour. Qu'il me suffise de citer les noms de Lobatchewsky,
Diels, d'abord Riemann, ensuite Dale (le grand analyste Dale). Actuellement,
les liens entre physique théorique et biologie se font à un niveau déjà bien connu,
qui est celui de la physique quantique, peut-être de la physique classique, ou
encore des équations aux dérivées partielles, dont nous a parlé M. Elsasser.
Nous voici maintenant les témoins d'une nouvelle branche des mathématiques,
la cybernétique, qui va engendrer des conceptions inédites, lesquelles à mon avis,
bouleverseront la biologie et la science exacte, c'est-à-dire la physique et maintes
disciplines. Que fait le physicien? Il applique tout ce qu'il sait à la biologie. C'est
très bien. Mais cela suffit-il? J'en doute. Veuillez excuser cette petite critique à
l'égard des physiciens.

Plusieurs orateurs, ont parlé de problèmes relativement nouveaux. C'est de la
cybernétique. Mais je n'ai pas entendu évoquer, en la matière, de nouvelles
découvertes qui pourraient féconder la biologie. Ce que nous attendons, nous
mathématiciens, quand nous en parlons entre nous.


R. WURMSER: J'ai l'impression que les biologistes sont moins pessimistes que
les physiciens dans leurs jugement sur la biologie. Par exemple, ils croient pouvoir
établir des équations. J'entends par là qu'ils connaissent de très nombreuses lois
quantitatives les unes statistiques, les autres portant sur des comportements
individuels.

M. Dalcq rappelait tout-à-l'heure la complexité de la division cellulaire. Néanmoins il sait — et beaucoup mieux que moi — que l'on peut prévoir à une minute près dans des conditions précises les premiers stades de l'évolution d'un oeuf d'oursin fécondé et, avec une bonne approximation, la suite de son développement jusqu'à l'état de larve. Les neurophysiologistes savent provoquer à coup sûr le comportement de la faim, de la soif, du sommeil, en stimulant à l'aide d'une microélectrode un endroit précis du cerveau. Si bien que la position des biologistes est généralement différente de celle des physiciens. Ils sont les premiers à s'étonner de leurs prévisions étant donné qu'ils connaissent bien l'hétérogénéité essentielle du matériel vivant. C'est pourquoi ils recherchent des modèles mécaniques. Ils expliquent les choses par des templates, des surfaces complémentaires, des déformations, d'une manière générale par des agencements mécaniques, ceux-ci éliminant au maximum l'effet des bruits dus au désordre thermique. Tout au moins ils croient que cela leur est permis. Ils sont donc inquiets quand on leur démontre — ou qu'on essaye de leur démontrer — qu'ils ne devraient pas pouvoir faire des prévisions.

A. LICHNEROWICZ: Quel est leur problème alors?

R. WURMSER: Les biologistes s'efforcent de trouver des modèles, généralement mécaniques—comme je l'ai dit—mais qui ne sont pas des analogies naïves en ce sens qu'ils sont fondés sur les propriétés physicochimiques de mieux en mieux connues des biomolécules. Certaines difficultés que soulève l'application de la mécanique quantique ne sont pas réservées à la biologie.

Sauf aux confins de la neurophysiologie — touchant à la psychologie — où malgré des tentatives audacieuses existe encore trop d'inconnu — et en ce qui concerne l'origine de la vie, les biologistes croient comprendre dans une certaine limite ce qui se passe et que leurs modèles — est-ce une illusion? — ne sont pas contraires à la physique connue.

L. ROSENFELD: Oui, nous sommes d'accord.

A. FESSARD: Je suis aussi physiologiste. Donc, je fais partie de la famille des biologistes. Et je voudrais répondre brièvement à ce qu'a dit M. Wurmser.

Je crois qu'il y a deux catégories d'attitudes, et que l'on adhère soit à l'une soit à l'autre. Il y a l'attitude "réductionniste" — il me semble, M. Wurmser, que vous faites partie de cette catégorie . . .

R. WURMSER: Un peu . . . .

A. FESSARD: Et il y a ceux qui pensent que la physique d'aujourd'hui n'est pas capable d'expliquer tous les phénomènes de la Vie. Mais la physique actuelle n'est pas une science terminée. Et si je ne suis pas moi-même réductionniste, c'est uniquement pour le présent. J'espère que la physique, qui a assez montré sa vitalité, se développera encore et nous révèlera de nouvelles lois.

S. L. Sobolev: Plutôt les mathématiques que la physique, je pense.

A. Fessard: Oui. Quand je dis "physique", bien entendu, je pense aux cadres abstraits qui permettent à l'esprit humain de comprendre les phénomènes physiques; et, finalement, je conçois la physiologie comme une physique de l'être vivant, mais une physique qui est loin d'être à l'état de maturité où en est la physique mathématisable des phénomènes inorganiques.

L. Rosenfeld: Je me range plutôt du côté de M. Wurmser. Il me semble que la physique actuelle, disons la mécanique quantique, puisqu'il s'agit après tout de phénomènes moléculaires, est suffisamment bien établie pour que l'on puisse risquer, en tant que physicien, l'affirmation qu'elle doit suffire à rendre compte des mécanismes biologiques.

On sait, par les biologistes, que l'organisme est fait de composés moléculaires. On connaît l'ordre de grandeur des énergies en jeu, énergies chimiques même assez faibles; les mouvements ne sont pas très rapides. Donc on est dans les conditions de validité de la mécanique quantique, en laquelle on a confiance, car jusqu'à présent, dans tous les cas où on a tenté de l'appliquer, les résultats ont été satisfaisants. De ce point de vue, donc, aucune question de principe n'est soulevée. On ne demande pas de faire appel à des principes qu'on ignorerait encore. Pour prendre une analogie tirée de l'histoire de la physique: on connaît, depuis 1911, la supraconductivité; on ne l'a comprise que tout récemment. Dans l'intervalle entre la connaissance de ce phénomène et la théorie qui enfin en rend compte, en remontant aux premiers principes de la théorie quantique, personne n'a jamais cru que l'on aurait à faire appel à un nouveau principe, encore inconnu.

Sommerfeld, je m'en souviens, citait la supraconductivité comme la honte des physiciens, puisqu'ils étaient incapables de rendre compte du phénomène alors qu'ils avaient en main tous les éléments pour le faire. La raison pour laquelle cela a duré si longtemps est — comme on le comprend maintenant — que le phénomène présentait une complication d'un genre très particulier: un effet "coopératif", faisant intervenir un nombre d'éléments (d'électrons, dans ce cas) assez grand, mais non très grand. Et ce sont justement les systèmes composés d'un nombre "intermédiaire" d'éléments qui sont les plus difficiles à traiter. Aux systèmes d'un très grand nombre d'éléments, on peut appliquer des méthodes statistiques bien développées. Les processus élémentaires, moléculaires, sont aussi accessibles, parce qu'ils ne comportent qu'un petit nombre de paramètres. Mais les phénomènes coopératifs dont nous parlons font intervenir un nombre de paramètres insuffisant pour l'emploi des méthodes statistiques, trop grand cependant pour que les méthodes détaillées applicables à de très petits systèmes puissent être utilisées.

C'est plutôt là que je verrais l'origine de la grosse difficulté que nous avons à comprendre les phénomènes biologiques. C'est ce que M. Dalcq rappelait très justement: il ne faut pas se hâter de faire des généralisations, alors qu'on sait que l'on a affaire à un nombre considérable de facteurs, qui, probablement, jouent tous un rôle essentiel.

M. MARGENAU: I have hardly the right to claim your attention because I am here as a mere philosopher of science, not as an expert in biophysics. However, during this day of listening to various speakers, some questions have arisen in my mind, and since they are likely to have occurred to others, I would like to ask them in the hope of obtaining clarifications.

As a preamble to my inquiry, let me make a few pronouncements which will suggest my premises or, perhaps my prejudices. The issue of reducibility has recurred in our discussions. May I add that I believe that the circumstance that the phenomena of biology are in a measure explicable in terms of theories well known from physics and chemistry does not imply that new principles may not be required to understand them fully in the end. To make this evident let us start with classical quantum mechanics. To a certain extent we can comprehend the behavior of an assemblage of particles in terms of these basic laws. But we shall never understand thermodynamics without introducing *additional principles* concerning random phases, irreversibility, and so forth. In other words, as we pass from one level of complexity to a higher level of complexity two things seem to become necessary. One is to introduce concepts like temperature, phase, entropy, pressure and so forth which have no relevance whatever for individual particles. They become relevant only in a higher, more complex domain. Secondly, it has usually happened in the history of science that in these higher domains new laws need to be formulated which are applicable only to these higher-level phenomena but are never contradictory or even applicable on the lower levels. This may well be the situation as we pass from physics to biology.

Next I would like to say a few words about the relation between quantum mechanics and classical physics, a topic which was discussed a great deal this morning. Can we not agree that no classical theory of physics is acceptable if it contradicts the laws of quantum mechanics? In other words, do we not agree that quantal mechanics is the more general theory and that classical physics is a special case of it? Thus the question that was asked this morning concerns the relations between a logical system of larger range to one of smaller range; it has nothing to do with a change in basic epistemological relations.

Now my question: what is the magic of phases? The concept of phases has been injected into our discussions and invoked quite frequently as though the meaning of this term were clear. Many problems have been said to be resolved by means of phases. The context has not always made clear to me whether the word stood for state of aggregation, for phase in the sense of phase space, or the renowned phases of the state function in quantum mechanics. My comments will refer only to the latter. There are clearly two ways of describing a complex biological system. One is in terms of pure states, of unique vectors in Hilbert space. Such a vector does, indeed, have a determinate phase, and this phase is, contrary to popular opinion, observable. But I see no relevance of it for biology at all. If it does have a significance, I should like to know what it is.

Secondly, composite phases appear when a pure state in Hilbert Space is compounded out of other states. A superposition of states in terms of a complete set of vectors involves phases of the individual vectors as well as resultant phases of the entire state. Here the individual phases are very important in determining physical effects, but again their significance for biology escapes me.

For in the first place the composition in question is entirely arbitrary and can be performed in an infinite number of ways. Hence I confront a difficulty. But there is a way of performing this composition which is based on the fact that an organism consists of a great number of individual parts. Each part has a quantum mechanical state, and the state of the total system may be written as a unique product of the Hilbert space of the individual parts. This is what Dr. Löwdin had in mind this afternoon, and the procedure is meaningful if one can suppose that the total organism is significantly made up of constitutent parts. It is a most fruitful approach in the simpler parts of chemistry.

The example chosen this morning was the superposition of the state of two hydrogen atoms to form what is called a bond. Now this is more or less all right. But, remember, this kind of superposition of individual parts into the state of a total system has meaning only if the total system is indeed in some sense separable. Separability does, in fact, prevail in the cases discussed this morning, in the superposition of hydrogen atoms as well as in the superposition of the states of assemblies of atoms and a light wave, i.e. the laser. These are proper examples only, however, because they permit us to start with an assumption of spatial separability as a good approximation. In a biological system the interactions between cells, and the parts of a cell, are largely unknown in their dynamical details, but are likely to be crucial and strong? Is this still a good approximation? It seems to me that there is a real likelihood that we may not describe an organism by means of a wave function that makes reference to individual constituent molecules and atoms.

There is one further point I should like to raise. It concerns phases in a complex state which in von Neumann terminology is represented by a mixture. Whether biological systems are mixtures or pure cases is, I think, a very interesting question, and it may indeed be relevant for our discussion. The issue here involves extremely difficult philosophical questions which touch upon such issues as the objectivity of quantum states, the theory of measurement, and indeed the theory of knowledge.

What troubles me is this. If the state of an organism is a quantum mixture, then the phases of the ingredient states are again not unique, for a mixture can be decomposed into pure states in a great many different ways (provided there is no degeneracy). The question as to which phases are important, therefore, raises its head again and I should be at a loss to answer it.

S. BRESLER: I have some questions connected with the lecture of Dr. Fröhlich. There were some statements, some predictions in this lecture which can be verified experimentally, but to try to find the phenomenon we need some estimation of its size, some estimation of the amplitude of oscillations of the electric field or the dipole moment. So I would ask Dr. Fröhlich if he can make an estimation of the order of magnitude of the effects.

A second thing which I find difficult to understand is what can be the generating system for such oscillations of frequency $10^{11}$ and even higher. The single source of energy in biology are chemical reactions and mainly the cleavage of ATP molecules. If you try to think of some specific times, say of the cleavage of a ATP molecule, you always come to frequencies which are six to seven powers less than $10^{11}$. So I find it difficult to understand the mechanism of generation of such kind of oscillations.

H. HAKEN: I would like to make a comment on Dr. Margenau's comment with respect to the phases. I agree with the largest part you have said. But nevertheless I think in respect to phases I would like to avoid a misunderstanding.

In the laser we are not at all interested in the absolute phase of the laser light. In other words I agree with what you said, if you have a space vector in Hilbert space, you can choose any other state vector which has the same meaning. But what we are interested in in real life are correlations. An example is the correlation function $\langle b^*(t')b(t)\rangle$. Say we measure the light amplitude at a time $t$ and later at a time $t'$. The absolute phase is unimportant but the relative phase is. That means we can choose any state vector here, but nevertheless the physics remains the same. Now comes a remark if the concept of coherence might be in some relation with biology or not. What does this mean?

If $b$ at the time $t'$ describes a process of very short memory then the $b$'s are uncorrelated at all. What we find in biology is that living things retain their form or their function over very long time. Dr. Fröhlich's point of view was that we have phenomena which have a very slowly decaying phase. I think what we must look for is a model which explains us how we can understand that on the microscopic scale the things decay very fast in $10^{-8}$s but on the macroscopic scale like the laser they decay only in minutes or hours. This is one of the problems. It may well be that what we are saying is completely irrelevant to biology but it also might direct biologists in further directions.

H. Fröhlich: I should like just to reply to Dr. Bresler. The first question is can we predict anything more definite as far as the frequency is concerned. We cannot, but we can notice that my range agrees with that mentioned by Dr. Careri for molecular frequencies. As far as the strength of the field is concerned one could indeed make a certain prediction but this would of course require some computation.

As I mentioned this morning a simple prediction can be made in the beginning when the processes are linear but when they become non-linear then one has to know quite a bit more about the surface tension of the cell and some other quantities. Since you were interested in these experiments I would be prepared to set up these calculations and give a definite answer. The second question I cannot answer, I know there must be a process which builds up just as for lasers but I do not know what it would be.

H. C. Longuet-Higgins: I would like to follow up Dr. Margenau's point with what I regard as an even stronger argument for the irrelevance of the quantum-mechanical phase to the main problems of biology. Namely, when one describes a physical system by a wave function, or even by a density matrix, one must have an absolutely clear idea as to the physical boundaries of the system. It is very unclear to me what are the physical boundaries of an organism. For example, in trying to describe me would you include in your description the air which is in my lungs? The food which is now in my stomach, being digested? Because if you are in any doubt at all on these questions, you ought to be in very grave doubt about the possibility of describing me by a wave function. That is my first point.

Another point: If we *must* attempt to formalise the relation between quantum mechanics and the properties of living systems, let us use a Kubo type formalism rather than an elementary dynamical description, because all we can in fact do is to determine *correlations* between macroscopically observable quantities.

Now may I refer again to something that Dr. Tisza said. I regard it as exceedingly important, so perhaps he will forgive me if I repeat his point that there is a real problem in maintaining structural stability in an organism against thermal and quantum-mechanical fluctuations. His point was that this problem is in fact solved for us by the fact that one can separate two kinds of degrees of freedom from one another very well indeed. The fact that one can talk about benzene as benzene, or even of a particular benzene molecule as having a constant identity — even when it is under bombardment by other molecules — arises from the fact that no *electronic* transitions occur inside the molecule when it is knocked around fairly gently, though of course energy is transferred to and from other degrees of freedom. This seems to be the answer, formulated much more clearly than I ever heard before, to the question of how we can have, at the same time, highly random correlations in some degrees of freedom and highly persistent correlations in the degrees of freedom that really matter, namely those that maintain the order of the atoms or groupings in large and elaborate biological molecules.

L. TISZA: It seems to me that the foregoing discussion suffers from an ambiguity surrounding the term "determinism" that is no less misleading for being sanctioned by general usage. The difficulty is closely related to the paradoxes of classical physics that I discussed. When people talk about determinism, they usually mean the temporal determinism of classical mechanics as it is actually observed in celestial mechanics. Yet in biology we have a determinism of a quite different sort; say, the development of the seed into the tree. A greatly simplified version of this process is found in the operation of a chemical factory. The two cases have very different qualitative properties. In the mechanical temporal determinism we have to rely on the exhaustive specification of the system, and this includes even the environment if there is any coupling between the two. Moreover, the system should not exhibit any dissipative mechanism leading to the loss of memory of the initial state. In contrast, in the chemical and the biological case the environment should satisfy only some very general requirements concerning the temperature, pressure and the proper supply of raw materials. Also the process is enhanced by a proper dissipative mechanism bringing about the loss of memory of the initial state: namely by stirring and other procedures that speed up the attainment of chemical equilibrium in each intermediate state of the chemical process. The lack of discrimination in the conceptual accounts of such different situations tends to produce conceptual confusion.

H. HAKEN: First of all I agree completely with you Dr. Longuet–Higgins that all we are saying from the physical side might be completely irrelevant for biology. What I would like to say is that in physics we have certain experiences and methods to explain them, for instance "feeding" lasers. One excites the laser all the time by a high flux of incoherent light that means you push the system all the time at random and nevertheless the laser light keeps its form. I think that's amazing, therefore I believe one should think a little bit of it.

L. ROSENFELD: I may perhaps say a few words in my defence. I may repeat, to begin with, what Dr. Haken has just said. It is quite possible that the models that we are discussing are not relevant. However we see general features in the biological phenomena which strongly remind us of those cooperative phenomena which we know in physics. Therefore it is natural to try and follow up this analogy. Now it is clear, as Dr. Fröhlich said, and well known to those who have looked into these phenomena, that the phase of the general wave function describing the system is of primary importance just for describing the correlations, and so it is natural to consider it.

It is quite possible that all the techniques that we have will prove insufficient if we try to apply them to realistic models of biological organisms, but it may also prove that they will work, and I think that from the pragmatic point of view it is a constructive proposal to start with those methods which we know after all do work and to try them on new cases which we do not yet understand. That is the background for my foolhardy expectations.

R. P. Dou: I would like to comment on the possibility of a theoretical biology. I am a mathematician and will follow the line of Drs. Rosenfeld and Sobolev. I am not so optimistic as Dr. Rosenfeld. I do not think that we already know all the fundamental phenomena to construct a theoretical biology. I am also confident that in a short time biology will be formalized, as physics and as quantum theory have been formalized. Therefore we think that, starting from few axioms, theorems will be proved; or at least problems will be set, and when they are set soon afterwards will be solved. Theoretical biology will grow up as theoretical physics has grown up.

The question is whether the axioms we have from physics and chemistry are enough for explaining the formalizing of a fundamental theory of biology. I would rather think this is not the case. We need very much new phenomena. Not only new concepts, new theories for explaining the phenomena that we already know; but we need few phenomena and probably global phenomena, as mathematicians say, global phenomena, which may be quite independent of the sum of the phenomena of its parts. These phenomena are very difficult. The only way to get them will be just to try again and again. To study the biological processes. I am also not so optimistic as Dr. Sobolev that this time again mathematical theory will come before; perhaps it will come later. I do not know which will be the process, but something is needed very much. I think, in order to construct theoretical biology, it is important to get new axioms, and in order to find them, we have to understand new phenomena.

P. O. Löwdin: I would like to say to Dr. Margenau that I whole-heartedly agree with him when he explains his difficulties in understanding the nature of the "phases". Unfortunately, we don't get away from the problem. In classical mechanics, one has a Liouville-space based on the coordinates $(x, p)$ and real functions, whereas, in quantum mechanics, one has an $x$-space (or a $p$-space) with complex wave functions which determine the probability distributions with respect to $x$ and $p$. The physical situation is not fully determined unless we know the "phase" of the complex wave function.

In quantum statistics, the discussions are based on density matrices $\Gamma = \Gamma(x|x')$ instead of wave functions $\Psi = \Psi(x)$, but the phase problem persists. The diagonal elements $\Gamma(x|x)$ of the density matrix are, of course, independent of the phase, but, unfortunately, one needs also the non-diagonal elements $\Gamma(x|x')$ for $x \neq x'$ in the theory, and they are difficult to understand. However, I don't agree with

Dr. Longuet–Higgins that the phases are irrelevant in biology, since it would mean that one would throw away all aspects of the chemical bond. One would not be able to distinguish between bonding orbitals, say $(a+b)$, and non-bonding orbitals, say $(a-b)$, and the entire molecular-orbital approach would have to be abandoned.

H. C. LONGUET–HIGGINS: I speak of very large systems.

P. O. LÖWDIN: For large systems of random-phase character, I would agree with you, but I believe that the situation is different for the highly organized systems in the cells. The phase rule

$$\alpha = \frac{2\pi}{h}(px - Et),$$

given by Louis de Broglie in 1924 forms a very good starting point for studying phases of free particles, the stationary states have usually well-determined phases (except for some irrelevant constant), but one has otherwise little definite instructions how to "measure" a phase for a manyparticle system.

B. B. LLOYD: I have a question for Dr. Fröhlich. He made a prediction from his proposed longitudinal oscillations in which he dealt with the question of the concentration of cells and the distance between the cells, and the effect that these might have on cell-growth. Now if I got it correctly, you suggested that the cells have to be far apart, and that when they divide and increase in number they come closer together, so that these special forces would cease to operate and cell division would tend to cease. As far as I know, the concentration of cells in a very young organism such as a foetus is the same as in you and me, but the cell of course divides quite happily in the foetus, and much more than in adults like you and me, although it is quite true that we both have cell systems which are dividing quite rapidly, for example in the skin and in the bone marrow. Once again both these sets of cells are highly concentrated and as concentrated as in those regions of the body where no division is taking place. The advantages of your model are that they do immediately imply some sort of observational or experimental test; but when the test is applied, it is on the whole against the model.

G. CARERI: As a physicist, I would like to ask the biologists whether there is any evidence of long range effects inside a cell between cell subunits. There are certainly chemical effects which I call first neighbour effects; one molecule acts on the next one, possibly this one on the next one again, and even feedback is possible if there is some circular arrangement of molecules, as is well known from enzymatic reactions for instance. So far for short range effects. Now the point is this. Is there any evidence pro and contra the existence of long range effects (effects that propagate at large distances)? If these do exist, I do not know any other way of explaining them than by the electromagnetic field, and in that case you must have a window in the polarizability of water. Whenever I ask a biochemist, he replies usually that short range effects explain everything. But chemists explain everything in terms of chemistry, I am afraid. I would really like to have this question settled experimentally by biologists: is there anything that cannot be explained purely by short range effects? I hope one day you may give us the answer; this would be an immense benefit for us.

M. DALCQ: A l'intéressante question de notre Collègue, je crois pouvoir répondre que, dans l'ontogenèse, on connaît certaines manifestations qui sont préparées très tôt et restent longuement latentes. Le cas le plus frappant concerne l'apparition de la lignée germinale. Les belles recherches de feu M. Bounoure sur les Amphibiens ont montré que les gonocytes primordiaux dérivent des cellules de la gastrula qui ont reçu un matériel, le déterminant génital, décelable au pôle végétatif de l'oeuf récemment fécondé. Cette notion vient encore de faire l'objet d'une confirmation basée sur l'instillation d'un isotope dans le déterminant. Entre l'apparition de celui-ci et celle des gonocytes, il s'écoule plusieurs dizaines d'heures. On pourrait encore invoquer d'autres faits relatifs d'une part aux Mammifères et d'autre part à des résultats de l'hybridation moléculaire, mais j'ai l'impression que ma réponse ne répond pas effectivement à la préoccupation de notre Collègue?

G. CARERI: Ce que je disais visait seulement la cellule et non un organisme où la chose peut être trop compliquée pour moi.

S. BRESLER: Une petite observation: quand vous avez beaucoup de chromosomes, séparés par des distances microscopiques de l'ordre de grandeur du micron, ils arrivent à se retrouver pendant la méiose et de façon très détaillée. C'est une juxtaposition qui se passe. Les gènes, même les points homologues, arrivent à se retrouver deux à deux. Ici, il doit y avoir un rapport, à mon avis, avec les forces à longue distance. Je ne sais pas d'autre forces moléculaires à longue portée que les forces de Coulomb. Et ces forces peuvent être specifiques pour des réseaux compliqués de charges. Je crois que c'est l'exemple le plus courant, indiquant que la cellule redispose de forces, qu'on peut expliquer, en détail, comme étant à longue portée.

# PHYSICOCHIMIE DE LA VIE: DONNEES ET PROBLEMES

*1ère séance*

PRÉSIDENT R. S. MULLIKEN

B. PULLMAN

Les Calculs Quantiques en Biologie

S. BRESLER

Molecular Forces and Molecular Recognition

B. CHANCE

The Time Domains of Biochemical Reactions in Living Cells

Discussions

*Theoretical Physics and Biology*, © 1969, *North-Holland Publ. Co., Amsterdam*

# LES CALCULS QUANTIQUES EN BIOLOGIE

BERNARD PULLMAN

*Institut de Biologie Physico-Chimique, Paris, France*

La rencontre de la physique théorique et de la biologie me paraît pouvoir se faire sur deux plans distincts:

1) Le plan des *principes généraux*, des développements fondamentaux sur le rôle possible de certains concepts de la mécanique quantique en biologie, sur l'unité ou la complémentarité (ou si certains le préfèrent la non complémentarité) de ces sciences, sur l'universalité des lois de base etc. Nous avons eu hier à l'occasion de l'exposé du M. Fröhlich et de la discussion qui a suivi l'exemple de la façon dont les événements se déroulent sur ce plan;

2) Le plan *d'application* de ces différentes professions de foi et qui est le plan des *contacts* effectifs entre les physiciens théoriciens et les biologistes sur des problèmes qui seront alors naturellement en général nettement *plus restreints* mais aussi *plus précis* et qui *préoccupent* directement ces derniers. Il comporte pour le théoricien la nécessité de relever le défi que lui lance l'expérimentateur de contribuer *effectivement* par ses méthodes et ses techniques à la solution de ces problèmes.

Je voudrais préciser immédiatement que mon exposé se place *entièrement* à ce deuxième niveau, sur ce deuxième plan. Mais je crois que malgré toutes les limites, les limitations qu'il comporte, et qu'il risque de comporter encore longtemps, ce mode d'approche est important par ce qu'il a de positif en ce qui concerne le fond même du problème. En effet, dans une réunion telle que celle-ci il y a toujours le danger de voir le sujet composite qui en est l'objet, surtout lorsque ce sujet se réfère à deux domaines aussi *différents* que la physique théorique et la biologie, de voir ce sujet se discuter sur deux plans parallèles, c'est-à-dire d'assister en fait à deux réunions superposées. L'autre danger est de voir le sujet planer sur les hauteurs olympiennes de grands principes sinon, ce qui me paraît bien pire encore, de hardies mais douteuses spéculations. C'est pour le moins ces dangers là que je voudrais éviter. Or l'un des domaines de recherches qui me paraissent les plus prometteurs pour les contacts utiles entre les physiciens ou les chimistes théoriciens et les sciences biologiques est *l'étude de la structure électronique des biomolécules par les méthodes de la mécanique ondulatoire* qui ont fait leur preuve en chimie quantique. Point n'est besoin de rappeler ici, je pense, les contributions essentielles de la chimie quantique à notre compréhension générale du monde des molécules, depuis le premier travail de Heitler et London, élucidant sur l'exemple de la molécule d'hydrogène la nature fondamentale de la liaison chimique, jusqu'aux travaux les plus récents traitant de la structure électronique des composés autrement complexes. Il suffit de mentionner les noms bien connus des pionniers de cette science tels que Mulliken, Pauling, Slater, Eyring, Hundt, Hückel, Löwdin, Kotani, Coulson, et j'en omets, pour que viennent à l'esprit immédiatement des travaux mémorables, qui ont ouvert la voie à une connaissance approfondie de la structure *électronique* des molécules, avec toutes les conséquences importantes qui en ont résulté pour notre compréhension de leurs propriétés chimiques et physicochimiques.

Dans ces conditions il peut paraître surprenant, à première vue, qu'il ait fallu attendre longtemps, jusqu'au milieu des années 1950, pour voir paraître les premiers travaux de *biochimie quantique*, dans le sens des calculs quantiques explicites de structure électronique des molécules d'intérêt biologique directe (des constituants fondamentaux de la cellule), effectués dans le but conscient de leurs applications aux problèmes biologiques. Comme raison principale de ce retard je vois essentiellement, personnellement, l'ignorance mutuelle de la biologie, de la biochimie par les chimistes théoriciens et de la chimie théorique par les biochimistes. Certes on peut aussi dire que la complexité relative des molécules biologiques, jointe à l'indiscutable faiblesse des méthodes de la chimie quantique (méthodes d'approximation) en particulier dans leurs applications aux grandes molécules ont pu également freiner le développement des recherches dans ce domaine. Mais je ne crois pas que ce soit là  la raison véritable car s'il est exact que les composés biologiques importants sont souvent des polymers qui même aujourd'hui encore échappent en grande partie aux méthodes de calcul, leurs constituants sont souvent des molécules de dimensions et même de catégorie (hétérocycles conjugués) tout à fait comparables aux molécules que les chimistes théoriciens des années 40 traitaient volontiers et pour lesquels malgré les imperfections, les déficiences très nettes des méthodes alors employées ils ont ammassé, à force de patience et d'ingéniosité, un nombre de résultats impressionnant.

En revanche, on peut très bien distinguer les causes qui vers le milieu des années 50, donc il y a une dizaine d'années aujourd'hui, ont provoqué le démarrage accéléré des calculs quantiques dans le domaine des molécules biologiques. Ce furent en premier lieu les découvertes fondamentales de la biologie moléculaire, en particulier la détermination de la structure des acides nucléiques, la mise en évidence du rôle central joué par les purines et pyrimidines, donc des composés tout à fait du type de ceux auxquels nous avons, en tant que chimistes théoriciens, souvent à faire. De là à s'apercevoir qu'il en était de même pour beaucoup d'autres constituants moléculaires fondamentaux de la cellule, il n'y avait qu'un pas à faire, qui consistait surtout à lire attentivement un traité de biochimie moderne. Et je dois ajouter immédiatement que bien que nous ayons été conscients, tout au moins dans notre laboratoire, dès le début, que les problèmes biochimiques et biologiques dépassaient nettement par leur complexité et rien que par le caractère polymérique de leurs agents principaux ce que l'on avait pu rencontrer auparavant en chimie quantique, il nous a paru évident qu'un intérêt appréciable s'attachait à la connaissance de la structure électronique des unités de base, des monomères. Par ailleurs, dans le cas de polymères ordonnés, comme les acides nucléiques, et surtout des polynucléotides, la possibilité est apparue très tôt de pouvoir déduire tout au moins certaines caractéristiques électroniques des polymères à partir de la connaissance de celles des monomères par un perfectionnement du calcul introduisant peu à peu les interactions entre les monomères.

Comme deuxième facteur favorisant l'épanouissement de la biochimie quantique il convient de signaler l'apparition et la propagation, à peu près vers la même période, tout au moins en Europe, d'ordinateurs électroniques simplifiant de beaucoup les travaux matériels de calculs et en augmentant notablement les possibilités.

Finalement, il convient encore de signaler parmi ces facteurs le perfectionnement notable, survenu également à peu près vers la même époque, des méthodes de calcul quantiques en particulier en ce qui concerne les systèmes conjugués, auxquels appartiennent un bon nombre de composés biochimiques fondamentaux et aussi l'apparition des méthodes de calcul appropriées aux molécules saturées. Il s'agit là essentiellement de la méthode des orbitales moléculaires, qui pratiquement, pour des raisons techniques, est depuis de nombreuses années la plus utilisée dans les calculs quantiques des structures électroniques des molécules, méthode qui doit tant au professeur Mulliken. Cette méthode, comme vous le savez tous, se présente comme toute méthode d'approximation, dans différents degrés de perfectionnement, à commencer par des approximations simples telle l'approximation dite de Hückel, et en allant jusqu'aux approximations nettement plus perfectionnées comme celle du champ moléculaire self consistant, en passant par des approximations intermédiaires telles que celles (dite P.P.P.) de Pariser-Parr-Pople, avec ou sans interactions de configurations, etc. . .

Initialement c'était naturellement, pour des raisons évidentes, l'approximation la plus simple qui fut essentiellement utilisée pour l'étude des molécules biologiques et il se fait que le champ d'action était tellement vaste et tellement vierge que même cette approximation simple a conduit à une abondance de résultats dont la majorité conserve d'ailleurs encore aujourd'hui pratiquement toute leur signification.

Depuis quelques années, les méthodes plus perfectionnées, en particulier la méthode dite P.P.P. convenablement paramétrisée prennent le relai, quand cela est possible, de sorte queaujourd'hui nous avons pratiquement refait presque tous les calculs plus anciens dans ces approximations plus perfectionnées. Ce factuer, à vrai dire, n'a donc pas joué dans la première période de l'application de théories quantiques à la biochimie. Mais aujourd'hui sa contribution, son poids se font sentir, d'une part par l'accroissement de la précision quantitative des résultats et, d'autre part, par l'extension du champ d'application de la méthode par exemple aux problèmes spectroscopiques, difficilement abordables à fond dans les approximations simples.

Quels sont alors les caractéristiques et si possible les avantages de l'utilisation des méthodes de la mécanique ondulatoire à l'étude de la structure électronique des molécules biologiques et naturellement quels sont les résultats, les types de résultats auxquels ces procédés conduisent et leur signification pour la biologie? 1) La première caractéristique et qui me paraît également être un des grands avantages de ce procédé de recherche réside dans la *généralité* de la méthode et dans son *amplitude*. Je m'explique. Toutes les méthodes physicochimiques d'étude de molécules ont pour objectif *une* ou de toutes façons un nombre limité de propriétés moléculaires. Elles sont spécialisées pour, faites pour, l'étude, disons, des moments dipolaires, ou des spectres d'absorption, ou des potentiels d'ionisation etc... La situation est toute différente en chimie ou en biochimie quantiques où un calcul *unique* relatif à une molécule nous fournit, *en principe*,

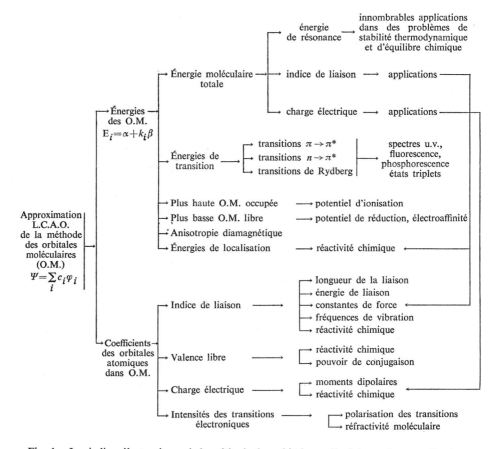

Fig. 1.   Les indices électroniques de la méthode des orbitales moléculaires et leurs applications.

toute une gamme d'indices de structure dont l'utilisation permet une étude *simultanée* de pratiquement toutes les propriétés électroniques des molécules.

Cette situation est illustrée schématiquement sur la fig. 1, correspondant à l'approximation habituelle, L.C.A.O., de la méthode des orbitales moléculaires, dans laquelle ces orbitales sont représentées sous forme d'une combinaison linéaire des orbitales atomiques. Le calcul direct, tel que nous le commandons à l'ordinateur nous fournit en premier lieu deux quantités fondamentales, les niveaux d'énergie des électrons (les orbitales moléculaires) et la forme de ces orbitales, déterminée par les coefficients des orbitales atomiques qui composent l'orbitale moléculaire. A partir de ces grandeurs fondamentales, nous pouvons définir alors toute une série d'indices de structure. Chacun de ces indices se rattache directement à une propriété physicochimique déterminée. Par exemple: la charge électrique se rattache au moment dipolaire et à la réactivité chimique; les énergies de transition, aux différents types de spectres, etc.

Un calcul unique fournit donc toute une gamme d'indices conduisant à une vue d'ensemble de la réalité moléculaire. Cela est particulièrement important en biologie, où très fréquemment les molécules interviennent, non point par une propriété déterminée, mais par l'ensemble de leurs propriétés.

2) Le deuxième avantage, évident, de l'utilisation des théories quantiques pour l'étude des molécules biologiques réside dans la possibilité qu'elles offrent et que dans beaucoup de cas elles sont les seules à offrir D'ÉLUCIDER dans des *termes appropriés*, c'est à dire en les rattachant aux caractéristiques structurales appropriées des propriétés et de comportements de ces substances. Je pense que cet avantage évident se passe de tout commentaire.

3) Finalement quant au 3e avantage, *il se rattache à la possibilité de devancer l'expérience par des prédictions*. Tout à l'heure j'ai utilisé le terme "élucider". Je l'ai fait à bon escient, en entendant par là à la fois *l'interprétation des propriétés connues* et la *prédiction des propriétés inconnues*, chaque fois que la théorie à la chance ou la possibilité de *précéder* l'expérience en ce qui concerne la connaissance de certaines caractéristiques des biomolécules. A ce propos, il est frappant et en fait étonnant même de constater jusqu'à quel point sont mal connues ou même totalement inconnues (parce que très difficilement accessibles à l'expérience, il faut l'admettre), certaines caractéristiques physicochimiques fondamentales des biomolécules, tel que par exemple leurs moments dipolaires, leurs potentiels d'ionisation, leurs affinités électroniques etc... dont la connaissance est pourtant indispensable pour la compréhension de certains aspects de leurs activités biologiques. La théorie peut palier tout au moins certaines de ces déficiences de l'expérience, avec naturellement plus ou moins de succès et de précision selon le degré de perfectionnement de la méthode de calcul employée, mais dans l'ensemble, aujourd'hui, avec une garantie d'exactitude raisonable. De toute façon, il est évident que devant l'absence des données expérimentales, des données théoriques même approximatives sont les bienvenues. J'appelle ces

types de prédictions, résultat du travail quotidien d'un biochimiste théoricien des *prédictions au niveau simple*, surtout pour les distinguer de ces *prédictions au niveau complexe* dont je parlerai plus tard et qui représentent des vastes théories, à propos des problèmes complexes, ou les calculs conduisent à une vision des choses dépassant de beaucoup l'acquit de l'expérience. Ce que je veux toutefois souligner immédiatement c'est que même dans le domaine d'interprétation ou de prédictions au niveau simple, l'apport de calcul peut avoir, en fait *a* souvent, pour conséquence, l'élaboration soit des généralisations appréciables, soit des théories explicatives, *sources d'expériences nouvelles* et peut même suggérer parfois une *correction*, une *rectification* aux résultats expérimentaux.

Ces affirmations méritent, je crois, que je les étaie par des exemples concrets. C'est ce que je vais faire maintenant.


*Aspects de la structure électronique des acides nucléiques*

Comme premier exemple considérons les acides nucléiques et leurs constituants, les bases puriques et pyrimidiques, composés dont je n'ai pas besoin de souligner, je pense, le rôle central en biologie moléculaire moderne. Il est encore impossible aujourd'hui bien sûr, malgré l'aide que peuvent nous apporter les énormes ordinatuers dont nous pouvons disposer, de calculer une molécule réelle d'un acide nucléique. En revanche un bon nombre de calculs ont été faits pour les bases et les paires de bases complémentaires et cela, aujourd'hui, par différents groupes d'auteurs, utilisant différentes variantes donc différents degrés d'approximations de la méthode. Tous ces différents calculs n'ont pas naturellement la même valeur mais ce qui est essentiel, c'est qu'ils convergent tous vers certaines conclusions d'une importance générale et fondamentale.

Ainsi, considérons, pour simplifier le problème et comme point de départ une sorte d'acide nucléique "miniature", schématisé sur la fig. 2, formé d'une paire adénine-thymine et d'une paire guanine-cytosine, liées ensemble par la chaîne sucres-phosphates.

Les calculs quantiques, comme je l'ai dit, sont effectués pour les bases puriques et pyrimidiques isolées et pour les paires de bases couplées par les liaisons hydrogènes. Dans les calculs simples l'interaction entre les paires de bases n'est pas introduite. Elle ne le sera qu'ultérieurement, comme nous le verrons. Le point de départ correspond, dans une certaine mesure, à la supposition que les bases gardent leur individualité dans les acides nucléiques, ou qu'éventuellement il y a de légères perturbations de leur structure du fait de leur interaction. A tous points de vue, c'est un point de départ intéressant, parcequ'il est toujours intéressant de voir jusqu'à quel point cette hypothèse de départ est vérifiée et jusqu'à quel point elle ne l'est pas; ce qui veut dire jusqu'à quel point les perturbations entre les bases modifient certaines caractéristiques locales.

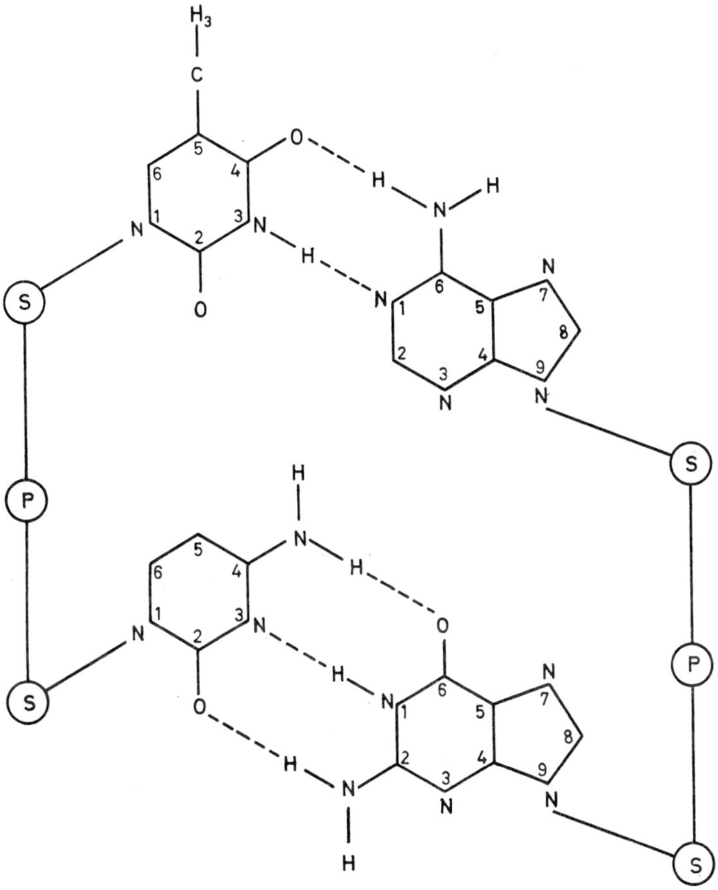

Fig. 2.  Acide nucléique "miniature".

Comment travaille-t-on alors?

On fait appel à l'une des approximations disponibles de la méthode. Naturelle-ment, plus l'approximation est perfectionnée, mieux cela vaut. Mais il se pose toujours des questions de temps disponible et de la complexité des calculs qui font que l'on doit souvent transiger sur la précision au profit de la rentabilité. Quoiqu'il en soit on évalue alors à l'aide de l'approximation choisie les différentes grandeurs électroniques caractéristiques de telles bases.

Vous avez, par exemple, sur la fig. 3 la répartition des électrons $\pi$, représentée dans une des approximations des calculs.

Si j'utilisais une autre approximation des calculs, j'obtiendrai quelque chose de semblable, mais tous les nombres figurant ici seraient modifiés (j'y reviendrai tout à l'heure).

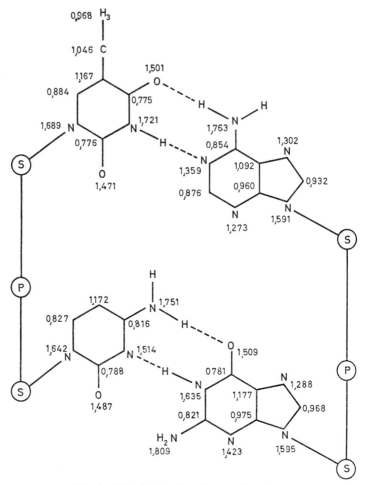

Fig. 3. Distribution des charges électriques.

On peut faire la même chose pour tous les indices électroniques énumérés il y a un instant. Vous avez sur la fig. 4 des calculs analogues, effectués pour un autre indice non pas des atomes mais des liaisons: c'est le célèbre indice de liaison mobile. On peut souligner l'image extrêmement nuancée obtenue par les calculs. Je vous rappelle que l'image purement chimique est une simple alternance des liaisons simples et doubles. Ici, chaque liaison est spécifiquement caractérisée par un nombre différent des autres et fractionnaire.

Mais, ce qui en réalité est essentiel c'est que fréquemment ce ne sont pas les nombres, les valeurs absolues des indices qui importent mais plutôt l'image générale de la structure des acides nucléiques qui ressort de ce genre de calculs. La raison en est évidente. Lorsque nous traitons un phénomène chimique, ce n'est pas tellement la grandeur explicite d'une charge électronique sur un carbone

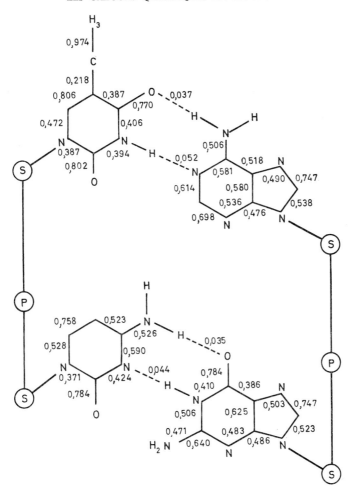

Fig. 4.   Distribution des indices de liaison.

ou un azote qui nous intéresse. Ce qui nous intéresse, c'est le site particulièrement frappant de cette grandeur. Autrement dit l'endroit où cette grandeur a sa valeur, minimum ou maximum, selon les circonstances, parceque c'est le site de localisation de la valeur maximum qui aura le plus de signification dans les phénomènes physiques et physicochimiques appropriés. Par exemple, là où il y aura les plus grandes charges électroniques se produiront certains types de réactions chimiques.

Or, ce qui est remarquable, c'est que, si l'on travaille correctement, alors, quel que soit le degré d'approximation de ces méthodes une même image générale est fournie, qui se conserve même lorsque l'on change d'approximation. Il n'y a pas de changement, de déplacement de ces sites essentiels de localisation d'indices, lorsque l'on varie l'approximation de calcul utilisé. C'est là un résultat pratique essentiel.

Ainsi, par exemple la fig. 5 indique le résultat d'un travail général pour notre acide nucléique miniature. Un grand nombre d'indices électroniques y figurent qui furent calculés par toute une gamme d'approximations. Vous y voyez certains indices soulignés en traits pleins avec une flêche pointant vers un certain site de localisation. Cela veut dire que l'indice en question a sa valeur la plus significative à l'endroit indiqué. Lorsque le même indice est souligné en pointillé cela signifie que le site correspondant est le deuxième dans l'ordre d'importance décroissant de cet indice. Ainsi par exemple, la fig. 5 indique que l'énergie de résonance

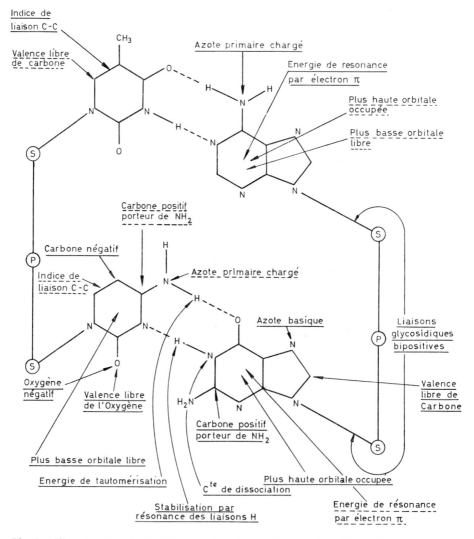

Fig. 5. Sièges les plus significatifs des indices électroniques et énergétiques des bases hétéro-cycliques des acides nucléiques.

par électron $\pi$ doit être la plus élevée dans l'adénine, suivie par la guanine. Comme l'énergie de résonance est une grandeur extrêmement importante pour l'étude de toute une série de propriétés moléculaires, comme d'autre part sa valeur expérimentale dans les purines et pyrimidines biologiques est totalement inconnue, on se rend compte immédiatement du parti important que l'on doit pouvoir tirer des prédictions théoriques dans ce domaine. De même par exemple la dénomination "azote basique" est soulignée en traits pleins pour l'azote numéro 7 de la guanine. Cela veut dire que les calculs permettent de prévoir que de tous les azotes, et il y en a beaucoup, présents dans les bases des acides nucléiques c'est cet azote là qui devrait être le plus basique. etc. . . etc. . .

Ce que je voudrais souligner tout particulièrement à propos de ce tableau, c'est le fait qu'il est *purement théorique. Nous n'avons besoin de rien savoir sur les acides nucléiques, sauf leur géométrie, pour arriver aux résultats figurant sur la fig. 5.* Cet ensemble de résultats constitue donc autant de prédictions théoriques. Bien sûr il peut arriver souvent que ce soit des prédictions "a postériori", l'expérience correspondante ayant déjà été effectuée auparavant. Mais il arrive aussi souvent que l'inverse soit vrai et que nous nous trouvions en présence des prédictions véritables. Quoi qu'il en soit, fondamentalement, les calculs peuvent être effectués indépendamment de toute connaissance expérimentale des propriétés étudiées, pourvu que la géométrie du système soit connue.

Les questions qui se posent alors naturellement maintenant sont:

Quelle est la valeur de ces prédictions? Comment utiliser ces différents résultats? Que pouvons-nous en sortir qui permette d'éclaircir les phénomènes biologiques ou biochimiques?

Nous ne pouvons pas discuter de toutes ces propriétés ou problèmes. Examinons en un d'une façon plus quantitative. Choisissons un problème fondamental et loin d'être complètement résolu: celui relatif à l'origine de la stabilité même de la structure en hélice double de l'ADN. C'est un problème à propos duquel il règne encore aujourd'hui souvent dans beaucoup de publications une grande confusion.

Ainsi par exemple lorsque fut établie cette structure en hélice double de l'ADN la majorité des biologistes ont cru que la stabilité de cet édifice provenait essentiellement des liaisons hydrogène entre les paires de bases complémentaires. Lorsque plus tard divers arguments, sur lesquels je n'insiste d'ailleurs pas ici, ont indiqué qu'une telle conception était defectueuse ou pour le moins insuffisante, beaucoup d'auteurs ont avancé que la stabilité des acides nucléiques était dûe surtout aux interactions Van der Waals-London entre les bases ou les paires de bases empilées. Encore aujourd'hui on voit souvent posée la question de savoir si la stabilité des acides nucléiques est dûe principalement aux liaisons hydrogènes entre les bases horizontales ou aux forces Van der Waals-London entre les bases superposées. Or posée de cette façon la question est surtout mal posée car elle laisse implicitement supposer qu'il existe une différence fondamentale entre les forces opérant entre les bases horizontales et celles opérant entre les bases verticales; elle laisse supposer en particulier que les forces de Van der Waals-London n'opèrent pas entre les bases horizontales mais seulement entre les bases verticales. Or c'est là une conception tout à fait erronée. On peut s'en rendre compte facilement en effectuant le calcul suivant: on fixe dans un plan un noyau d'adénine et l'on tourne autour de cette base (et autour de son propre axe) un noyau d'uracile. On calcule la variation d'énergie potentielle en utilisant tout simplement la théorie classique des interactions de Van der Waals-London, modifiée un peu du point de vue technique: on remplace les dipôles par des monopôles. On constate qu'il apparaît, dans la courbe d'énergie potentielle, des minima, et que, chaque fois que ces minima apparaissent, la configuration est celle qui correspond aux "liaisons hydrogène". Il est très facile de voir que cette situation est dûe à la très forte contribution à l'énergie electrostatique des interactions entre les atomes rapprochés des "liaisons hydrogène".

Ceci montre immédiatement que la liaison hydrogène, telle qu'on peut la concevoir, est un phénomène essentiellement électrostatique déterminé par les forces de Van der Waals-London.

D'autre part ce problème se rattache étroitement à celui du mécanisme des processus de reconnaissance tels qu'ils se produisent lors de la replication des acides nucléiques ou lors de transcriptions ou lors des réactions modèles entre des polynucléotides artificiels etc. Les recherches effectuées dans ce domaine ont mis en évidence et posé toute une série de problèmes connexes. Parmi ceux-ci quelques un des plus frappants me paraissent associés aux observations suivantes:

1) La découverte par Hoogsteen que la cocristallisation de l'adénine et de la thymine, substituées toutes deux sur leurs azotes glycosyliques, conduit à une association par liaison hydrogène, qui n'est pas conforme au modèle de Watson-Crick: la thymine est liée à $N_7$ de l'adénine et non pas à $N_1$. Ce type de cristaux mixtes s'est montré assez général dans les associations entre les dérivés de l'adénine et de l'uracil. En revanche c'est la configuration Watson-Crick qui paraît la seule observée dans les cocrystallisations des dérivés de la guanine et de la cytosine (toujours substitués sur leurs azotes glycosidiques).

2) L'exclusivité des associations par liaisons hydrogène (que ce soit dans des cocrystallisations ou en solution dans des solvants non aqueux) entre les bases complémentaires dans le sens de Watson-Crick (A–T ou A–U et G–C), aucune association ne paraissant s'établir entre les bases non complémentaires en ce sens (A–G, C–T, A–C ou G–T). C'est un phénomène qui, à première vue, a une allure un peu magique, car *chimiquement* rien ne paraît s'opposer à ce que de telles associations non complémentaires se forment en dehors des acides nucléiques.

3) Lorsqu'au lieu d'utiliser les bases substituées sur leur azote glycosylique on fait appel à des bases entièrement libres *aucune* association ne paraît plus s'établir entre elles, etc., etc. Je pourrais citer beaucoup d'autres exemples ex-

trêmement frappants de la diversité de la situation, diversité difficilement explicable dans le langage classique, dont les utilisateurs se borneraient à compter le nombre de liaisons hydrogène entre les partenaires liés.

On peut se demander alors si ce vaste ensemble d'observations est susceptible d'une interprétation homogène? Les calculs quantiques paraissent fournir une réponse positive à cette question en montrant qu'elles s'interprètent toutes en termes du jeu des forces de Van der Waals-London. L'intéressant est que pour y parvenir ils ont dû faire appel à toute une série d'évaluations théoriques des moments dipolaires, charges électroniques, potentiels d'ionisation etc. quantités intervenant dans l'expression de ces forces mais dont les valeurs sont inconnues pour la majeure partie.

Ainsi par exemple, si l'on connaît le moment dipolaire de certains dérivés simples de l'adénine et de l'uracil, on ignore totalement le moment de la guanine et de la cytosine. On ignore totalement la direction du moment dans toutes ces molécules. Or, comme l'indique le tableau I, la théorie arrive à reproduire très correctement les moments connus et prévoit, avec par conséquent une bonne garantie de réussite, les moments inconnus. On peut remarquer le caractère significatif de la prédiction, qui indique un moment de l'ordre de 7–8 D pour la guanine et la cytosine, à la différence d'un moment de 3–4 D connu pour l'adénine et la thymine.

De même, en ce qui concerne les potentiels d'ionisation, on ignorait totalement jusqu'à cette dernière année leurs valeurs dans les purines et les pyrimidines biologiques. On était donc dans ce domaine entièrement dépendant de la théorie qui a pu, elle, prévoir leurs valeurs. En ce qui concerne les bases des acides nucléiques, ces valeurs sont, 7,6 eV. pour la guanine, 7,9 eV. pour l'adénine, 8,2 eV. pour la cytosine et 9,2 eV. pour l'uracile. Très récemment, Bergmann de l'Université Hébraïque de Jérusalem a réussi à mesurer ces potentiels à l'aide de spectrographie de masse et ses résultats expérimentaux, en voie de publication, apportent une très bonne confirmation des prédictions théoriques.

C'est parceque la théorie a permis d'évaluer ces différentes grandeurs physicochimiques qu'une explication a pu être apportée aux différents problèmes mentionnés plus haut à propos des interactions entre les purines et pyrimidines.

Ainsi, le tableau II résume tout un ensemble de résultats de calculs sur les interactions horizontales entre les purines et pyrimidines nucléiques correspondant à la formation des associations par liaisons hydrogènes. Les résultats figurant sur ce tableau correspondent aux calculs effectués dans l'approximation des monopôles, le nombre figurant sur le tableau représentant l'énergie totale de l'interaction, somme des énergies électrostatiques, d'induction et de dispersion. Ces nombres correspondent dans chaque cas à l'association la plus forte obtenue dans l'hypothèse où les azotes glycosidiques ne sont pas disponibles pour l'association. Comme nous le verrons dans le tableau suivant ça sera, par exemple, la configuration de Hoogsteen pour la paire A–T et celle de Watson-Crick pour

TABLEAU I

Moments dipolaires de purines et pyrimidines

| Direction de localisation | Moment théorique (en D) | Moment expérimental (en D) |
|---|---|---|
| Purine | 4.15 | 4,3 dans 9-méthyl-purine |
| Adénine | 3,16 | 3,0 dans 9-méthyl-adénine |
| Guanine | 6,76 | |
| Uracile | 3,86 | 3,9 dans 1,3-diméthyl-uracile |
| Thymine | 3,58 | |
| Cytosine | 7,10 | |

TABLEAU II

Energies d'interaction dans les paires de bases liées par liaisons hydrogènes

| (Kcal/mole) | | | |
|---|---|---|---|
| A — A | — 5,8 | A — T | — 7,0 |
| T — T | — 5,2 | | |
| G — G | — 14,5 | G — C | — 19,2 |
| C — C | — 13 | | |
| A — A | — 5,8 | A — C | — 7,8 |
| C — C | — 13 | | |
| G — G | — 14,5 | G — T | — 7,4 |
| T — T | — 5,2 | | |
| C — C | — 13 | C — T | — 6,5 |
| T — T | — 5,2 | | |
| A — A | — 5,8 | A — G | — 7,5 |
| G — G | — 14,5 | | |

$$A - T \quad > \quad \begin{matrix} A - A \\ T - T \end{matrix}$$

$$G - C \quad > \quad \begin{matrix} G - G \\ C - C \end{matrix}$$

$$A - C \quad < \quad C - C$$
$$G - T \quad < \quad G - G$$
$$C - T \quad < \quad C - C$$
$$A - G \quad < \quad G - G$$

la paire G–C. Comme vous pouvez le remarquer nos calculs couvrent à la fois les associations existantes et celles qui n'existent pas. C'est évidemment un avantage de théoricien sur l'expérimentateur que de pouvoir étudier ce qui n'existe pas.

Rermarquons tout d'abord, et cela sera le point de départ de la réponse à certains problèmes qui nous intéressent, qu'en ce qui concerne les autoassociations de bases on peut diviser celles-ci en deux groupes: d'une part les autoassociations G–G et C–C correspondant aux énergies d'interactions élevées et d'autre part A–A et T–T auxquelles correspondent des énergies d'intéraction nettement plus faibles.

Si nous regardons maintenant les différentes associations mixtes que l'on peut construire à partir de ces mêmes bases on constate que celles-ci aussi se divisent en deux groupes: d'une part les paires complémentaires A–T et G–C dont les énergies d'interaction sont supérieures aux énergies d'autoassociation de leur deux constituants (ou à la moyenne de ces énergies) et, d'autre part, toutes les paires non complémentaires pour lesquelles les énergies d'interaction seraient

## TABLEAU III

Energies d'interaction dans différentes configurations des paires A–T et G–C

| Paire | Configuration | $E_{\rho\rho}$ | $E_{\rho\alpha}$ | $E_L$ | $E_M$ |
|-------|---------------|------|------|------|------|
| A–T | Watson–Crick | −4,64 | −0,25 | −0,69 | −5,58 |
| A–T | Watson–Crick inversé | −3,98 | −0,25 | −0,71 | −4,94 |
| A–T | Hoogsteen | −5,86 | −0,22 | −0,88 | −6,96 |
| A–T | Hoogsteen inversé | −5,63 | −0,17 | −0,94 | −6,74 |
| G–C | Watson–Crick | −15,91 | −2,02 | −1,25 | −19,18 |
| G–C | Type "Hoogsteen" | −3,98 | −1,33 | −0,44 | −5,75 |

en elles-mêmes appréciables mais toujours inférieures à l'une des énergies d'auto-association de l'un de leur constituant (G ou C). Cette situation suggère par elle-même l'explication de l'exclusivité "magique" des associations complémentaires A–T et G–C, seules suffisamment stables par rapport aux autoassociations pour pouvoir se former à leur dépens.

Le tableau III indique avec un peu plus de détails les résultats de calculs pour les différents modes de couplage possibles dans les associations A–T et G–C. Soulignons la prépondérance du couplage Hoogsteen pour la paire A–T et du couplage Watson-Crick pour la paire G–C. Remarquons également que la partie essentielle de l'énergie d'interaction provient dans ces couplages de la composante électrostatique. Signalons aussi que presque toutes les prédictions de détails contenues dans les calculs se trouvent vérifiées actuellement par l'expérience. Signalons aussi que ces calculs ont permis de rendre compte de la structure cristalline de différentes purines et pyrimidines.

Jusqu'ici nous avons considéré, en accord avec la réalité expérimentale la plus courante des interactions entre les bases substituées sur les azotes glycosyliques. La question peut être posée de savoir ce qui se passerait si l'on mettait en présence des bases entièrement libres. Dans ce cas il convient d'envisager un type complémentaire d'auto-association et d'association mixte de bases mettant en jeu le proton attaché à l'azote glycosylique. Les résultats de calculs obtenus dans cette nouvelle hypothèse ou plutôt les modifications que cette nouvelle hypothèse entraîne pour les résultats antérieurs est illustré sur le tableau IV. La modification essentielle concerne l'énergie (maximum) d'autoassociation de l'adénine qui dans ce nouveau mode d'interaction est supérieure à toutes celles prévues pour les associations mixtes possibles entre l'adénine et la thymine. Par conséquent, en accord avec la règle précédente et à la différence de ce qui se produit pour les dérivés substitués, l'adénine et la thymine libres ne devraient pas s'associer. En revanche, le nouveau mode d'autoassociation de la guanine et de la cytosine correspond toujours à une énergie d'interaction inférieure à celle des associations mixtes G–C et ces deux bases libres devraient pouvoir s'associer. L'expérience indique que jusqu'ici aucune association entre les bases nucléiques libres n'a pu être mise en évidence. Toutefois, l'expérimentation avec le guanine et la cytosine ne saurait être considérée comme décisive du fait de l'insolubilité presque totale de la guanine dans les solvants utilisés.

On peut se demander maintenant quelle lumière ce genre de calcul est-il susceptible de jeter sur la structure des acides nucléiques eux-mêmes et, en particulier, sur l'origine de la stabilité de l'arrangement en hélice double. La réponse à cette question est contenue dans le tableau V de l'examen duquel il apparaît nettement que les interactions horizontales et les interactions verticales contribuent toutes deux et cela d'une façon assez comparable à la stabilité de la structure à double hélice. De plus, on constate que les différentes combinaisons de paires de base se divisent, au point de vue de l'énergie totale d'interaction, en trois groupes: les

TABLEAU IV

Energies d'interaction entre les bases non substituées

| (Kcal/mole) | | | |
|---|---|---|---|
| A — A | — 8,13 | | |
| T — T | — 5,2 | A — T | — 7,0 |
| G — G | —14,5 | | |
| C — C | —13 | G — C | —19,2 |
| A — A | — 8,13 | | |
| C — C | —13 | A — C | — 7,8 |
| G — C | —14,5 | | |
| T — T | — 5,2 | G — T | — 7,4 |
| C — C | —13 | | |
| T — T | — 5,2 | C — T | — 6,5 |
| A — A | — 8,13 | | |
| G — G | —14,5 | A — G | — 7,5 |

$$A-T \begin{array}{c} < \\ > \end{array} \begin{array}{c} A-A \\ T-T \end{array}$$

$$G-C > \begin{array}{c} G-G \\ C-C \end{array}$$

$$\begin{array}{c} A-C \\ G-T \end{array} \begin{array}{c} < \\ < \end{array} \begin{array}{c} C-C \\ G-G \end{array}$$

$$\begin{array}{c} C-T \\ A-G \end{array} \begin{array}{c} < \\ < \end{array} \begin{array}{c} C-C \\ G-G \end{array}$$

combinaisons les plus stables s'établissent entre deux paires G-C, les moins efficaces ont lieu entre deux paires A–T, et les différentes combinaisons $\frac{G-C}{A-T}$ correspondent à une stabilité intermédiaire. On peut remarquer que si ces résultats, qui de tout évidence peuvent être rapprochés de l'accroissement de la stabilité thermique des acides nucléiques en fonction de leur contenu en G-C, mise en évidence par Marmur et Doty, correspondent également à l'ordre d'interactions horizontales, ils ne correspondent pas au seul ordre des interactions verticales.

Ainsi, ce premier exemple d'utilisation de la théorie, illustre un certain aspect des possibilités qu'elle offre pour l'évaluation de propriétés partiellement inconnues et l'utilisation de ces données pour des théories explicatives concernant un vaste domaine d'observations fondamentales. On peut naturellement étendre ce mode d'investigations à toute une série d'autres problèmes concernant les acides nucléiques. La fig. 6 illustre le résultat d'un tel travail pour toute une série de propriétés. Je ne peux malheureusement pas m'étendre ici sur les détails.

TABLEAU V

Energies d'interaction entre paires de bases voisines dans DNA
(dans le vide)

| Paires adjacentes (*) | Interactions verticales | | | Energie totale d'empile-ment | Contribution moyenne des interactions horizontales | Energie totale d'inter-action |
|---|---|---|---|---|---|---|
| | $E_{\varrho\varrho}$ | $E_{\varrho\alpha}$ | $E_L$ | | | |
| ↑ C—G / G—C ↓ | + 0,9 | — 2,0 | — 10,2 | — 11,3 | — 19,2 | — 30,5 |
| ↑ G—C / C—G ↓ | — 1,6 | — 2,5 | — 4,0 | — 8,5 | — 19,2 | — 27,7 |
| ↑ G—C / G—C ↓ | + 2,6 | — 2,0 | — 8,3 | — 7,7 | — 19,2 | — 26,9 |
| ↑ A—T / G—C ↓ | + 1,2 | — 0,8 | — 10,3 | — 9,9 | — 12,2 | — 22,1 |
| ↑ A—T / C—G ↓ | — 0,6 | — 1,7 | — 4,9 | — 7,2 | — 12,2 | — 19,4 |
| ↑ T—A / C—G ↓ | — 0,1 | — 1,7 | — 5,2 | — 7,0 | — 12,2 | — 19,2 |
| ↑ T—A / G—C ↓ | + 1,8 | — 1,0 | — 7,8 | — 7,0 | — 12,2 | — 19,2 |
| ↑ A—T / A—T ↓ | + 0,5 | — 0,5 | — 7,4 | — 7,4 | — 5,5 | — 12,9 |
| ↑ A—T / T—A ↓ | + 0,4 | — 0,3 | — 6,2 | — 6,1 | — 5,5 | — 11,6 |
| ↑ T—A / A—T ↓ | + 1,5 | — 0,7 | — 5,8 | — 5,0 | — 5,5 | — 10,5 |

(*) Les flèches qui désignent la direction de la chaîne sont dirigées du carbone 3′ sur un sucre vers le carbone 5′ sur le sucre adjacent. exemple: T A représente: T-sucre-3′-phosphate-5′-sucre-A.
$\longrightarrow$

## Pouvoir donneur et accepteur d'électrons des molécules

Passons maintenant à un autre type de problème qui en fait n'est qu'un développement d'une question que j'ai déjà mentionnée. Ainsi, j'ai parlé tout à l'heure des potentiels d'ionisation qui, avec l'affinité électronique, sont une caractéristique physico-chimique très importante des molécules biologiques, intervenant dans tous les phénomènes de transfert d'électrons, dans les complexes de transfert de charges, dans les problèmes de la semiconductivité des macro-molécules biologiques etc. Or ces grandeurs physicochimiques sont pratiquement encore tout à fait inconnues expérimentalement à l'heure actuelle et elles étaient totalement inconnues il y a dix ans lorsque nous avons fait nos premiers travaux

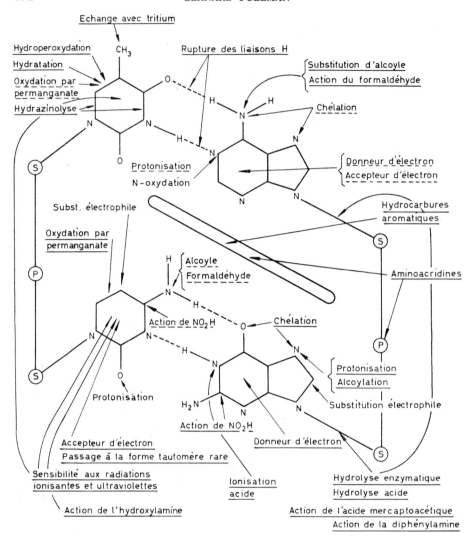

Fig. 6.  Localisation des centres essentiels de propriétés chimiques et physico-chimiques des
bases hétérocycliques des acides nucléiques.

de biochimie quantique au laboratoire du Albert Szent-Györgyi à Woods-Hole.
A ce moment Szent-Györgyi était très intéressé par le rôle possible des complexes de transfert de charges en biologie et avait donc besoin d'avoir des
renseignement sur le pouvoir donneur et accepteur d'électrons des composés
biologiques fondamentaux. Nous avons effectué à ce moment, par une méthode
très simple, la méthode de Hückel, des calculs quantiques qui nous ont permis
de déterminer une échelle de pouvoir donneur et accepteur d'électrons pour
la grande majorité des molécules biologiques. Nous avons pu prédire lesquelles

de ces molécules devraient fonctionner comme des donneurs d'électrons et lesquelles devraient fonctionner comme des accepteurs et indiquer l'ordre relatif de leur activité. L'application de ces résultats était d'une grande utilité pour toute une série de problèmes biochimiques, sur lesquels je ne peux pas insister ici. En revanche ce que je voudrais indiquer ici c'est que ces calculs bien qu'effectués il y a 10 ans et par une méthode approchée ont néanmoins été confirmés, tout au moins indirectement (en l'absence de confirmation directe), par l'expérience, comme l'indique le tableau VI.

TABLEAU VI

Pouvoir donneur et accepteur d'électrons des purines et pyrimidines

| Composé | La plus haute orbitale moléculaire occupée | Oxidabilité Polarographique | La plus basse orbitale moléculaire libre | Réducibilité Polarographique |
|---|---|---|---|---|
| Purine | 0.69 | — | — 0.74 | + |
| Adénine | 0.49 | + | — 0.87 | + |
| Guanine | 0.30 | + | — 1.05 | — |
| Hypoxanthine | 0.40 | + | — 0.88 | + |
| Xanthine | 0.44 | + | — 1.01 | — |
| Acide Urique | 0.17 | + | — 1.19 | — |
| Uracil | 0.70 | — | — 0.96 | — |
| Thymine | 0.51 | — | — 0.96 | — |
| Cytosine | 0.60 | — | — 0.80 | + |
| Acide barbiturique | 1.03 | | — 1.30 | |
| Alloxane | 1.03 | | — 0.76 | + |
| 2,6-Diaminopurine | | | — 0,92 | + |

Une application particulière mérite peut-être d'être mentionnée, tout spécialement dans ce domaine. Au cours de nos recherches sur ce sujet du pouvoir donneur d'électron, nous avons trouvé à propos d'un type de molécules, les phénothiazines, un résultat très spécial. La théorie prédisait que la plus haute orbitale occupée de ces molécules devrait être antiliante, ce qui signifiait que ce type de composés devait constituer des donneurs d'électrons exceptionnellement bons, (potentiel d'ionisation particulièrement bas). La prédiction était tellement frappante qu'elle a immédiatement suscité des travaux expérimentaux (Calvin à Berkeley, Lyons en Australie) qui en ont confirmé la justesse. De plus, Szent-Gyorgyi a eu l'idée de proposer que l'activité psychotrope des phénothiazines (chlorpromazine) pourrait être liée à cette qualité particulière de "bon donneur d'électrons", en favorisant par exemple l'établissement de complexes de transferts de charges entre ces composés et les récepteurs cellulaires appropriés.

Je dois dire qu'en fait l'hypothèse est beaucoup plus vaste. Parce qu'on peut montrer également qu'on certain nombre de composés indoliques, qui se comportent comme des composés psychomimétiques, comme le célèbre acide lysergique, sont aussi de très bons donneurs d'électrons.

Je ne vais pas m'appesantir sur la justesse ou non d'une telle hypothèse. Ce que je peux dire, c'est qu'elle a démarré toute une branche de recherche en pharmacologie et qu'il y a maintenant constamment, des travaux effectués, en partant de cette idée fondamentale.

## Structure électronique et activité cancérogène des substances chimiques

Finalement, pour terminer, un autre exemple, très particulier celui-ci. Il concerne l'un des problèmes les plus passionants de la biochimie électronique, et aussi l'un des plus importants. Un problème étudié d'ailleurs par beaucoup d'autres disciplines, mais dont malheureusement la solution nous échappe encore toujours. Il s'agit du problème de la carcinogénèse chimique et je voudrais vous montrer ce qui, dans un problème tel que celui-ci, peut être obtenu par l'utilisation des procédés de la mécanique quantique.

Nous allons aborder le problème sous l'angle de la relation entre la structure électronique et l'activité cancérogène des substances chimiques. Nous ne pouvons qu'en résumer ici quelques aspects. En particulier, nous bornerons notre exposé au cas des hydrocarbures aromatiques. Ce groupe de composés représente d'ailleurs, certainement, l'exemple le plus important et le plus abondamment étudié.

La grande difficulté qui domine toute tentative d'établir une corrélation entre la structure et l'activité cancérogène des substances aromatiques provient de l'extrême sensibilité du phénomène. En effet, il suffit d'une très petite modification structurale pour faire apparaître ou disparaître le pouvoir cancérogène ou pour en modifier l'intensité.

Ainsi, par exemple, dans les hydrocarbures fondamentaux considérés ici, le pouvoir cancérogène ne se manifeste qu'à partir de quatre noyaux benzéniques accolés. Les composés I, II, III et IV sont totalement inactifs. Parmi les six molécules possibles à quatre noyaux, une seule est certainement active, le benzo-3, 4-phénanthrène VIII. On a quelques doutes sur une activité éventuelle du benzanthracène VI. De même, parmi les quinze hydrocarbures possibles à cinq noyaux benzéniques accolés, qui furent tous étudiés en vue d'une action cancérogène éventuelle, cinq seulement se sont avérés actifs. Ce sont, approximativement dans l'ordre d'activité décroissante: le benzo-3, 4-pyrène XIII très cancérogène (alors que le pyrène IX est totalement inactif), le dibenzo-1, 2, 5, 6-anthracène XI moyennement cancérogène, le dibenzo-1, 2, 5, 6-phénanthrène XIV et le dibenzo-1, 2, 3, 4-phénanthrène XV faiblement actifs et le dibenzo-1, 2, 7, 8-anthracène XII très faiblement cancérogène. En revanche, des isomères strictement apparentés, comme, par exemple, le benzo-1, 2-pyrène XVI ou le dibenzo-1, 2, 3, 4-anthracène XIX sont totalement inactifs.

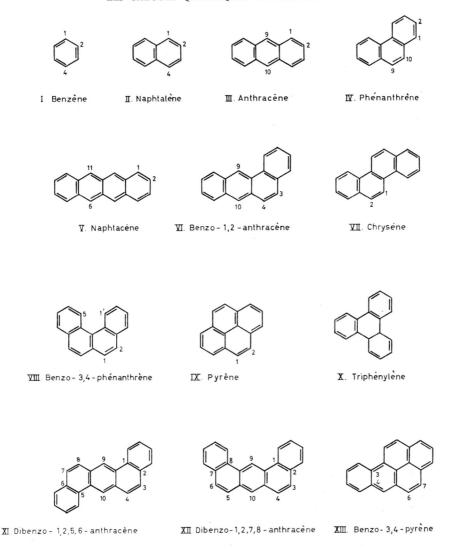

I Benzène    II. Naphtalène    III. Anthracène    IV. Phénanthrène

V. Naphtacène    VI. Benzo-1,2-anthracène    VII. Chrysène

VIII. Benzo-3,4-phénanthrène    IX. Pyrène    X. Triphénylène

XI. Dibenzo-1,2,5,6-anthracène    XII. Dibenzo-1,2,7,8-anthracène    XIII. Benzo-3,4-pyrène

Seul un nombre limité d'hydrocarbures à six noyaux benzéniques accolés a été étudié. Parmi ces composés, trois seulement, le dibenzo-1, 2, 3, 4-pyrène XXXVI, le dibenzo-3, 4, 8, 9-pyrène XXXVII et le dibenzo-3, 4, 9, 10-pyrène XCI se sont révélés cancérogènes (d'ailleurs fortement). En revanche, des molécules analogues comme le dibenzo-1, 2, 6, 7-pyrène XXVIII, le dibenzo-3, 4, 6, 7-pyrène XXIX, le naphto-2′, 3′, 3, 4-pyrène XXXII, le dibenzo-1, 2, 9, 10-naphtacène XXXIV et le dibenzo-1, 2, 7, 8-naphtacène XXXV sont totalement inactifs.

Finalement, peu de composés à plus de six noyaux benzéniques accolés ont été expérimentés au point de vue de leur activité pathogénique. Néanmoins, la majeure partie de ceux qui furent étudiés sont inactifs.

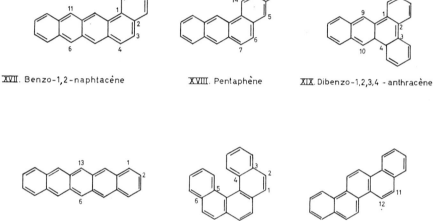

XIV. Dibenzo-1,2,5,6-phénanthrène    XV. Dibenzo-1 2 3 4-phénanthrène    XVI. Benzo-1,2-pyrène

XVII. Benzo-1,2-naphtacène    XVIII. Pentaphène    XIX. Dibenzo-1,2,3,4 - anthracène

XX. Pentacène    XXI. Dibenzo- 3,4,5,6-phénanthrène    XXII. Picène

Ainsi, dans l'ensemble, sur une cinquantaine d'hydrocarbures fondamentaux étudiés, neuf seulement sont certainement cancérogènes. Ces composés actifs peuvent être divisés en trois groupes:

a)  le benzo-3, 4-phénanthrène VIII et ses deux dérivés benzéniques XIV et XV;

b)  les deux dibenzanthracènes XVI et XII;

c)  le benzo-3, 4-pyrène XIII et ses trois dérivés dibenzéniques XXXVI, XXXVII et XXXVIII.

Cet aperçu rapide montre bien, de toute évidence, que le problème est un problème de structure fine et que c'est seulement en atteignant la constitution intime de ces composés que l'on peut espérer déceler éventuellement les facteurs responsables de la présence ou de l'absence du pouvoir pathogène. Il s'agit, en premier lieu, de trouver des règles de sélection qui permettraient de distinguer, parmi tous les hydrocarbures précités, les neuf molécules actives. De telles règles

XXIII. Dibenzo - 2,3,7,8 - phénanthrène    XXIV. Dibenzo - 2,3,5,6 - phénanthrène    XXV. Pérylène

XXVI. Anthanthrène    XXVII. Tribenzo - 1,2,3,4,5,6 - anthracène    XXVIII. Dibenzo - 1,2,6,7 - pyrène

XXIX. Dibenzo - 3,4,6,7 - pyrène    XXX. (Anthra - 1',2')-1,2 - anthracene    XXXI. (Anthra - 2',1')-1,2 - anthracène

ont pu être obtenues après une longue série de recherches et elles se rattachent à certaines aptitudes réactionnelles de molécules mises en jeu. Ainsi, en étudiant la structure électronique de l'ensemble des hydrocarbures, actifs ou non, on constate qu'ils sont caractérisés par deux régions de propriétés électroniques particulières et qui sont d'une importance spéciale pour leur réactivité chimique: ce sont d'une part la région du type méso-phénanthrénique (liaison 9–10 du phénanthrène), d'autre part celle du type mésoanthracène (9–10 de l'anthracène), que j'appellerai respectivement régions K et L (fig. 7).

Région L

Région K

Fig. 7.

XXXII. (Naphto-2', 3')-3,4 - pyrène

XXXIII. (Phénanthra-2', 3')-1,2-anthracène

XXXIV. Dibenzo-1,2,9,10 - naphtacène

XXXV. Dibenzo-1,2,7,8 - naphtacène

XXXVI. Dibenzo-1,2,3,4-pyrène    XXXVII. Dibenzo-3,4,8,9-pyrène    XXXVIII. Dibenzo-3,4,9,10-pyrène

La région K est celle qui est douée de la plus faible énergie d'ortholocalisation, c'est-à-dire qu'elle est particulièrement apte aux additions en ortho.

La région L, de son côté, correspond aux plus faibles valeurs de l'énergie de para-localisation; elle est donc particulièrement apte aux additions en para.

Ensuite, la recherche d'une relation entre la structure électronique et l'activité cancérogène de l'ensemble des hydrocarbures aromatiques a conduit aux deux propositions fondamentales suivantes:

1°) Pour être cancérogène, un hydrocarbure polybenzénique doit posséder une région K "active", c'est-à-dire d'indice égal ou inférieur à un certain seuil déterminé comme devant être égal à 3,31 $\beta$. ($\beta$ étant une unité de mesure, caractéristique de la méthode des orbitales moléculaires).

2°) Cette condition est nécessaire mais non suffisante. En effet, si la molécule contient aussi une région L, il faut alors que cette région soit au contraire peu active, en fait qu'elle ait un indice égal ou supérieur à 5,66 $\beta$.

On peut interpréter l'ensemble de ces deux propositions de la façon suivante:

supposons qu'une étape essentielle de la carcinogénèse soit constituée par une réaction entre l'hydrocarbure et un récepteur cellulaire. Cette réaction dont nous ne précisons pas la nature a priori se ferait sur ou par l'intermédiaire de la région K et, pour que la réaction se produise, il faut que cette région soit suffisamment active. Mais cette réactivité de la région K ne conduit à la cancérisation que si elle n'est pas concurrencée par la présence d'une région L très active qui risquerait d'engager la molécule dans une réaction qui, elle, est incompatible avec la manifestation d l'activité cancérogène.

L'ensemble des résultats numériques relatifs aux indices des régions K et L dans les hydrocarbures aromatiques qui doivent permettre de préciser le bien-fondé de ces propositions fondamentales sont résumés dans le tableau VII.

Voyons le sens de ces résultats sur quelques exemples particulièrement frappants.

L'anthracène III, non cancérogène, n'a pas de région K proprement dite et possède d'ailleurs une région L fortement active. Lorsque l'on accole à l'anthracène un autre noyau benzénique linéairement, V, on accentue énormément la réactivité de la région L et l'on rend par conséquent encore plus improbable l'apparition de l'activité. Cette observation explique immédiatement le fait fondamental que l'activité cancérogène ne se manifeste jamais dans des hydrocarbures possédant une longue chaîne linéaire de noyaux.

Lorsque, en revanche, on accole à ce squelette anthracénique un noyau latéral, VI, on produit deux effets différents: d'une part on crée une région K d'indice favorable à la manifestation de l'activité, et d'autre part on diminue l'activité de la région L (pas assez toutefois pour atteindre le seuil à partir duquel cette activité n'est plus défavorable à la manifestation de celle de la région K). Si l'on accole au benzo-1, 2-anthracène un second noyau latéral en 7–8, XII, ou en 5–6, XI, on modifie peu la région K, mais on désactive la région L de telle sorte que le pouvoir cancérogène apparaisse. Enfin, si l'on passe au benzo-5, 4-pyrène, XIII, on s'aperçoit que l'on a considérablement activé la région K, et que l'on a supprimé la région L; on se trouve donc dans les meilleures conditions possibles pour avoir une forte activité et effectivement le benzo-3, 4-pyrène est très fortement cancérogène.

La différence frappante entre l'activité et l'inactivité de quelques autres isomères apparentés, cités au début de ce chapitre, se comprend aussi immédiatement à l'aide des conditions précitées. Ainsi, par exemple, les dibenzopyrènes XXXVI, XXXVII et XXXVIII sont tous trois cancérogènes puisqu'ils sont dépourvus de région L et que leur région K a un indice convenable. En revanche, les dibenzopyrènes XXVIII et XXIX ne sont pas cancérogènes, bien que dépourvus de la région L car leur région K est bloquée. De même, le napthopygène XXXII est inactif car tout en possédant une région K aussi propice à l'action cancérogène que celle du benzo-3, 4-pyrène XIII, il possède toutefois une région L trop réactive. Exactement la même situation se présente dans les dibenzonaphtacènes XXXIV

TABLEAU VII

Réactivités des régions K et L

| Composé | Région K | | Région L | | Activité cancérogène |
|---|---|---|---|---|---|
| | Liaison | Indice | Carbones | Indice | |
| I | (1–2) | (4,07) | (1–4) | (6,54) | — |
| II | (1–2) | (3,56) | (1–4) | (5,98) | — |
| III | (1–2) | (3,43) | 9–10 | (5,33) | — |
| IV | 9–10 | (3,37) | | | — |
| V | 1–2 | 3,33 | 6–11 | 5,26 | — |
| VI | 3–4 | 3,29 | 9–10 | 5,53 | + |
| VII | 1–2 | 3,38 | | | — |
| VIII | 1–2 | 3,41 | | | + |
| IX | 1–2 | 3,33 | | | — |
| X | (1–2) | (3,81) | | | — |
| XI | 3–4 | 3,31 | 9–10 | 5,69 | + + |
| XII | 3–4 | 3,31 | 9–10 | 5,66 | + |
| XIII | 6–7 | 3,23 | | | + + + + |
| XIV | 7–8 | 3,41 | | | + |
| XV | 9–10 | 3,42 | | | + |
| XVI | 6–7 | 3,37 | | | — |
| XVII | 3–4 | 3,20 | 6–11 | 5.25 | — |
| XVIII | 6–7 | 3,23 | 5–14 | 5,56 | — |
| XIX | (5–6) | (3,51) | 9–10 | 5,67 | — |
| XX | (1–2) | (3,27) | 6–13 | 5,03 | — |
| XXI | 1–2 | (3,38) | | | — |
| XXII | 11–12 | (3,37) | | | — |
| XXIII | 9–10 | (3,27) | 1–4 | 5,47 | — |
| XXIV | 9–10 | (3,30) | 1–4 | 5,48 | — |
| XXV | | | | | — |
| XXVI | 1–2 | 3,20 | | | — |
| XXVII | 7–8 | 3,33 | 9–10 | 5,80 | — |
| XXVIII | | | | | |
| XXIX | 1–2 | 3,35 | | | — |
| XXX | 3–4 | 3,20 | 9–10 | 5,45 | — |
| XXXI | 3–4 | 3,24 | 9–10 | 5,44 | — |
| XXXII | 6–7 | 3,14 | 1'–4' | 5,30 | — |
| XXXIII | 3–4 | 3,24 | 9–10 | 5,54 | — |
| XXXIV | 3–4 | 3,25 | 5–12 | 5,40 | — |
| XXXV | 3–4 | 3,24 | 5–12 | 5,42 | — |
| XXXVI | 6–7 | 3,24 | | | + + + |
| XXXVII | 6–7 | 3,18 | | | + + + + |
| XXXVIII | (6–7) | (3,16) | | | + + + + |

et XXXV où l'allongement de la chaîne centrale des noyaux benzéniques crée des régions L trop fortes.

Les mêmes considérations peuvent s'appliquer avec un égal succès aux autres composés figurant dans le tableau. Les exceptions à nos propositions fondamentales sont rares. En fait, il en existe deux:

a) La première est constituée par le groupe du benzo-3, 4-phénanthrène VIII et de deux dibenzophénanthrènes apparentés XIV et XV, molécules cancérogènes qui sont dépourvues de région L mais dont la région K n'est pas, en principe, assez réactive. Cette exception a pu toutefois être expliquée par des facteurs d'encombrement stérique, provoquant une destruction partielle de la planéité de la molécule et un accroissement de la réactivité de sa région K au-delà de la valeur théorique.

b) La seconde exception est offerte par l'anthanthrène XXVI, qui, selon les calculs, devrait être cancérogène et qui ne l'est pas. C'est une exception qui n'a pas encore trouvé d'explication entièrement satisfaisante. Il se peut que l'absence d'activité de la part de ce corps soit due à des causes physiques. Il se peut aussi qu'elle puisse être liée au fait que, bien que dépourvu de région L, l'anthanthrène possède deux carbones très réactifs (11 et 12) situés relativement près l'un de l'autre.

Ajoutons que les résultats relatifs aux molécules à cinq et six noyaux permettent de comprendre facilement la raison de l'absence d'activité dans les molécules plus complexes. Celles-ci risquent en effet le plus souvent de posséder une région L défavorable (cas des molécules allongées) et quelquefois de manquer de région K favorable (cas des molécules trop condensées). Il est difficile d'imaginer des molécules volumineuses satisfaisant aux conditions que nous avons imposées aux régions K et L.

Ainsi, nous sommes donc arrivés à obtenir une relation précise entre les caractéristiques électroniques bien définies et quantitativement évaluées des hydrocarbures polybenzéniques et leur activité cancérogène. A part trois composés dans lesquels se manifestent des effets stériques particuliers, la théorie ne laisse subsister sur tous les corps expérimentés qu'une seule exception. Ajoutons que pour les hydrocarbures cancérogènes exempts d'effets stériques, la théorie reproduit quantitativement l'ordre relatif de leurs activités cancérogènes.

Nous allons voir maintenant que la théorie est susceptible aussi d'une vérification expérimentale chimique, plus précisément, la théorie entraîne implicitement certaines conséquences pour la réactivité chimique de ces composés, conséquences que l'expérience doit permettre de vérifier si la théorie est correcte. Ces vérifications sont nombreuses et nuancées, et nous n'en indiquerons ici que les grandes lignes. Ainsi, nous venons de voir que pour qu'une molécule soit cancérogène la théorie demande une région K active et une région L peu active. Or, le choix même des indices qui nous ont servi pour définir cette "activité" des régions implique des conséquences directes sur la réactivité chimique puique ce sont les indices qui rendent compte de l'aptitude de ces régions aux additions. Il s'ensuit que les molécules cancérogènes devraient être réactives vis-à-vis des additions se faisant sur la région K et, au contraire, peu ou pas réactives vis-à-vis des additions se faisant sur la région L. En outre, la réactivité vis-à-vis des substitutions, qui, elles, mettent en jeu non pas des régions mais des centres actifs des molécules ne figurant pas dans nos postulats, ne devrait pas être reliée de façon monotone avec le pouvoir cancérogène.

L'expérience chimique confirme entièrement ce point de vue. Cette expérience n'a pas toujours été effectuée pour tous les composés qui nous intéressent mais, heureusement, elle l'a néanmoins été pour des composés assez représentatifs pour nous permettre d'en tirer des conclusions générales.

*1. Réactivité de la région L.* Les additions moléculaires les mieux connues qui se font sur la région L sont:

a)   la fixation de l'anhydride maléique;
b)   la photoxydation;

Or l'expérience montre que:

a)   La facilité de fixation d'anhydride maléique décroît dans la série anthracène, benzo-1, 2-anthracène, dibenzo-1, 2, 5, 6-anthracène, benzo-3, 4-pyrène. Ceci est conforme à la théorie. En revanche le naphtacène et le pentacène, non cancérogènes, sont très actifs vis-à-vis du même réactif (plus que l'anthracène).
b)   La photoxydation est très facile dans les acènes (anthracène, naphtacène) non cancérogènes, plus difficile lorsqu'on accole des noyaux latéraux: le dibenzanthracène ne réagit plus. Le benzopyrène ne réagit pas du tout non plus. Ces faits sont encore conformes aux prévisions théoriques.

*2. Réactivité de la région K.* C'est le développement de la théorie qui a attiré l'attention sur la réactivité de la région K et la réaction qui a été la plus étudiée est l'addition de tétraoxyde d'osmium (réaction du type addition électrophile). L'ordre croissant de réactivité vis-à-vis de cet agent est, parmi les composés étudiés: phénanthrène, pyrène, benzo-1, 2-anthracène, dibenzo-1, 2, 5, 6-anthracène, benzo-3, 4-pyrène. Cet ordre est parallèle à l'ordre d'activité cancérogène croissante. Ainsi, grâce à la suggestion de la théorie, on a mis en évidence, pour la première fois, une réaction chimique dont la facilité évolue parallèlement à l'activité cancérogène.

*3. Réactivité vis-à-vis des substitutions.* Les substitutions se font non sur des régions mais sur des centres actifs, c'est-à-dire sur des carbones isolés. En accord avec les principes de la théorie aucune relation homogène (parallélisme ou antiparallélisme) n'existe entre le pouvoir cancérogène des molécules et leur aptitude aux substitutions (couplage diazoïque, thiocyanuration, action du monochlorure de soufre, du péroxyde de benzoyle, etc.). Ainsi, par exemple, l'anthracène, non cancérogène, et le benzo-3, 4-pyrène, très cancérogène, sont tous deux en général très réactifs vis-à-vis des substitutions précitées.

Un intérêt plus grand encore se rattache toutefois aux découvertes expérimentales récentes qui apportent des confirmations expérimentales directes des hypothèses fondamentales de la théorie. Des travaux de nombreux chercheurs ont naturellement contribué à l'éclosion de ces résultats mais une mention toute particulière doit être donnée au remarquable groupe de chercheurs réunis à l'Université de Madison (Wisconsin) aux Etats-Unis, sous la direction de Heidelberger. Ainsi ces savants ont, en premier lieu, réussi à isoler le composé d'addition hydrocarbure-tissus cancéreux, apportant ainsi une preuve de l'idée essentielle de la théorie qui repose entièrement sur la nécessité d'existence de tels complexes. Des études soignées ont permis de montrer que la formation de ce complexe, bien qu'elle ne mette en jeu qu'une faible partie du composé administré, joue un rôle essentiel pour le développement des tumeurs. Ces études ont comporté la démonstration (à l'aide des composés marqués avec du carbone radioactif) d'un parallélisme presque complet entre l'incidence des tumeurs et la quantité du complexe formé, la démonstration, dans le cas des cancérogènes n'exerçant leur action que sur certains organes spécifiques, que ce complexe ne se formait qu'avec ces organes, la démonstration du fait que l'abondance de ce complexe varie parallèlement à la susceptibilité des différentes espèces animales vis-à-vis du cancérogène en question, la démonstration du fait que la formation du complexe diminue lorsque l'animal était soumis à un régime ayant pour effet de diminuer l'activité cancérogène etc.

De plus Heidelberger et ses collaborateurs ont démontré sur l'exemple du 1, 2, 5, 6-dibenzanthracène qu'en accord avec l'idée centrale de la théorie la formation du complexe cancérogène-récepteur cellulaire avait lieu effectivement par l'intermédiaire de la région K des hydrocarbures. En effet, cet auteur a réussi à former puis à dissocier le complexe 1, 2, 5, 6-dibenzanthracène-récepteur cellulaire. La difficulté consiste à dissocier le complexe sans tout démolir, ce qui n'est pas chose facile car ce complexe une fois formé est stable, ce qui indique la formation de véritables liaisons chimiques. Le produit isolé par la dissociation est l'acide dicarboxylique:

dont l'existence prouve quasi directement l'intervention de la région K dans la formation du complexe.

Finalement, le même savant a réussi à prouver que non seulement la jonction

TABLEAU VIII
Théorie et expérience en carcinogense chimique

| Expérience | Théorie |
|---|---|
| 1. "Binding of chemical carcinogens to tissue proteins is obligatory for the initiation of cancer" (Heidelberger, 1959). | .... L'idée fondamentale ... est ... la nécessité d'une fixation élective de la molécule sur un élément cellulaire defini ... Le problème est, selon toute probabilité, un problème de réactivité chimique (A. Pullman and B. Pullman, 1955). |
| 2. "In order for binding to occur, the intact ring system is required" (Heidelberger, 1959). "There is an apparent necessity for a completely aromatic system for binding" (Oliverio and Heidelberger, 1958). | Le problème de l'activité cancérogène met en jeu, selon toute probabilité, la molécule initiale et entière (A. Pullman and B. Pullman, 1955). |
| 3. "Binding takes place largely through the K region. The isolation of PDA (2-phenyl-phenanthrene-3′-2-dicarboxylic acid) ... was the first experimental verification of the Pullman's concept of the interaction of a carcinogen with cellular constituents at the K region" (Oliverio and Heidelberger, 1958). | The idea of the K region (a reactive bond) as the active center of carcinogens, through which the bonding to a cellular receiver occurs, the fundamental idea of the theory, postulated by one of us as early as in 1945 (A. Pullman). |
| 4. "The binding through the K region involves an addition reaction and is subject to steric hindrance. When one K region is blocked there is a possibility of binding to the second K region" (Oliverio and Heidelberger, 1958). | These are complete confirmations of the constant predictions of the theory, e.g., "The mechanism of action of carcinogenic molecules involves ... a chemical reaction between the carcinogen and cellular receiver ... The reaction probably consists in the formation of an addition product or complex.... The cellular receiver is probably of an electrophilic nature ..." (A. Pullman and B. Pullman, 1955). |
| 5. "The work reported here demonstrates that PDA is bound to the proteins through its carboxyl groups in amide linkage" (Bhargava and Heidelberger, 1956). | "Our scheme is in accordance with the theoretical speculations of the Pullman's with respect to a quinonoid bond between the hydrocarbon and the tissue" (Bhargava and Heidelberger, 1956). |

entre l'hydrocarbure et le récepteur cellulaire avait lieu à travers la région K mais, que, en parfait accord avec les prévisions de la théorie, cette jonction mettait en jeu des liaisons du type quinonique. Il montrait en fait que l'hydrocarbure était lié à la cellule par l'intermédiaire des fonctions amine et qu'il était doublement lié à ces fonctions.

L'extraordinaire accord entre les conceptions théoriques et les résultats expérimentaux est visible dans les tableaux VIII contenant quelques citations essentielles des articles de Heidelberger et de mes propres travaux. On peut remarquer que dans chaque cas la proposition théorique est antérieure à la constatation expérimentale.

Selon les hypothèses de Heidelberger l'attaque du cancérogène se ferait probablement sur les protéines cellulaires et leur action aurait peut-être pour effet l'inactivation ou la destruction d'un enzyme essentiel pour le fonctionnement normal de la division cellulaire. Il convient d'ajouter que d'autres auteurs considèrent que ce sont plutôt les acides nucléiques qui sont les sites des actions directes des cancérogènes. C'est un problème essentiel mais complexe et sur lequel je ne peux pas m'étendre ici.

**Conclusions**

En fait les domaines du sujet qui m'a été confié sont si vastes que je n'ai pu pratiquement vous en indiquer que quelques aspects très limités. En particulier j'aurais pu, j'aurais voulu vous donner encore de nombreux autres exemples d'applications de calcul à toute une gamme d'autres problèmes biochimiques ou biophysiques.

Mais enfin il faut que je me contente de ce que j'ai pu faire dans ce laps de temps limité et je dois conclure.

Pour cette conclusion je voudrais souligner auprès de vous le fait que quelque soit cette abondance de résultats déjà obtenus il convient *naturellement* de bien se rendre compte que l'utilisation des calculs quantiques en biologie en est au tout premier stade de son développement, stade qui a été comparé au traçage d'une piste à travers la jungle, et j'ajouterai à cette comparaison, cela avec des outils primitifs et des pionniers peu nombreux. En particulier *les méthodes* dont nous disposons à l'heure actuelle pour l'étude des molécules et des problèmes de dimensions de celles et de ceux que l'on rencontre en biologie sont *naturellement* très approximatives et très imparfaites encore. Combien de fois n'ai-je donc pas entendu des collègues chimistes théoriciens dire: comment peut-on essayer de comprendre la structure et les propriétés des grandes molécules alors que nous n'arrivons même pas à comprendre encore valablement celles des toutes petites molécules? Bien qu'une réponse appropriée à une telle question se trouve surtout dans la différence de la nature de résultats que l'on désire obtenir dans ces deux cas nettement différents, je préfère en général, me borner à leur faire remarquer que pour ma part, je trouve tellement plus excitant et surtout tellement moins frustrant de ne pas comprendre les grandes molécules plutôt que de ne pas comprendre les petites. Quoiqu'il en soit malgré cela, malgré ces limites, l'utilisation appropriée des méthodes *disponibles* permet de toute évidence d'interpréter, de faire des corrélations, des généralisations, des théories, de prévoir et parfois de corriger même l'expérience. C'est plus que l'on aurait pu espérer et cela indique que la voie choisie est bonne et riche en promesses.

# References

## General References:

M. Kasha and B. Pullman Eds., *Horizons in Biochemistry*, (Academic Press, 1962).
B. Pullman Ed., *Electronic Aspects of Biochemistry*, (Academic Press, 1964).
M. Weissbluth and B. Pullman Eds., *Molecular Biophysics*, (Academic Press, 1965).
A. Szent-Gyorgyi, *Introduction to a Submolecular Biology*, (Academic Press, 1960).
J. Duchesne Ed., *The Structure and Properties of Biomolecules and Biological Systems*, (Interscience, 1964).
M. Weissbluth Ed., *Quantum Aspects of Polypeptides and Polynucleotides*, (Biopolymers Symposia n° 1, 1963).
A. Pullman and B. Pullman, *Cancérisation par les Substances Chimiques et Structure Moléculaire*, (Masson, 1955).

## Purines and pyrimidines:

A. Pullman and B. Pullman, *Bull. Soc. Chim. France*, (1958) 766; (1959) 594.
T. Nakajima and B. Pullman, *Bull. Soc. Chim. France*, (1958) 1502; (1959) 663.
J. I. Fernandez-Alonso in *The Structure and Properties of Biomolecules and Biological Systems*, Ed. J. Duchesne, (Wiley, 1964) p. 3.
H. Berthod and A. Pullman, *J. Chim. Phys.* **62** (1965) 942.
D. F. Bradley, S. Lifson and B. Honig in *Electronic Aspects of Biochemistry*, Ed. B. Pullman, (Academic Press, 1964) p. 77.
A. Veillard and B. Pullman, *J. Theoret. Biol.* **4** (1963) 37.
C. Nagata, A. Imamura, Y. Tagashira and M. Kodama, *Bull. Chem. Soc. Japan* **38** (1965) 1638.
H. Berthod, C. Giessner-Prettre and A. Pullman, *Theoret. Chim. Acta* **5** (1966) 53; *Intern. J. Quantum Chem.* **1** (1967).
R. K. Nesbet, *Biopolymers Symp.* **1** (1964) 129.
J. Ladik and K. Appel, *Theoret. Chim. Acta* **4** (1966) 132.
S. Kwiatkowski, *Acta Phys. Polon.* **29** (1966) 573.
M. Tanaka and S. Nagakura, *Theoret. Chim. Acta* **6** (1966) 320.

## Purine-pyrimidine base pairs:

B. Pullman and A. Pullman, *Biochim. Biophys. Acta* **36** (1959) 343.
A. Pullman, *Compt. Rend. Acad. Sci. Paris* **256** (1963) 5435.
J. Ladik in *Electronic Aspects of Biochemistry*, Ed. B. Pullman, (Academic Press, 1964) p. 203.
P. O. Löwdin in *Electronic Aspects of Biochemistry*, Ed. B. Pullman, (Academic Press, 1964) p. 167; *Adv. Quantum Chem.* **2** (1965) 213.
R. Rein and F. E. Harris, *J. Chem. Phys.* **41** (1964) 3393; **42** (1965) 2277; **43** (1966) 4415.
H. De Voe and I. Tinoco Jr., *J. Mol. Biol.* **4** (1962) 500.
D. F. Bradley, S. Lifson and B. Honig in *Electronic Aspects of Biochemistry*, Ed. B. Pullman, (Academic Press, 1964) p. 77.
H. A. Nash and D. F. Bradley, *Biopolymers* **3** (1965) 261; *J. Chem. Phys.* **45** (1966) 1380.
B. Pullman, P. Claverie and J. Caillet, *Proc. Natl. Acad. Sci. U.S.* **55** (1966) 905; *J. Mol. Biol.* **22** (1966) 373; *Compt. Rend. Acad. Sci. Paris* **263** (1966) 2006.
M. Polak and R. Rein, *J. Theoret. Biol.* **11** (1966) 490.
Yu A. Kruglyak, V. I. Danilov and O. V. Shramko, *Biofizika* **10** (1965) 399, 561.

*Purine and pyrimidine analogs:*

B. Pullman and A. Pullman, *Bull. Soc. Chim.* (1958) 973.
A. Pullman and B. Pullman, *Compt. Rend. Acad. Sci. Paris* **246** (1958) 611.
H. Berthod and A. Pullman, *Compt. Rend. Acad. Sci. Paris* **262** (1966) 76.
B. Pullman and N. Dupuy, *Compt. Rend. Acad. Sci. Paris* **262** (1966) 2773; *J. Chim. Phys.* **64** (1967) 708.
J. I. Fernandez-Alonzo in *The structure and Properties of Biomolecules and Biological Systems,* Ed. J. Duchesne, (Interscience, 1964) p. 3.
P. Pithova, A. Piskala, J. Pitha and F. Sorm, *Coll. Czech. Chem. Comm.* **30** (1965) 90, 1626, 2801.
J. Pitha, P. Fiedler and J. Gut, *Coll. Czech. Chem. Comm.* **31** (1966) 1864.
H. G. Mautner and G. Bergson, *Acta Chem. Scand.* **17** (1963) 1694.


*Pteridines:*

J. I. Fernandez-Alonso in *The Structure and Properties of Biomolecules and Biological Systems,* Ed. J. Duchesne, (Wiley, 1964) p. 3.
A. Bopst, *Helv. Chim. Acta* **50** (1967) 1480.


*α-amino acids and proteins:*

G. Del Re, B. Pullman and T. Yonezawa, *Biochem. Biophys. Acta* **75** (1963) 153.
T. Yonezawa, G. Del Re and B. Pullman, *Bull. Chem. Soc. Japan* **37** (1964) 985.
M. G. Evans and I. Gergely, *Biochim. Biophys. Acta* **3** (1949) 188.
M. Suard, G. Berthier and B. Pullman, *Biochim. Biophys. Acta* **52** (1961) 254.
M. Suard, *Biochim. Biophys. Acta* **59** (1962) 227; **64** (1962) 400; *J. Chim. Phys.* **61** (1964) 79, 89.
A. Pullman, *Biopolymers Symp.* **1** (1964) 29; *Modern Quantum Chemistry*, Ed. O. Sinanoglu, (Academic Press, 1965) vol. III, p. 283.
R. Itoh, *Ann. Rept. Res. Group Biophys. Japan* **1** (1961) 11.
S. Yomosa, *Biopolymers Symp.* **1** (1964) 1; *Ann. Rept. Res. Group. Biophys., Japan* **5** (1965) 39; *J. Phys. Soc. Japan* **18** (1963) 1494.
A. Imamura, M. Kodama, Y. Tagashira and C. Nagata, *J. Theoret. Biol.* **10** (1966) 356.


*Porphyrins, Iron porphyrins:*

B. Pullman, C. Spanjaard and G. Berthier, *Proc. Natl. Acad. Sci. U.S.* **46** (1960) 1011.
A. Veillard and B. Pullman, *J. Theoret. Biol.* **8** (1965) 317.
K. Ohno, Y. Tanabe and F. Sasaki, *Theoret. Chim. Acta* **1** (1963) 378.
M. Kotani, *Rev. Mod. Phys.* **35** (1963) 717.
M. Kotani in *The Structure and Properties of Biomolecules and Biological Systems*, Ed. J. Duchesne, (Interscience, 1964) p. 159.
M. Zerner and M. Gouterman, *Theoret. Chim. Acta* **4** (1966) 44.
J. S. Griffith, *Biopolymers Symp.* **1** (1964) 35.
P. Day, G. Scregg and R. J. P. Williams, *Biopolymers Symp.* **1** (1964) 271.


*Carotenoids and retinenes:*

A. Pullman and B. Pullman, *Proc. Natl. Acad. Sci. U.S.* **47** (1961) 7.
H. Berthod and A. Pullman, *Compt. Rend. Acad. Sci. Paris* **251** (1960) 808.

*Energy-rich phosphates:*

B. Grabe, *Biochim. Biophys. Acta* **30** (1958) 560.
B. Grabe, *Arkiv. Fysik* **15** (1959) 207.
B. Pullman and A. Pullman, *Radiation Res. Suppl.* **2** (1960) 160.
K. Fukui, A. Imamura and Ch. Nagata, *Bull. Chem. Soc. Japan* **36** (1963) 1450.
R. L. Collin, *J. Am. Chem. Soc.* **88** (1966) 3281.


*Oxidation-reduction coenzymes:*

B. Pullman and A. Pullman, *Proc. Natl. Acad. Sci. U.S.* **45** (1959) 136.
G. Karreman, *Bull. Math. Biophys.* **23** (1961) 55, 135.
G. Karreman, *Ann. (New York) Acad. Sci.* **96** (1962) 1029.
B. Grabe, *Biopolymers Symp.* **1** (1964) 283.
J. P. Malrieu and B. Pullman, *Theoret. Chim. Acta* **2** (1964) 302.
A. Pullman, *Tetrahedron* **19**, suppl. 2 (1963) 441.
J. L. Fox, K. Nishimoto and L. S. Forster, *Biochim. Biophys. Acta* **109** (1965) 626.
J. L. Fox, S. P. La Berge, K. Nishimoto and L. S. Forster, *Biochim. Biophys. Acta* **136** (1967) 544.
J. Koziol, *Photochem. Photobiol.* **5** (1966) 41.


*Folic acid coenzymes:*

A. M. Perault and B. Pullman, *Biochim. Biophys. Acta* **44** (1960) 251; **52** (1961) 266.
R. Collin and B. Pullman, *Biochim. Biophys. Acta* **232** (1964) 232.
W. B. Neely, *J. Mol. Pharmac.* **3** (1967) 108.


*Pyridoxal phosphate:*

A. M. Perault, B. Pullman and C. Valdemoro, *Biochim. Biophys. Acta* **46** (1961) 555.
B. Pullman in *Chemical and Biological Aspects of Pyridoxal Catalysis*, Ed. E. E. Snell, P. M. Fasella, A. E. Braunstein and A. Rossi-Fanelli, (Pergamon, 1963) p. 103.


*Thiamine pyrophosphate:*

B. Pullman and C. Spanjaard, *Biochim. Biophys. Acta* **46** (1961) 576.
R. Collin and B. Pullman, *Arch. Biochem. Biophys.* **108** (1964) 535.


*Vitamine B$_{12}$:*

A. Veillard and B. Pullman, *J. Theoret. Biol.* **8** (1965) 307.


*Hydrogen bonding:*

B. Pullman and A. Pullman, *Biochim. Biophys. Acta* **36** (1959) 343.
A. Pullman, *Compt. Rend. Acad. Sci. Paris* **256** (1963) 5435.
P. O. Löwdin in *Electronic Aspects of Biochemistry*, Ed. B. Pullman, (Academic Press, 1964) p. 167; *Biopolymers Symposia* **1** (1964) 161, 293; *Advan. Quantum Chem.* **2** (1965) 213.
R. Rein and F. E. Harris, *J. Chem. Phys.* **41** (1964) 3393; **42** (1965) 2177; **43**, 4415; **45**, 1797.
H. A. Nash and D. F. Bradley, *Biopolymers* **3** (1965) 261; *J. Chem. Phys.* **45** (1966) 1380.
B. Pullman, P. Claverie and J. Caillet, *Proc. Natl. Acad. Sci. U.S.* **55** (1966) 905; *J. Mol. Biol.* **22** (1966) 373; *Compt. Rend. Acad. Sci. Paris* **263** (1966) 2006.

*Intermolecular forces:*

H. DeVoe and I. Tinoco, Jr., *J. Mol. Biol.* **4** (1962) 500.

D. F. Bradley, S. Lifson and B. Honig in *Electronic Aspects of Biochemistry*, Ed. B. Pullman, (Academic Press, 1964) p. 77.

H. A. Nash and D. F. Bradley, *Biopolymers* **3** (1965) 261; *J. Chem. Phys.* **45** (1966) 1380.

B. Pullman, P. Claverie and J. Caillet, *Proc. Natl. Acad. Sci. U.S.* **55** (1966) 905; *J. Theoret. Biol.* **12** (1966) 419; *J. Mol. Biol.* **22** (1966) 373; *Science* **147** (1965) 1305.

O. Sinanoglu and S. Abdulnur, *Photochem. Photobiol.* **3** (1964) 333; *Federation Proc.* **24** (1965) S 12.

N. F. Gersch and D. O. Jordan, *J. Mol. Biol.* **13** (1965) 138.

H. Jehle, W. C. Parke and A. Salyers in *Electronic Aspects of Biochemistry*, Ed. B. Pullman, (Academic Press, 1964) p. 313; *Biopolymers* **4** (1965) 433.

H. Jehle, W. C. Parke, R. M. Shiven and D. K. Aein, *Biopolymers Symp.* **1** (1964) 209.

L. Salem, *Nature* **193** (1962) 476.

L. Salem, *Canadian J. Biochem. Physiol.* **40** (1962) 1287.


*Electron donor and acceptor properties, charge transfer:*

B. Pullman and A. Pullman, *Proc. Natl. Acad. Sci. U.S.* **44** (1956) 1197; *Rev. Modern Phys.* **32** (1960) 428.

B. Pullman, in abstracts of the *Sixth International Congress of Biochemistry*, New York **10** (1964) p. 737.

M. Rossi and A. Pullman, *Biochim. Biophys. Acta* **88** (1964) 211.

B. Pullman, *Biochim. Biophys. Acta* **88** (1964) 440.

G. Karreman, *Bull. Math. Biophys.* **23** (1961) 135; *Ann. New York Acad. Sci.* **96** (1962) 1029; in *Data Acquisition and Processing in Biology and Medicine*, (Pergamon, 1962) p. 51.

A. Szent-Gyorgyi, *Introduction to a Submolecular Biology*, (Academic Press, 1960).

L. Brillouin in *Horizons in Biochemistry*, Eds. M. Kasha and B. Pullman, (Academic Press, 1962) p. 295.

M. J. Mantione and B. Pullman, *Compt. Rend. Acad. Sci. Paris* **262** (1966) 1492.

A. Pullman and B. Pullman in *Quantum Theory of atoms, Molecules and the Solid State*, Ed. P. O. Löwdin, (Academic Press, 1966) p. 345.


*Free radicals in biology:*

B. Pullman and A. Pullman, *Proc. Natl. Acad. Sci. U.S.* **45** (1959) 136.

A. Pullman, *J. Chim. Phys.* **61** (1964) 1666.

A. V. Guzzo and G. Tollin, *Arch. Biochim. Biophys.* **103** (1963) 231, 244.

J. P. Malrieu and B. Pullman, *Theoret. Chim. Acta* **2** (1964) 293, 302.

J. N. Herak and W. Gordy, *Proc. Natl. Acad. Sci. U.S.* **54** (1965) 1287; **56** (1966) 1354.

B. Pullman and M. J. Mantione, *Compt. Rend. Acad. Sci. Paris* **260** (1965) 5643; **261** (1965) 5679.


*Semiconductivity:*

M. G. Evans and I. Gergely, *Biochim. Biophys. Acta* **3** (1949) 188.

M. Suard, G. Berthier and B. Pullman, *Biochim. Biophys. Acta* **52** (1961) 254.

J. Ladik in *Electronic Aspects of Biochemistry*, Ed. B. Pullman, (Academic Press, 1964) p. 203.

T. A. Hoffmann and J. Ladik in *The Structure and Properties of Biomolecules and Biological Systems*, Ed. J. Duchesne, (Interscience, 1964) p. 84.

D. D. Eley in *Horizons in Biochemistry*, Eds. M. Kasha and B. Pullman, (Academic Press, 1962) p. 341.

D. D. Eley and R. B. Leslie in *Electronic Aspects of Biochemistry*, Ed. B. Pullman, (Academic Press, 1964) p. 105.

A. Pullman in *Molecular Biophysics*, Eds. B. Pullman and M. Weissbluth, (Academic Press, 1965), p. 81.

*Radiation effects and the mechanism of radioresistance:*

B. Pullman and A. Pullman in *Comparative Effects of Radiation*, Eds. M. Burton, J. S. Kirby-Smith and J. L. Magee, (Wiley, 1960) p. 105.

B. Pullman, *Mem. Acad. Roy. Belg.* **33** (1961) 174.

J. Duchesne and A. Van de Vorst, *Bull. Classe Sci. Acad. Roy. Belgique* **51** (1965) 778.

A. van de Vorst, M. Richer and K. V. Rajalakshmi, *Bull. Classe Sci. Acad. Roy. Belgique* **52** (1966) 276.

*Mechanisms of photobiology:*

A. Pullman and B. Pullman, *Proc. Natl. Acad. Sci. U.S.* **47** (1961) 7; *Biochim. Biophys. Acta* **75** (1963) 269.

M. J. Mantione and B. Pullman, *Biochim. Biophys. Acta* **91** (1964) 387; **95** (1965) 668.

C. Nagata, A. Imamura, Y. Tagashira, M. Kodama et N. Fukuda, *J. Theoret. Biol.* **9** (1965) 357.

*Mechanisms of mutagenesis:*

A. Pullman in *Electronic Aspects of Biochemistry*, Ed. B. Pullman, (Academic Press, 1964) p. 135; *Biochim. Biophys. Acta* **87** (1964) 365.

P. O. Löwdin in *Electronic Aspects of Biochemistry*, Ed. B. Pullman, (Academic Press, 1964) p. 167; *Advan. Quantum Chem.* **2** (1965) 213.

R. Rein and F. E. Harris, *J. Chem. Phys.* **41** (1964) 3393; **42** (1965) 2177; **43** (1965) 4415.

Ch. Nagata, A. Imamura, H. Salto and K. Fukui, *Gann* **54** (1963) 109.

*Mechanisms of carcinogenesis:*

A. Pullman and B. Pullman, *Cancérisation par les substances Chimiques et Structure Moléculaire*, (Masson, 1955); *Advan. Cancer Res.* **3** (1955) 117.

A. Pullman, *Biopolymers Symp.* **1** (1964) 47.

B. Pullman, *Biopolymers Symp.* **1** (1964) 141.

B. Pullman, *J. Cell. Comp. Physiol.* **64** suppl. 1 (1964) 91.

B. Pullman, P. Claverie and J. Caillet, *Science* **147** (1965) 1305.

*Biochemical evolution and the origin of life:*

B. Pullman and A. Pullman, *Nature* **196** (1962) 1137.

B. Pullman in *Molecular Orbitals in Chemistry, Physics and Biology*, Eds. P. O. Löwdin and B. Pullman, (Academic Press, 1964) p. 547.

*Optical properties of biopolymers:*

W. Rhodes, *J. Am. Chem. Soc.* **83** (1961) 3609; *J. Chem. Phys.* **37** (1962) 2433.

I. Tinoco, *J. Am. Chem. Soc.* **82** (1960) 4785; **84** (1961) 5047.

H. DeVoe and I. Tinoco, *J. Mol. Biol.* **4** (1962) 518.

R. K. Nesbet, *Mol. Phys.* **7** (1964) 211.

I. Tinoco in *Molecular Biophysics*, Eds. B. Pullman and M. Weissbluth, (Academic Press, 1965) p. 262.

D. F. Bradley, S. Lifson and B. Honig in *Electronic Aspects oᶠ Biochemistry*, Ed. B. Pullman, (Academic Press, 1964) p. 77.

K. Rosenheck and B. Sommer, *J. Chem. Phys.* **46** (1967) 532.

C. A. Bush and I. Tinoco Jr., *J. Mol. Biol.* **23** (1967) 601.


*Electronic aspects of pharmacology:*

B. Pullman in *Electronic Aspects of Biochemistry*, Ed. B. Pullman, (Academic Press, 1964) p. 559.

J. P. Malrieu and B. Pullman, *Theoret. Chim. Acta* **2** (1964) 293.

G. Karreman in *Data Acquisition and Processing in Biology and Medicine*, (Pergamon, 1962) p. 51.

G. Karreman, I. Isenberg and A. Szent-Gyorgyi, *Science* **130** (1959) 1191.

K. Fukui, Ch. Nagata and T. Yonezawa, *J. Am. Chem. Soc.* **80** (1958) 2267.

K. Fukui, Ch. Nagata and A. Amamura, *Science* **132** (1960) 87.

A. Inouye and Y. Shinagawa, *Bull. Chem. Soc. Japan* **35** (1962) 701.

J. P. Green and J. P. Malrieu, *Proc. Natl. Acad. Sci. U.S.* **54** (1965) 659.

W. B. Neely, *Mol. Pharmacol.* **1** (1965) 137.

D. Agin, L. Hersch and D. Holtzman, *Proc. Natl. Acad. Sci. U.S.* **53** (1965) 952.

M. Cocordano and J. Ricard, *Physiol. Veg.* **1** (1963) 129.

T. Ban, Jap. *J. Pharmacol.* **12** (1962) 72.

S. Snyder and C. R. Merril, *Proc. Natl. Acad. Sci. U.S.* **54** (1965) 258.

S. Snyder, C. R. Merril and D. F. Bradley, *Biochim. Biophys. Acta* **118** (1966) 316.

E. Yeargers and L. Angensten, *Nature* **212** (1966) 251.

M. Cocordano, J. Ricard and A. Julg, *J. Theoret. Biol.* **12** (1966) 291.

*Theoretical Physics and Biology,* © 1969, North-Holland Publ. Co., Amsterdam

# MOLECULAR FORCES AND MOLECULAR RECOGNITION

S. BRESLER

*Institute for High Molecular CPDS, Academy of Sciences, Leningrad, U.S.S.R.*

I want to discuss in very general terms the problem which is in the centre of interest of molecular biology. This is specificity of interaction or molecular recognition. The interaction of an enzyme with its co-enzyme and its substrate or an antigen with its antibody are well-known instances. The recent interest in information problems in biology has revealed some new aspects. The coding of a polypeptide by means of a polynucleotide template gives a rather well-known example. There is recognition of anticodons by codons and recognition of aminoacids and t-RNA's by the corresponding enzymes (aminoacyl t-RNA synthetases). The diversity of aminoacids is coded by triplet codons and by some twenty enzymes. What are the molecular forces involved in this specific interaction?

In the case of codon–anticodon recognition there are few doubts that these are hydrogen bonds regulating the complementarity of bases. We have one case where this complementarity was proved by direct analysis by comparing the primary structure of a t-RNA in a suppressor strain with a corresponding t-RNA of a wild strain by Smith et al. What is important is the possibility to predict not only the regular structure of complementary anticodons, but some possibility of ambiguity for anticodons recognizing different codons. These ambiguities are revealed because some species of t-RNA's with obviously fixed anti-codons are stuck sometimes on different codons. Crick in his wobble hypothesis gave a succesfull explanation of this case of incomplete complementarity. And this has very much in common with Dr. Pullman's talk about the ambiguity in the matching of bases.

We have no experimental data for the energy and entropy of codon–anticodon interaction. We don't know if the structure of ribosomes is important and influences the codon–anticodon interaction. We have no quantitative data on the influence of bivalent or trivalent ions like magnesium or streptomycine. We are trying now in our laboratory to accumulate such data which seem important for the physical features of the coding machine. Nevertheless we can make some simple theoretical estimations. If we take for the average number of proteins in a bacterial cell the number $10^3$, for the average polymerisation degree something like $10^3$ and for the average number of molecules per enzyme also $10^3$, we obtain a reasonable estimate of the cell weight. Now we consider the probability of mistakes in the process of decoding of genetic information. Let there be one mistake in a hundred molecules. I think this is a reasonable figure. It gives us a probability of a wrong codon–anticodon interaction of the order of $10^{-5}$. The free-energy difference between the right and the wrong codon–anticodon combination we shall denote $\Delta F$. Then $\exp(-\Delta F/RT) = 10^{-5}$. This gives us for $\Delta F$ some 5000 calories per mol. This difference is provided by the hydrogen bond between 3 nucleotides which make 1700 cal per mol for one pair. This is a reasonable figure for 2.3 H bonds in water. We suppose of course that the whole anticodon will not fit if one of the 3 nucleotides does not match according to Watson–Crick complementarity. So the codon–anticodon fit must be a cooperative phenomenon. In case of enzyme substrate recognition, many different kinds of molecular forces may be involved. Electrostatic interaction can be quite specific because of specific patterns of positive and negative charges on the surface of protein molecules.

This was demonstrated in a model experiment with adsorption of peptides or proteins on ion-exchanger resins with − or + alternating charges and different distances between charged groups. Experiments started in our laboratory by Samsonof showed that extreme specificity of adsorption can be obtained for every peptide or protein species if complementary charges on the surface are separated by an appropriate distance. You can easily get equilibrium constants of the order of $10^5$ in this case and as you know the equilibrium constants for simple ion-exchanger resins are never higher than 10 to 100. So the electrostatic interaction or ionic forces can be very specific if you have complementary charges at appropriate distances on both surfaces. I think this may be the answer to questions that were discussed about long range forces in cells. An instance of this is meiosis. The juxtaposition (synapse) of chromosomes during meiosis can be explained by ionic forces of complementary charges forming a specific pattern. Hydrogen bonds and coordination bonds, or weak covalent bonds are also quite usual. This explains the widespread participation of transition metal ions in enzymes. Even Van der Waals dispersion forces may be specific, and a model experiment demonstrating this specificity was done by means of silica gel formation in the presence of molecules of methyl orange. The resulting gel was specific for the absorption of the same dye. This is readily explained by the formation of specific holes in the structure of the gel, where the molecules of the dye fit exactly.

So all kinds of molecular forces, ionic, covalent, hydrogen bonds and even dispersion forces participate in enzyme substrate, enzyme–coenzyme, antibody-antigene recognition. Of course this recognition is not ideal. The possibility to design antimetabolites or analogues of a substrate shows that the enzyme can be cheated in the recognition of its substrate. All mutations which make the microorganism resistant to metabolic poisons are some minor changes of the protein structure which make the enzyme more selective, more discriminating in the process of substrate recognition.

A fundamental question which remains up to now unsolved is the mechanism of recognition of polynucleotides sequences by proteins. We find this problem in case of aminoacyl t-RNA synthesis, but the main objects are the repressors and inducers which regulate the transcription of genes. There are many proofs that the repressors in bacteria postulated by Monod and Jacob are proteins. The best evidence is the finding of temperature sensitive and amber mutants for the corresponding bacteria. The repressor must recognize the starting point for transcription or the operator which is probably nothing else than a sequence of nucleotides.

The situation is even more complex in a higher organism. During onthogenesis, differentiation occurs. The information contained in the nucleus of the zygote is transferred to differentiated cells. Every species of differentiated cells has its characteristic pattern of enzymes. We can estimate the order of magnitude of those for every specific cell as 200–300. In a developing organism the number of different cell species both in the embryo and in the adult organism can be estimated of the order of 1000.

What is the organization of genetic material? How is realized the repression of 999/1000 of all these genes in every cell and the release of information only in one of 1000 genes for every particular kind of differentiated cells? This is a general question of great complexity, if you compare it with the question of coding 20 aminoacids which was already solved in molecular biology. We can imagine that the genetic material is organized in sections for every species of cells rather than in independent small operons. From an informational point of view, this is the most economic way. A whole section can be switched on by a chemical signal and all the proteins and the nucleic acids of a particular species of cells are synthesized simultaneously. This situation is possible for a somatic cell of a complicated organism because it has not the problem to adapt to external conditions as has a microorganism. On the other hand, the amount of genetic material in the nucleus is enough for providing more than 300000 cistrons (or elementary genetic units). If the genetic material is organized this way, the question arises how to recognize 1000 different points on the chromosome by specific protein inducers. We know from the experiments with puffs that sections of chromosomes which become active during differentiation are placed at random, so the only possibility for switching on of genes is molecular recognition.

Now here comes the question how do inducers recognize one from a thousand of specific polynucleotide sequences. To have such a diversity of possibilities, the piece of the chain which effects recognition must be an oligonucleotide composed at least of 5 monomeric units because we have 4 different kinds of nucleotides which gives us $4^5 = 1024$ different combinations. It is very improbable that a protein can be specifically fit for the recognition of a sequence of 5 nucleotides.

Aminoacids and nucleotides are much too different in structure to give specific interaction. We are forced to suppose that the inducer in differentiation, if it is a protein, must contain as prosthetic group a polynucleotide, at least a pentanucleo-tide complementary to the polynucleotide which opens a section of genes used in this particular differentiated cell. Up to now these are pure speculations based on logic. But in case of bacterial repressors, there are some data indicating that the repressor is really a protein which contains a polynucleotide as prosthetic group. These are findings of Wacker in Germany.

Anyway there seems to be no other issue, and therefore we expect the same Watson–Crick principle to be potent in this most important problem to be solved in the years to come.

*Theoretical Physics and Biology*, © 1969, *North-Holland Publ. Co., Amsterdam*

# THE TIME DOMAINS OF BIOCHEMICAL REACTIONS
# IN LIVING CELLS

BRITTON CHANCE

*Johnson Research Foundation, University of Pennsylvania, Philadelphia, U.S.A.*

Three physical methods for perturbing biological systems have now been applied *in extenso* to enzymes isolated from living cells, to subcellular structures of living cells (particularly mitochondria), to suspensions of intact living cells (ascites, yeast, photosynthetic bacteria, and algae), and, in preliminary form, to solid tissue (the perfused, beating, heart, and plant leaves). The fastest and as yet the most specialized methods are those involving photolysis, especially in the activation of electron transfer reactions in chlorophyll-cytochrome systems following 30 nsec illumination by laser flashes. Temperature perturbations such as those employed in studies of isolated enzyme systems by Eigen and his colleagues at Göttingen are now being applied in relatively primitive form to suspensions of cells and to cell extracts [1]. But the workhorse of fast reaction methods, covering the widest range of time and of materials, is the rapid flow method of Hartridge and Roughton, particularly as adapted to studies of intracellular reactions [2]. These three methods now allow us to assign time domains to the various types of intracellular reactions, and to indicate those in which physics or chemistry might predominate.

Laser flash activation of cytochrome-chlorophyll electron transfer in the photosynthetic bacterium *Chromatium* provides a unique system for the detailed study of electron transfer mechanisms *in vivo*. The nature of the reaction studied in detail with steady state illumination [3] comes more sharply into focus when laser flashes are employed to give the time sequence of electron transfer reactions in a living cell. The fastest reaction is the donati on of an electron from a cytochrome of high oxidation-reduction potential ($c_{422}$) in a temperature-dependent reaction having a half-time of 2 $\mu$sec at room temperature and an activation energy of 3.3 Kcal [4]. This reaction apparently requires thermal motion of the cytochrome relative to the chlorophyll in order to achieve such high velocity [5]. As the cells are frozen and temperatures near those of liquid nitrogen are reached, the reaction velocity is further slowed and reaches a constant half-time of 2 msec, which, in experiments of DeVault and Parkes, is maintained down to a temperature of 4.5° K [6]. The application of quantum mechanical electron tunneling as an explanation of the temperature insensitivity of this reaction affords a unique opportunity to employ a purely physical theory for a biochemical reaction which has been studied in detail experimentally [4]. Barrier widths of 30 Å and barrier height of 1 eV are consistent with the observed speed of the reaction and with the expected frequency factor of electrons under these conditions. While the "fit" of the tunneling mechanism to the experimental data is satisfactory, it does not eliminate other possibilities from consideration [5].

The kinetics of rapid electron transfer reactions have recently been studied by a modified form of the rapid flow apparatus which solves boundary layer problems in the mixer and extends the time range of the apparatus to the detection of time differences of 30 $\mu$sec [7]. Observations of the overall activity of the enzyme, catalase, have suggested that the lifetime of the rate-limiting intermediate was as short as $10^{-8}$ sec [8]. Studies of a peroxidase specific for the oxidation of cytochrome $c$ reveal an intramolecular oxidation-reduction reaction with a computed half-time of 30 $\mu$sec in the following sequence:

$$E + S \xrightarrow[t_{\frac{1}{2}}=170\ \mu sec]{k_1} ES_I \xrightarrow[t_{\frac{1}{2}}=30\ \mu sec]{k_2} ES_{II} \rightarrow E + P.$$

Thus, the red Complex ES of Yonetani is preceded by a prior complex involving an extremely rapid intramolecular oxidation-reduction reaction.

The kinetics of electron transfer reactions involving cytochrome $c$ as a bound component of the electron transport chain of mitochondria and submitochondrial particles can also be measured by the rapid flow apparatus, and half-times of 1 to 2 msec are observed. It is apparent that reactions which involve energy coupling do not proceed as rapidly as those of pure electron transfer. For this reason, a point of greatest interest from the chemical and physical standpoints is the degree to which the energy coupling mechanism may actually control the rate of electron transfer into cytochrome $c$; experiments on mitochondria, and more recently on submitochondrial particles [9] indicate that control ratios of over 10-fold may be exerted in this manner. The nature of such control mechanisms is of the greatest interest, and a number of possibilities are under consideration at the present time [5].

One mechanism proposes that interaction between the spin state of the iron porphyrin of cytochrome $c$ and secondary and tertiary structural features of the molecule lead to both energy conservation and control of electron transport by the energy coupling apparatus. While crystallographic data on cytochrome $c$ are not yet of adequate resolution to provide detailed support for this hypothesis, parallel crystallographic studies on myoglobin now clearly identify secondary and tertiary structural feature changes associated with the conversion of the high spin aquo form into the low spin species formed in the presence of cyanide or alkali [10]. Xenon binding to the hydrophobic interior of the molecules not only causes additional conformation changes [11], but also alters the reactivity towards cyanide and hydroxide [12].

Perturbations of metabolizing cells and tissue extracts have emphasized rate-controlling steps in the multi-enzyme system of glycolysis. Here, not only is direct biochemical readout necessary, but rapid sampling techniques for the evaluation of changes in the metabolite pattern are essential, so that the time resolution is diminished to a few tenths of a second [1]. Nevertheless, the time required for transitions in the glycolytic system of the metabolizing cell appear much longer than those of the structure-bound enzymes, such as the cytochrome chains of respiration, or of photosynthesis, mentioned above. The application of such perturbation methods to metabolic control phenomena is a developing topic of the greatest interest.

## References

[1]   P. K. Maitra, I. Y. Lee and G. F. Williamson, *Fed. Proc.* **26** (1967) 564.

[2]   B. Chance, *Faraday Soc. Disc.* **17** (1954) 365.

[3]   J. M. Olson and B. Chance, *Arch. Biochem. Biophys.* **88** (1960) 26.

[4]   D. DeVault and B. Chance, *Biophys. J.* **6** (1966) 825.

[5]   B. Chance, *Biochem. J.* **103** (1967) 1.

[6]   D. De Vault, J. Parkes and B. Chance, *Nature* **215** (1967) 642.

[7]   B. Chance, D. De Vault, V. Legallais, L. Mela and T. Yonetani, in: *Fast Reactions and Primary Processes in Chemical Kinetics*, Ed. S. Claesson, (Almqvist & Wiksell, 1967) 437.

[8]   B. Chance in *Currents in Biochemical Research*, Ed. D. E. Green, (Interscience, 1956) 308.

[9]   C. P. Lee and L. Ernster, *Fed. Proc.* **26** (1957) 610.

[10]  H. C. Watson and B. Chance, in: *Hemes and Hemoproteins*, Eds. B. Chance, R. W. Estabrook, and T. Yonetani, (Academic Press, 1966) 149.

[11]  B. P. Schoenborn, in: *Abstracts, Division of Biological Chemistry*, (Am. Chem. Soc., 1968) 30.

[12]  A. S. Mildvan, N. Rumen and B. Chance, in: *Abstracts, Division of Biological Chemistry*, (Am. Chem. Soc., 1968) 32.

*Theoretical Physics and Biology*, © 1969, *North-Holland Publ. Co., Amsterdam*

# DISCUSSIONS

M. Eigen: I want to talk about some experiments on codon–anticodon interactions. Especially three types of problems have been or are being studied: (1) The elementary step and specificity of recognition between single complementary base pairs [1]. (2) The cooperativity of base pair interaction in oligo- and polynucleotides [2]. (3) The enzymic mechanism of replication.

## 1. Pairing of single bases

The specificity of base pairing was investigated in nonpolar solvents where these processes occur to a quite pronounced extent. A new technique was introduced with these studies [3] which should be discussed here briefly. The single bases have an appreciable electrical moment, which is reduced considerably upon pairing. As a consequence the pairing equilibrium depends on the electric field strength ($E$) according to

$$\frac{\eth \ln K}{\eth |E|} = \frac{\varDelta M}{RT},$$

where $\varDelta M$ is the difference in partial molar moments for reactants and products. This quantity contains essentially—due to Langevin's theory—the squares of the dipole moments ($\mu$) of each species and the electrical field strength (e.g. for a pairing between the bases A and U, $\varDelta M$ is proportional to $(E/kT)(\mu_A^2 + \mu_U^2 - \mu_{AU}^2)$). If we integrate the above relation we would find a quadratic dependence of $K$ on field strength. This is the reason why chemical contributions to dielectric loss never had been observed. At small field strengths (necessary for the usual dielectric measurements) the relaxation of the chemical equilibrium does not contribute measurably to the dielectric loss. However, if we superimpose a large D.C. field ($\sim 200$ kV/cm) on the low amplitude (high frequency) A.C. field, the contribution becomes detectable. In the frequency range (around $1/\tau$) where chemical equilibration lags behind the oscillating field we observe an uptake of energy (dielectric loss) characteristic of the chemical effect. It is measured by means of a line broadening technique and yields information about the relaxation process in the micro- to nanosecond region.

By means of this technique the following problem was studied. We know that the recognition of nucleotides is based on complementarity. The two complementary base pairs are AU (or AT resp.) and GC. However, almost any other combination involving two hydrogen bonds is quite possible. There must have been a very early stage in evolution where enzymes did not yet exist which were sufficiently adapted to control this process. Thus the complementarity observed

159

must have been established by a preferred stability of these complexes. This preferred stability of, for instance, the AU complex compared to any other combination of A and U is indeed found. These findings are in agreement with spectroscopic results obtained by other authors [4, 5]. The dielectric studies in addition also yield information about rates. It turns out that base pair formation is a very rapid, diffusion controlled process and that lifetimes of the pairs are in the neigborhood of $10^{-7}$ sec (which is important for a rapid reading process). On the other hand, the stability of the complementary pair is by far not high enough to account for the low error rate observed in replication, which must result then from cooperativity and especially from enzyme recognition.

## 2.  Cooperativity of base pairing and code reading

If one would look for the pairing of single bases in aqueous media none would be found due to the competition of hydrogen bond formation with water molecules. The stability of any double-helical structure of nucleic acids is due to 'stacking' interaction and 'structure freezing' (entropic effect) which comes into play when several bases can pair cooperatively. The build up of cooperativity has been studied with oligo-nucleotides (oligo A + oligo A at pH ⩽ 4 and oligo A + oligo U in the neutral region) using the temperature jump relaxation technique [2]. For this purpose oligo compounds containing 3 to 10 residues of A and U have been prepared. Their saturation curves become steeper with increasing degree of polymerization as would be expected for a cooperative process. A perturbation of the binding equilibrium by a temperature jump results in a single (second order) relaxation process even for compounds containing as many as 10 residues. If we look at a system where 50% of the bases are paired ('melting point') we expect a single relaxation time only if an 'all or none' binding process is present. This would mean that at the melting point the oligo-nucleotides are present either as single unpaired chains or as double helices with the maximum number of bases paired. All intermediates (i.e. complexes with less than the maximum number of bases paired) must be present at undetectably small concentration. This is a result of the very strong cooperativity which gives an increase of stability by a factor $10^2$ to $10^4$ for the second base pair formed next to an already existing one. Only at higher degrees of polymerization is a distribution of the number of base pairs per complex found, resulting in a continuous spectrum of time constants (unpublished results, obtaining by Günther Maass). Here suitable averaging procedures for describing the relaxation spectrum have been introduced [6, 7].

If we consider the kinetics we find much lower recombination rates ($k \sim 10^6 M^{-1}sec^{-1}$) than for single base pair formation. This is due to the requirement of nucleation demonstrated by a negative 'apparent' activation energy.

We may draw the following conclusions about the mechanism:

(1) The rate of base pair formation in a polymeric strand could proceed as fast as $10^7$ pairs/sec. Loops in double-stranded nucleic acid molecules therefore exhibit a very high mobility allowing a very high 'reading rate'.

(2) The 'nucleation length' includes only about three base pairs. Strong cooperativity is introduced by addition of the first and a second neighbor. Any sequence containing more than three base pairs might get quite 'sticky'. The life times of the double-strands at the melting point were found in the time range of 1 to 0.1 sec. Thus only small code units would show the required high dynamic lability.

(3) These studies resemble an important principle of nature, the combination of specificity with high dynamic lability by introduction of several different levels of recognition. The stability of the single pair just allows distinction between 'right' and 'wrong'. If one would reduce the error rate by increasing the stability one would at the same time slow down the performance of the system. Instead by introduction of a new level, e.g. an enzymic recognition site, one may further increase the ability of distinction using the unique geometry of the complementary pair. Further levels, such as enzymic code checking devices may be found in the highly organized in vivo systems. The mechanisms of enzymic code checking processes are being studied by coupled T-jump-flow devices.

[1]  T. Funck, Göttingen, unpublished results.
[2]  D. Pörschke, Diplom-Thesis, Göttingen (1967).
[3]  K. Bergmann, M. Eigen and L. De Maeyer, Ber. Bunsenges. phys. Chem. **67** (1963) 819.
[4]  Y. Kyogoku, R. C. Lord and A. Rich, J. Am. Chem. Soc. **89** (1967) 496.
[5]  E. Küchler, J. Derkosch, Z. Naturforsch. **21b** (1966) 209.
[6]  D. Crothers, J. Mol. Biol. **9** (1964) 712.
[7]  G. Schwarz, Habilitation-Thesis, Göttingen, (1966) to be published.

S. Bresler: Nous sommes arrivés à peu près aux mêmes conclusions que M. Eigen. Nous avons mesuré l'interaction d'une chaîne polynucléotidique et des oligomères très courts. Nous l'avons fait en mesurant l'isotherme d'absorption à températures différentes. L'idée était d'étudier les forces en prenant des triplets, qui ne sont pas tout à fait complémentaires, il faut voir s'ils peuvent s'accorder avec des types polymères, comme prévu par Crick. Nous sommes arrivés à des conclusions très semblables à celles que M. Eigen a énoncées. Vous avez une forte coopérativité. L'interaction mesurable, dans un milieu aqueux, commence au niveau des trinucléotides. Ce qui est important, c'est que, dans les constituants ioniques du milieu, la concentration du magnésium est cruciale pour l'interaction des chaînes oligonucléotides et polynucléotides. La streptomycine aussi peut intervenir. Nous n'avons pas encore eu de données définitives. Mais je crois que l'on pourra constater que la streptomycine, qui est un ion trivalent, devra jouer un rôle important dans l'interaction.

E. D. BERGMANN: I would like to elaborate on some of the things which Dr. Pullman said and some of the things which he did not say but which he has indicated in his résumé. I hope to show by my remarks where is the place of organic chemistry in this dialogue between theoretical physicists and biologists. As so far organic chemistry has not been mentioned, I would like to say that there is good "interaction" between organic chemistry and quantum chemistry, although from our experience in cooperating with Dr. Pullman and Mrs. Pullman we know that inspiration has so far come very much from the side of quantum chemistry and not from the organic chemist. If I may, I would like to give two or three examples.

Firstly, I would like to say a few words on one point which Dr. Pullman mentioned, the ionisation potential of purines and pyrimidines. Together with the electron affinity of other compounds, they give us a good idea to what extent it is correct to assume that these other compounds can give charge transfer complexes with purines and pyrimidines.

We have done some work on these ionisation potentials by a simple but I think efficient method, using a mass spectrometer and employing an internal standard, a rare gas, the ionisation potential of which is known; thus one can very easily determine the ionisation potential of the compound to be studied. It is remarkable to what extent the sequence of the measured values of the ionisation potentials parallels the sequence Dr. and Mrs Pullman have calculated. We intend to measure the electron affinities of compounds of biological interest and also their dipole moments, not by the elegant method of which we have just heard, but by a more conventional method which will permit us to work in the necessary concentrations in non-polar solvents. In order to achieve this, we will have to modify somewhat the structure of the natural purines and pyrimidines.

May I refer to another contribution which perhaps organic chemistry can make to our knowledge of these compounds? A great deal of interest has been aroused by the fact that it is possible to make adenine from hydrocyanic acid under conditions which are similar to "prebiotic" ones. It may perhaps interest you to know that if one studies the fragmentation of purines and pyrimidines by electron impact and uses adenine, one can show that it simply decomposes into five molecules of hydrocyanic acid. Thus we have the exact reversal of the pathway which has been considered to be the primordial synthesis of adenine.

The second point which I would like to mention is the importance of free radicals in biological systems. Dr. Pullman has not mentioned this in his statement but has given some literature about it in his printed résumé. Perhaps I could say a few words about an observation which we have made and which is interesting from the point of view of the correlation between physical and chemical properties and biological effects. It has been known for a very long time that if one lyophilises bacteria and then tries to "revive" them, they lose 90 to 95% of their viability. Now one can show that if one carries out this operation under absolute exclusion of oxygen, one can revive the bacteria almost completely. Thus it is clear that for the revival process oxygen is a poison which has to be avoided. There are certain inorganic and organic compounds of great specificity which can protect bacteria against the influence of oxygen in the revival of the lyophilised material. I do not want to go into details; of some of these compounds we shall hear some more from Dr. Szent-Gyorgyi. This specificity has to do with an important observation which has been made by one of my younger colleagues. One can show by electron spin resonance the presence of radicals which have an extremely long life time, of the order of magnitude of many months, and which disappear when oxygen is admitted. We do not know what these radicals are but it is interesting that we have these very long-lived radicals play an important part in a biological process.

The third point which I would like to mention has to do with carcinogenic hydrocarbons. Every chemist and probably every theoretician has been surprised by the fact that hydrocarbons have some biological activity at all and moreover, that if one introduces polar substituents, one destroys, contrary to all our accepted notions, immediately the activity of the compounds. We have found one substitution which does not abolish, but indeed, in many cases, increases the ability to cause cancerous growth, and this is the substitution by fluorine in certain areas of the molecule. Obviously, the introduction of the fluorine changes the electron density; indeed the insistence of Dr. and Mrs Pullman on the electron density as a determining factor, has been of importance in encouraging this kind of experiment. I may add that the small atomic radius of the fluorine is undoubtedly responsible for this effect, because one cannot make the same experiment with halogens other than fluorine, which has a van der Waals radius most similar to that of the hydrogen atom.

These are the kind of things which a chemist can do under the influence of quantum-chemical calculations and which in the end may perhaps also induce the quantum chemists to reach out into new directions.

In conclusion, if you permit me to say something more general, I think the role of the organic chemist in this dialogue between the biologists and theoretical physicists lies in his ability to explain the phenomenon of specificity which has been very little mentioned here so far. All the theories of which we have heard yesterday do not give us an indication of what biological specificity means. And we know there is such a thing; we have only to think of all the work which has been done on the active sites of enzymes, on the mechanism by which anti-metabolites act, on the mode of action of certain drugs on chemoreceptors. It is obvious that these receptors cannot be large molecules. The large molecules are only the matrix for the receptors. The chemical action which explains specificity must be a reaction on small molecules or at least on small site of a macromolecule. I believe the importance of the macromolecules, fascinating as they are, has been exaggerated; they cannot explain specificity.

R. S. MULLIKEN: We are getting contributions from theoretical chemistry and practical chemistry and I think this rounds out the complete picture from theoretical physics to biology.

J. DUCHESNE: L'un des résultats remarquables de l'exposé de M. Pullman est de montrer, sur la base de la chimie théorique, que de petits changements dans la structure électronique d'une molécule sont tout à fait susceptibles d'entraîner, du point de vue biologique, des modifications de nature qualitative. Un des exemples les plus frappants qu'il ait donnés est sans doute le fait d'une substance fortement cancérogène, perdant sa cancérogénicité lorsqu'on supprime sur celle-ci un radical méthyle.

Ceci est aussi en accord avec ce que vient de souligner le M. Bergmann en ce qui concerne la spécificité biologique. Sans doute nous sommes armés, du point de vue de la physique moléculaire, pour reconnaître, à côté des déterminations de nature purement théorique, par nos méthodes expérimentales, des variations petites dans toute une série de grandeurs physiques moléculaires, comme par exemple les potentiels d'ionisation. C'est la cas aussi des moments dipolaires électriques. Mais il est une grandeur physique, comme la densité des charges électroniques, souvent calculée par les Pullman, et dont personne ne doute de l'efficacité, mais qui ne souffre guère de contrôle experimental. Pour le réaliser, il faudrait être capable de disséquer la molécule au point de pouvoir déterminer, à un niveau quelconque de celle-ci, sur un quelconque des atomes de carbone, par exemple, quelle est la structure électronique. Eh bien, pour cela, il faudrait faire usage d'une sonde adéquate. Et c'est ce que l'on est en train de faire à l'heure actuelle dans mon laboratoire (J. Mol. Structure, Elsevier, 1967). Cette sonde doit être aussi générale que possible et elle nu peut être, par conséquent, qu'un atome d'hydrogène ou un atome de deutérium. Par les méthodes de la résonance nucléaire magnétique, il est possible de déduire, au niveau d'un atome de deutérium fixé en un point quelconque d'une molécule, de l'analyse de la structure fine des spectres, le couplage nucléaire quadripolaire associé. Comme cette grandeur, surtout au niveau d'atomes aussi simples que l'atome d'hydrogène, ou son isotope, est une représentation remarquable de l'état électronique, on voit tout de suite l'usage que l'on peut faire, en principe, de cette méthode. Par exemple, dans le cas de la cancérogénicité, il est parfaitement possible de fixer, sur la molécule responsable, un atome de deutérium et de prospecter, par conséquent, sa structure électronique, selon le lieu de fixation.

Il était utile d'indiquer cette voie nouvelle, qui s'ouvre en ce moment, pour creuser davantage la question si admirablement posée par M. Pullman.

Je voudrais aussi, comme le M. Bergmann, dire quelques mots sur des problèmes considérés par M. Pullman, mais dont il n'a pas pu, naturellement, parler aujourd'hui. Il s'agit, en particulier, de ce problème fondamental—qui intéresse tous les biologistes, et, en fait, tout le monde—: les mutations.

Nous avons étudié, par la méthode de transfert de charge, qui commence à nous devenir familière (il faut cependant user de prudence quand on la manie, pour route une série de raisons) les agents de mutation du type acridine, qui présentent un intérêt tout particulier, parce que leur interaction au niveau des ADN semble aujourd'hui assez bien connue, suite aux travaux de Lerman, aux Etats-Unis. Nous avons pu en déduire, pour les potentiels d'ionisation de ces substances remarquables, des valeurs en accord avec les prédictions de Pullman. Mais nous sommes allés plus loin, et nous avons voulu rechercher si les mécanismes de transfert de charge étaient susceptibles de jouer un rôle au niveau des ADN, c'est-à-dire dans le déterminisme des interactions entre les acridines et les ADN eux-mêmes. Nous avons effectivement trouvé que, dans certaines conditions, lorsqu'on met en présence—non pas l'ADN: la chose n'a pas été faite jusqu'ici— des nucléosides et des acridines, comme la proflavine, etc...., il apparaît des phénomènes de ce type qui pourraient peut-être jouer un rôle et expliquer complémentairement aux forces de Van der Waals, la stabilité de ces édifices complexes.

Je pourrais dire aussi un mot au sujet d'une question qui vient d'être évoquée par M. Bergmann, qui faisait allusion à l'efficacité des méthodes de radiofréquence pour l'analyse de toute une série de phénomènes d'intérêt biologique. Nous avons, quant à nous, étudié la protection, dans le cadre de nos études sur les photomutations, c'est-à-dire sur les perturbations que l'on peut engendrer chez un être vivant, comme un virus, par exemple, au moyen de la lumière visible, lorsque celle-ci agit par l'intermédiaire d'un colorant. Nous avons montré * que des radicaux libres, en grand nombre, sont induits dans l'ADN de Virus et qu'ils sont en relation étroite avec les mutations. Nous nous sommes alors demandé s'il est possible de protéger le virus, en d'autres termes s'il est possible d'éviter la genèse des radicaux libres là où nous ne désirons pas qu'ils apparaissent. Et nous avons ajouté à notre système un agent protecteur comme la cystamine (à laquelle faisait allusion, tout à l'heure, M. Bergmann). Nous avons eu la surprise d'observer, par résonance électronique paramagnétique, que, pour des concentrations relativement faibles, les radicaux libres ne se forment plus dans les phages que nous avons analysés, mais apparaissent seulement dans la cystamine elle-même, qui devient en quelque sorte, et par un mécanisme que nous ne connaissons pas encore, un véritable "dévoreur de dommages".

Et, maintenant, une question que je voudrais me permettre de poser à M. Pullman. Si l'on fait une perturbation d'une nature un peu différente de celle qui consiste à ajouter ou à supprimer un radical méthyle, à savoir si l'on excite électroniquement, par exemple, une molécule comme l'anthracène, qui, en son état fondamental, n'est pas cancérogène, on voit apparaître des phénomènes de cancérogénicité (D. Rondia, A. Van de Vorst et J. Duchesne, Comptes

---

* Mutation Research **6** (1968) 15.

Rendus, Paris, 1967, 264, série D, 3053). Serait-il possible de vérifier si la corrélation mettant en jeu les régions K et L des molécules cancérigènes dans leur état fondamental peut s'étendre aux états excités?

R. S. MULLIKEN: I would like just to say something about accurate calculations in the approximation of the SCF (self-consistent-field) including *all electrons*. As soon as one has an accurate self-consistent-field wave function, one can compute various properties which give one a more intimate understanding of the structure and of the possibilities of the molecule. Accurate SCF all-electron calculations on diatomic molecules are already becoming rather familiar. But recently several people have been making calculations on *larger* molecules. Among others, Enrico Clementi at the IBM Research Laboratory has calculated molecules like pyrrole, pyridine, and pyrimidine and he says he can calculate in the same way the structure of any of the nucleotides, given a little more machine time. He has some very ingenious methods. This work gives great promise for a more complete understanding of these molecules.

He has also made a calculation about bringing together one molecule of ammonia with one of hydrogen chloride, with the idea that this would be a prototype for hydrogen bonding.

Given the wave function, one can calculate the charge distribution on the atoms as it changes with the distance, and Clementi has obtained some very interesting results.

And other calculations can be made which should throw light on hydrogen bonding. The calculation can be made for various distances and orientations or relative distances of the parts.

P. O. LÖWDIN: In connection with Dr. Pullman's excellent lecture, I would like to comment on the method he is using. To many biologists, Dr. Pullman and his wife seem like magicians who look at the structure of a molecule in biochemistry and come out with a large number of interesting data, and the question is then how reliable these numbers are in reality. If a biologist asks a theoretical physicist about his opinion, he will probably say that the methods utilized are rather primitive and that it would be safer to wait for better calculations. In spite of the development of the electronic computers, however, better methods for treating large molecules quantum-mechanically are only slowly maturing, so, in such a case, the biologists would have to wait for many years to come.

The truth seems to be that the Pullmans have found something like a "golden rule" in biochemistry in the form of a simple connection between the topological structure of the molecules under consideration and some important quantum-mechanical concepts, which is revealed by the simple Hückel-method. There is little doubt that the calculations later have to be refined, but, for the biologist, they represent a useful first rough approximation.

Some of the most fundamental concepts are related to the symmetry of the molecules. The conjugated systems under consideration are all planar molecules, where the single bonds correspond to $\sigma$-electrons having wave functions symmetric with respect to reflections in the plane, whereas the classical double bonds correspond to mobile $\pi$-electrons having wave functions which are antisymmetric with respect to reflections in the plane, i.e. they change sign under reflection. The signs of the wave functions, i.e. their "phases", are hence here of basic importance.

In treating the $\pi$-electrons, the Pullmans have used the fundamental ideas introduced by Mulliken, Hund and Hückel. The molecular-orbital method so successfully developed by Mulliken belongs to the independent-particle model, in which the total wave function may be approximated by a single Slater-determinant formed by the molecular orbitals involved. In the early 1930's, it was proven by Fock and Dirac that such a determinant is uniquely described by a simple density matrix:

$$\varrho(x_1, x_2) = \sum_{k=1}^{N} \Psi_k(x_1)\, \Psi_k^*(x_2),$$

where one should sum over the occupied molecular orbitals $\Psi_k(x)$, and one can then show that all measurable properties of the system in the specific state under consideration are determined by $\varrho$. A theorem by Koopmans says that the ionization potentials are given by the orbital energies $\varepsilon_k$, and that, with some modifications, they may also be used to calculate excitation energies and these results have been utilized by the Pullmans.

In the MO-LCAO-scheme, one may build the molecular orbitals $\Psi_k$ by linear combinations of the atomic orbitals $\phi_\mu$:

$$\Psi_k = \sum_\mu \phi_\mu\, C_{\mu k},$$

and substitution into the formula for $\varrho$ gives:

$$\varrho(x_1,x_2) = \sum_{\mu\nu} \phi_\mu(x_i) R_{\mu\nu}\, \phi_\nu^*(x_2),$$

where

$$R_{\mu\nu} = \sum_k C_{\mu k} C_{\nu k}^*$$

is the "charge and bond order matrix" introduced by Coulson and Longuet-Higgins more than twenty years ago. This matrix of "molecular indices" has been extensively used by the Pullmans in their broad study of conjugated systems of importance in biochemistry.

Instead of using the full Hartree-Fock scheme representing the best realization of the independent-particle-model the Pullmans have usually used the so called naive Hückel-method in evaluating the charge orders $q_\mu = R_{\mu\mu}$, the bond orders $p_{\mu\nu} = 1/2(R_{\mu\nu} + R_{\nu\mu})$, and the orbital energies. This approach is probably rather well justified by the fact that in many cases, the charge and bond order matrix $R_{\mu\nu}$ is simply determined by symmetry. Even when this is not the case, the mathematical theorem of the inertia of quadratic forms tells us that the general features of the matrix $R_{\mu\nu}$ found in a simplified approach may remain intact in a more sophisticated theory. The relative results of a topological nature found by the Hückel-method are hence probably better than expected, and, if some theoretical data are fitted to experimental parameters, one obtains a semi-empirical theory which can predict certain results with a much higher degree of accuracy than originally anticipated.

Some recent large-scale calculations by Clementi indicate that certain of the basic assumptions in the independent-particle-model and the Hückel scheme are not fully valid in a more strict quantum-mechanical treatment, but this does not diminish the value of the simplified theories in obtaining, for biologists, a rough approximation of the electronic structure of the conjugated systems of importance in biology.

J. POLONSKY: Je n'ai pas la prétention d'intervenir dans cette discussion, n'étant pas spécialiste des membranes. Je voudrais seulement poser une question concernant le modèle simulant le rôle des protéines spécifiques dans une membrane. Y a-t-il une certaine analogie entre le modèle d'impuretés dans un cristal et celui des protéines spécifiques dans une membrane constituée d'assemblées macromoléculaires ordonnées?

Pour modifier les caractéristiques électriques d'un cristal, il suffit d'introduire une quantité infime ($10^{-6}$) de donneurs ou d'accepteurs d'électrons qui créent des états permis à l'intérieur d'une bande interdite du cristal. Ce rôle quasi magique des impuretés n'est possible que grâce à la présence de l'ordre périodique et uniforme du cristal. Dans un modèle analogue, des protéines spécifiques en petit nombre, placées à l'intérieur d'une matrice membranaire, joueraient le rôle d'une information protégée par un milieu à basse température statistique. Un autre modèle analogue est celui des plateaux basiques dans la double hélice de l'ADN.

H. C. LONGUET-HIGGINS: I was very much interested in Dr. Pullman's paper, and personally agree with Dr. Löwdin's comment on it. I would like to ask a technical question about the matter of carcinogenesis. One of the things which baffles the non-biologist is the extraordinary variety of different kinds of substance which can be powerfully carcinogenic. But can one claim to have a satisfactory picture of the process until one knows the shape of the locks into which these different keys actually fit?

I introduce this familiar analogy because shape is obviously a matter of some importance and I know that you, Dr. Pullman, have on previous occasions discussed this question. To be more precise: when one draws a picture of the most carcinogenic of your planar hydrocarbons one does see a resemblance to a pair of DNA bases—which is also a planar structure. I wonder whether you would care to comment on the experimental evidence which bears on this point?

B. PULLMAN: Just a few words to answer the questions that have been asked and in order to make also a few complementary comments.

I agree, of course, entirely and I think that everybody has to agree with the general philosophy developed by Dr. Löwdin about the significance of the different quantum-mechanical methods and in particular of the Hückel-type molecular orbital method in quantum chemistry and in particular in quantum biochemistry. What I would like to stress is that in problems of the nature of those encountered in quantum biochemistry, the value and the success of a given method depend very largely on the manner in which the method is being utilized.

One may use the Hückel method in this respect in a catastrophic way, one may also use it in a way in which it gives extremely satisfactory and significant results whose essence will be preserved in all more refined treatments. One must adapt one's exigencies to the method he is using and in particular one must be perfectly aware of the way in which one can interpret the results of the calculations obtained within a given procedure.

You might have noticed that when using the Hückel approximation of the molecular orbital method we have nearly constantly tried to compare the results on a *relative scale* rather than to give indications of the absolute values of the different quantities involved. The fact that we made no significant errors in any of the numerous predictions that we have made in many fields of biochemistry and biophysics is to a large extent due to this caution and appropriate way of utilizing the method.

I do not think that anybody can question the logic of having applied in the first place the classical Hückel procedure to problems of biochemistry. In the late fifties this was in fact the only method practically utilized even in quantum chemistry in so far as problems connected with somewhat large molecules were concerned. There was so much to be done in this original field of research that even the simplest methods were necessarily leading to an abundent harvest.

It must, however, be stressed that since a few years, all the fundamental results obtained by the Hückel method have been recalculated and checked by more refined methods based essentially on the self-consistent field approach. This has been done to a very large extent in our own laboratory. People who appear to believe still today that something like a glorification of the Hückel procedure was our utmost goal in quantum biochemistry are simply out of bounds. This procedure was never considered by ourselves and probably by anybody working in this field as anything else than a first step on a long route. A very convenient and useful step however, I would say, essential, and what is the most important, extremely fruitful step.

I must nevertheless add also a few words of caution about the sense of "theoretical progress". Thus there is no possible doubt that procedures based on the self-consistent field method are basically technically superior to those based on the Hückel technique. But it must also be remembered that in principle more is being asked from this refined techniques. In particular they are expected to be much more precise and useful as concerns the absolute values of the quantities calculated. Now, in this respect, the situation is far from being as simple as it may seem to be. When dealing with large molecules all these "refined" methods are, at one stage or another, semi-empirical in the sense that they always introduce, at one stage or another, some *parametrization* into the procedure. Because of the nature of the techniques this parametrization plays a capital role for the exactitude or even significance of the results obtained. I am sorry to say that, because of inadequate, superficially done parametrization a number of so-called "refined calculations" carried out on biomolecules give far worse results than do the simple calculations. Of course, it is not the method, but its careless utilization which is to be blamed. But I advise you very strongly to keep in mind that the mere label of a refined technique is in no way a guarantee by itself of a higher standard of work or of more significant results. To a large extent this is true even in connection with all-valence electrons or even all-electron calculations to which Dr. Mulliken referred in his remarks.

This does not mean, of course, that more and more refined methods must not be used continuously for the study of the electronic structure of biomolecules. Of course, they have to be used and in fact, this is being done constantly in our laboratory. But care must be taken, and is very much taken by us, that the methods should be used in such a way as to really correspond to a constant progress.

Now, I would like to answer the question raised by Dr. Duchesne. We did not study particularly the effect of light on the carcinogenic activity of molecules partly because the methods used originally for establishing the correlations between structure and activity were better suited for studies of the ground state of molecules than for studies on their excited states. Today there are no overwhelming difficulties in extending this type of studies to excited states and, recently, we have made such a study in connection with the problem of chemical mutagenesis. We have studied both the factors responsible for the mutagenic activity of a number of chemicals in their ground state and the influence of light in enhancing the rate of mutagenesis. One may, for instance, consider as a possible cause for spontaneous mutations the tautomeric shifts of the purine and pyrimidine bases of the nucleic acids and investigate the effect of electronic excitation on the rate of this tautomerisation. Such a study enables to pinpoint the active site for this mechanism of mutagenesis.

I would finally like to answer the important and exciting questions raised by Dr. Longuet-Higgins.

The first one concerns the big variety of different types of chemicals, all of them capable of producing cancer. It is, of course, evident that we cannot reduce all of them to the same kind of explanation, say to a K and L regions theory. Some of them are completely different from the aromatic hydrocarbons and don't have any region of this type.

I shall give an indirect answer to this question by going back to the discussion of mutagenesis which raises the same type of problem but the answer to which is easier to visualize. The reason for it is very simple. The great difficulty in discussing the mechanism of carcinogenesis—I really mean now the biological mechanism and not the correlation between structure and activity which is to a large extent a chemical problem—is the absence of a precise physico-chemical, one could say *molecular*, definition of carcinogenesis. This results in the difficulty of not being able yet to discuss the problem at the molecular level. The situation is quite different as concerns mutagenesis because in this case we have a precise physico-chemical definition of this phenomenon.

Namely, we know that the genetic code is embodied in the sequence of base-pairs along the principal axis of the nucleic acid so that mutagenesis is, *by definition*, a perturbation of this sequence. This is a molecular, a physico-chemical definition. Now, although, we have a great variety of mutagenic agents, as big as that of the carcinogenic ones, we have no difficulty in understanding the unicity of their action. All these mutagenic agents are capable of interacting chemically with DNA, they all act in different ways but they are all able to bring about the same essential result namely to perturb, in one way or another, the natural sequence of the base-pairs. I have a slide here which shows the site and the type of chemical action of the principal mutagens. (See fig. 1). As you can see they belong to very different types of chemicals so we understand easily that the modification of the sequence of the bases is produced by them in a variety of different ways and at different sites. Some of them increase the tendancy of the bases to shift to rare tautomeric forms and increase thus the probability of mispairings, others introduce a chemical modification in a base which again increases the chances of mispairings, still others eliminate a pair of bases from the helix etc. They have different reactions but they all lead to the same result namely the alteration of the original order and thus a mutation.

Unfortunately we cannot discuss yet carcinogenesis at the same level because we do not know yet the nature of the chemical, the molecular event responsible for this phenomenon. The most popular definition of carcinogenesis is that it represents a perturbation in the cellular division mechanism. This is not enough to be able to speak about the phenomenon in a molecular language. We do not know even yet whether the essential primordial site of action of the carcinogens resides in the nucleic acids or in the proteins or in both. Let's imagine for a

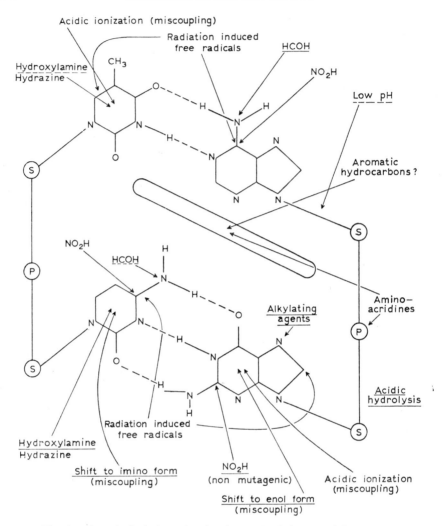

Fig. 1.  The principal sites of action in DNA of the essential mutagens.

while that this site is in the nucleic acids and that it consists in some kind of a very important mutation or a series of mutations. Then, of course, following the previous considerations, we have no difficulty in understanding that very different agents may bring about the same biological result. The same point of view can be preserved if the phenomenon implies primarily a protein rather the nucleic acids.

As to the second question raised by Dr. Longuet-Higgins, concerning the role of shape in chemical carcinogenesis, I must affirm that I do not think the shape of molecules to be of an essential role in carcinogenesis. This is in fact obvious from the very multiplicity of chemical agents of carcinogenesis previously referred to, which have very different shapes. But even in the limited field of the aromatic hydrocarbons I do not see any reason to believe the shape to be of a fundamental importance. In the electronic theory of carcinogenic activity which I described previously, the shape does not need to be introduced at all.

Besides, shape is something which cannot be defined properly. When some people affirm that the carcinogenic hydrocarbons have a shape very similar to that of a base-pair of the nucleic acids or to some fundamental oestrogens I can very easily indicate to a formula of a non carcinogen whose shape can be considered to be much closer to that of a base-pair or of the oestrogen than is that of the carcinogen. All statements about such analogies of shape which were supposed to involve the carcinogenic hydrocarbons are in fact completely arbitrary. The reason is simple. There are relatively few carcinogenic hydrocarbons, as opposed to many more non-carcinogenic ones. Those being built up of the same or close number of fused benzene rings have frequently very similar shapes. The distinction between carcinogens and non-carcinogens can only be obtained through a much more subtler analysis then that based on shape, or volume, or area.

Of course, one may consider that there is something deeper in Longuet-Higgins question because the shape of the molecule, its topology, has an essential influence on its electronic indices. In this way shape is important indirectly. But, I do not think that Dr. Longuet-Higgins had in mind this aspect of the problem.

E. D. BERGMANN: May I say a few words in connection with the question of Dr. Longuet-Higgins and the answer of Dr. Pullman. If we leave out of consideration some small molecules which can cause cancer and for which one knows how they react with DNA, I think there are mainly three types of carcinogenic compounds which have interested the chemists:

1. Aromatic polycyclic amines: 2-aminofluorene, or $\beta$-naphthylamine.
2. Azo-dyes, which all contain dimethylamino-groups.
3. Polycyclic aromatic hydrocarbons.

As to the first type, the aromatic amines, it is to-day quite obvious that their action is based on the oxydation of an hydrogen atom bound to nitrogen, to hydroxyl. Thus we get the same mechanism which we know for the various simple hydroxylamine derivatives and which has been studied by Berenblum at the Weizmann Institute.

As to the second type of carcinogenic agents, the azo-dyes, we know that they are demethylated in vivo. This is the first observable step; it is not unreasonable to assume (though it has not been proven experimentally) that these compounds methylate DNA. We have heard from Dr. Pullman that this process may change completely the structure of the double-stranded DNA, because it affects the intermolecular bonding between purines and pyrimidines. In addition, it has been shown in America that the methylation of DNA is species-specific and therefore can discriminate between various DNA's. Thus the mode of methylation of DNA after ingestion of the azo-dyes may give an insight into the mechanism of their action.

To the organic chemist the most surprising fact is the carcinogeneity of polycyclic aromatic hydrocarbons because one does not see how those hydrocarbons, devoid as they are of any "haptophoric" groups, can act. As Dr. Pullman very rightly says, one has to differentiate between the mode of carcinogenic action which must be some chemical reaction, and the question how these hydrocarbons arrive at the point where they are acting. There is one observation which has been made some years ago—however, it has not been followed up—namely that 3,4-benzopyrene reacts with guanine and gives a new chemical compound which can be isolated and has been more or less identified as to its structure. Thus it is possible to assume that, in fact, the *chemical* reaction takes place on some spot in the DNA and forms a defined chemical compound. However, this does not solve the problem how the hydrocarbons reach this spot. Here, I do not agree with Dr. Pullman. I believe that the shape, the geometrical shape and the size of the molecule play a part in making a certain hydrocarbon carcinogenic and a very similar one not, if this means that the first reaches the site of action on the DNA, the other not. It is not possible to explain by the assumption of hydrogen bonds, charge transfer or dispersion forces the phenomenon of specificity. If the shape of the hydrocarbon corresponds to that of an "active site", it is possible for the hydrocarbon to attach itself to the active site and then undergo the chemical reaction which represents the carcinogenic effect.

M. KOTANI: I give a very short comment to Dr. Pullman's lecture. Dr. Pullman emphasized that parametrisation is very useful and important in carrying out the practical calculations at the empirical level. I completely agree to this statement and I would like to give another example of parametrisation in quantal chemistry or quantal biophysics.

This is the Ligand field theory which can be possibly applied to the study of electronic structure of transition metals, for instance iron atoms in proteins. We know that there are many proteins which contain iron ions—haemoglobin, myoglobin, cytochromes, peroxydes, etc.

Fe ion is just in the center of the active spot of these enzymes and proteins. And their electronic structure can be studied by the application of the Ligand field theory in which we have parameters. The d-electrons have the same energy in the free ions. But in the case of ions in Ligand field orbital degeneracy is removed, and according to the symmetry of the environment of this metallic ion, we have 1, 2, 3 or 4 parameters, which represent orbital energy differences. With the use of these parameters in addition to the parameters which take account of inter electronic interaction, we can reproduce the electronic states and eigen functions very well through rather simple calculations.

In this connection I like to call your attention to the fact that Fe ions in these proteins have very singular electronic structures. I like to give an example to underline my statement. Hemoglobin can take oxygen and become oxygenated. The spin of the hemoglobin without oxygen molecule is $S = 2$, that means quintet, and we know that the ground state of the oxygen molecule has $S = 1$. But once hemoglobin captures oxygen, this becomes dimagnetic $S = 0$.

This is a very strange case and this situation arises because both hemoglobin and oxygen have several rather low excited states of the order of one electron volt or even less.

Dr. MULLIKEN: How does it help?

M. KOTANI: Deoxygenated hemoglobin whose ground state is a quintet has a very low excited singlet state, and, the triplet excited state is also not very high. On the other hand an oxygen molecule has low excited states with $S = 0$.

Thus the Fe ions in proteins have a lot of low excited states, some of which may be populated appreciably even at the room temperature. This means that iron in hemoglobin or in protein is a very flexible, adaptable entity and not like irons in inorganic compounds. This unique nature of iron atoms in proteins must be closely related to the biological activities of these molecules.

L. ROSENFELD: Je voudrais faire une remarque très élémentaire, mais qui n'est peut-être pas entièrement superflue, si elle peut enlever aux calculs du M. Pullman le caractère "magique" qu'ils semblent avoir aux yeux des biologistes. Le M. Pullman se limite aux états stationnaires des systèmes moléculaires qu'il étudie. La connaissance de ces états et de leurs propriétés est certes très importante, mais ce sont évidemment les transformations des systèmes moléculaires qui intéressent le plus directement les biologistes. L'étude de ces transformations pose des problèmes théoriques difficiles, mais non sans espoir. Nous avons une théorie déjà très avancée des réactions moléculaires, et je voudrais exprimer le souhait que le M. Pullman se décide à aborder ce problème plus large et à y concentrer ses puissants moyens de calcul.

# PHYSICOCHIMIE DE LA VIE:
# DONNEES ET PROBLEMES

*2ème séance*

PRÉSIDENT M. KOTANI

---

### H. H. USSING

Active Transport

### A. KATCHALSKY

Non-Equilibrium Thermodynamics of Bio-Membrane Processes

### A. SZENT-GYORGYI

Charge, Charge-Transfer and Cellular Activity

Discussions

*Theoretical Physics and Biology*, © 1969, *North-Holland Publ. Co., Amsterdam*

# ACTIVE TRANSPORT

HANS H. USSING

*Institute of Biological Chemistry, University of Copenhagen, Copenhagen, Denmark*

All living organisms as well as the cells of which they are composed are faced with two conflicting problems: In order to retain their integrity they must confine their components within a definite space, but at the same time they require an exchange of matter with their surroundings. The maintenance of a cellular composition which deviates from that of the surroundings thus only to a limited extent can be explained as the result of impermeability of the cell surface. In many cases adsorption phenomena may account for the accumulation of certain substances, and so far as charged particles are concerned, the electric potential difference between cells and their surroundings may bring about the uneven distribution of the substances in question. There remains, however, a number of cases where chemical processes in the cells provide all or most of the free energy required for the transport of specific substances. Such transports are called active.

A more practical definition is that the term active transport is used to characterize the transports which cannot be explained on the bases of observable physical forces. Clearly this means that as our knowledge advances we will have to remove some cases of apparent active transport from the list.

Active transport processes seem to occur in all living organisms from bacteria to man, and they are not restricted to the plasma membrane of cells but appear also to be integral parts of the function of mitochondria, vacuoles and other cellular organells. The substances which can be actively transported in one cell type or another is legio and comprises most of the low molecular buildingstones of cells and many substances which are listed as excretion products.

My alotted time does not permit me to discuss even shortly the different types of active transport which have been described. Suffice it to say that the mechanisms described are usually highly specific so that they transport only a single species or they handle a limited number of chemically related species which often inhibit each others transport in a competitive way. In this way they seem to resemble enzymes. They also often resemble enzymes in that they show saturation kinetics: As the concentration of the transported species increases, the transport rate approaches assymptotically an upper limiting value. They differ from enzymes in solution in that the reaction is directed in space. The difference between active transport systems and enzymes becomes less apparent when we consider enzymes arranged spatially like it is apparently the case in mitochondria. As a matter

179

of fact it is easy to visualize cases where the difference vanishes completely.

A large fraction of the work on active transport has been concerned with the behaviour of the alkali metal ions potassium and sodium in living systems. One reason is the well known fact that practically all living cells maintain a high potassium and a low sodium concentration. This fact has always been a challenge to biochemists and biophysicists due to the chemical similarity between the two ionic species. Also it has played a role that the assumption of adsorption as the reason for accumulation of potassium and exclusion of sodium seems unattractive due to the limited tendency for these ions to form complexes.

The role of these two ions in the electric activity of muscle and nerve is well established and has been beautifully expressed in the Hodgkin-Huxley theory. The action potential itself is created by a sequence of permeability changes to sodium and potassium, but since each action potential leads to loss of potassium and gain of sodium in the cell, it is an integral part of the theory that the cellular potassium sodium ratio should be maintained by active transport.

Even more direct evidence for the existence of active transport is obtained from the study of certain epithelial organs like amphibian skin and urinary bladder, intestine, kidney tubule etcetera, where large amounts of certain substances undergo undirectional transport. Good examples are the reabsorption of glucose from the blood ultrafiltrate in the kidney tubules which reduces the concentration of glucose in the normal urine to almost zero, and the active sodium transport through the skin of frogs which can take place even when the outside medium is more dilute than tap water with respect to sodium. The capacity to transport sodium from the outside to the inside medium is retained by the frog skin even when it is isolated, and this property has made the isolated surviving frog skin one of the favorite objects for the study of the transport of sodium. Since a large part of my work in recent years has been devoted to the study of this object I should like to take the transport of sodium in the frog skin as an example and discuss in some length the evidence for the active nature of this process.

In 1848 Du Bois-Reymond [2] observed that the frog skin maintains an electric potential difference between the inside and outside bathing solutions, the inside solution being positive relative to the outside solution. The dependence of the frog skin potential upon the composition of the bathing solutions was first studied by Galeotti [3]. Among his most important findings was that the potential is only developed in the presence of either sodium or lithium. In order to explain his results he proposed that the diffusion coefficient for these two ions were greater in the direction outside-in than in the direction inside-out. Although, at the time his explanation was rejected because it seemed to violate the second law of thermodynamics, there was an element of truth in his description as we shall see later. Numerous other explanations for the frog skin potential were advanced in the following years. Thus Meyer and Bernfeld [13] proposed that the potential arose due to the fact that hydrogen ions diffuse faster from epithelium cells to the inside bathing solution than do bicarbonate ions, so that the formation of $CO_2$ was the real source of the potential.

In the meantime it had been discovered [8] that the isolated frog skin when in contact with Ringer solution on both sides, will transport chloride ions from the outside to the inside medium. Although he did not analyze for sodium, he assumed that sodium chloride was being transported actively through the skin. Shortly afterwards it was shown by Krogh [11] that both chloride and sodium were taken up through the skin by live frogs, even from solutions as dilute with respect to NaCl as $10^{-5}$ molar. Since the blood of the animals is about 100 millimolar with respect to NaCl, the active nature of the uptake is obvious.

Our initial experiments on isolated skins, using isotopes to measure transport rates [17] indicated that even the isolated skin can take up NaCl from rather dilute solutions, Furthermore, it became apparent that the inward transport of chloride might be the consequence of the electric potential difference across the skins since the potential was usually sufficient to raise the electrochemical potential of the chloride ion in the outside solution over that in the inside solution. That is: $\bar{\mu}_{Cl(o)} > \bar{\mu}_{Cl(i)}$. This means that the chloride transport *might* be passive. For sodium, on the other hand we found $\bar{\mu}_{Na(o)} \ll \bar{\mu}_{Na(i)}$, indicating that the inward transport of sodium was taking place against the electrochemical potential. This illustrates the criterion proposed by Rosenberg [14] to characterize active transport, viz. that active transport is a transport against the electrochemical potential.

As far as the frog skin is concerned, one might argue that if the transport

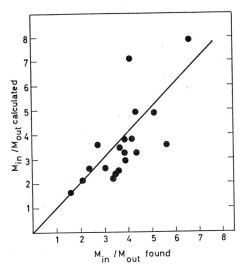

Fig. 1.   Calculated versus found flux ratios ($M_{in}/M_{out}$) for chloride ions in the isolated surviving frog skin.

TABLE I

The influence of DNP on in- and out-flux of Na through the frog skin. Solution bathing the outside 1/10 Ringer. Solution bathing inside Ringer

| Sample | $M_{in}$ | $M_{out}$ | $E$ | $M_{in}/M_{out}$ | $M_{in}/M_{out}$ |
|--------|----------|-----------|-----|------------------|------------------|
|        | $\mu$mole | cm$^{-2}$ h$^{-1}$ | m V | found | calculated |
| Control | 0.34 | 0.093 | 62 | 3.66 | 0.011 |
| DNP | 0.25 | 1.57 | −11 | 0.16 | 0.15 |
| Control | 0.445 | 0.145 | 82 | 3.07 | 0.025 |
| DNP | 0.049 | 0.56 | − 6 | 0.087 | 0.126 |
| Control | 0.228 | 0.008 | 72 | 28.5 | 0.017 |
| DNP | 0.032 | 0.216 | − 7 | 0.148 | 0.132 |

Note: active transport of sodium in control periods and passive transport under influence of Dinitrophenol (DNP).

took place as small droplets of solution or as ion pairs of NaCl, it might be meaningless to consider the transport of one ion as active and the other as passive. Therefore it was necessary to work out methods which would enable us to demonstrate that sodium and chloride move independently in the frog skin. It can be shown that for an ionic species which passes a membrane without interacting with other moving particles, the flux ratio is independent of the membrane structure and is determined only by the difference between its electrochemical potentials in the bathing solutions:

$$RTln(M_{in}/M_{out}) = \bar{\mu}_o - \bar{\mu}_i, \text{ or}$$

$$M_{in}/M_{out} = (a_o/a_i) \cdot \exp (zFE/RT),$$

[18, 16] where $M_{in}$ is the influx, $M_{out}$ the outflux, both measured with isotopes, $a_o$, the activity of the ion in question in the outside solution, $a_i$, its activity in the inside solution and $E$ the electric potential difference between the outside and inside solutions. $z$ is the charge of the ion and $F$, $R$ and $T$ have their usual meanings.

This equation obviously describes an idealized situation since there will always be some interaction between the ionic fluxes. Nevertheless the equation describes the behaviour of the chloride ion in the frog skin with reasonable accuracy [9]. Thus the "flux ratio analysis" supports the assumption that chloride is not subject to active transport in the frog skin but is transported inward due to the electric potential difference.

If, on the other hand, a similar analysis is carried through in the case of sodium it turns out that the flux ratio is often more than 100 times larger than predicted for passive behaviour, indicating that only an insignificant fraction of the sodium ions are passively transported. It is also significant that if the skin is poisoned with dinitrophenol which uncouples the metabolism from the formation of ATP, the flux ratio for sodium becomes very closely that for a passively transported ion.

The flux ratio analysis has the advantage that it can be used for systems which are not in a steady state, but strictly speaking, a deviation from the equation does not prove active transport, but only shows that there is strong interaction between the ion in question and other moving particles.

One type of interaction is of special importance, namely that between solute and solvent. If, for instance, the frog skin were a sive or pore membrane through which the solution was forced by some mechanism, both sodium and chloride ions would be dragged along.

The flow of solvent may be considered to create a "drag potential" which for each solute species is a function of the shape and size of the solvent molecules, the rate of flow of the solvent and the shape of the pores through which the solution flows. The drag effect on, say, the sodium ion can be estimated from the drag upon uncharged molecules of about the same size as the sodium ion. An approximate treatment [10] shows that the flux equation in the presence of solvent flow takes the shape:

$$ln\ (M_{in}/M_{out}) = \ln\ (a_o/a_i) + zFE/RT + k\Delta w/D,$$

where $k$ is a constant characteristic for the membrane but independent of the solute as long as it can permeate the membrane, $\Delta_w$ is the volume rate of flow of solvent, and $D$ is the diffusion coefficient for the solute in question. The last term is the drag potential. Since for a given membrane and a given rate of flow it depends only on the diffusion coefficient, one can estimate the drag upon an ion by measuring the flux ratio for an uncharged molecule of the same size.

Flux ratios were determined at different rates of osmotic flow through toad skin of the test substances thiourea and acetamide. Influx and outflux of thiourea can be performed by using the pair 14-C thiourea and 35-S thiourea. Whereas for acetamide we used two batches, one labelled in the 1-position and the other in the 2-position with 14-C. Toad skin rather than frog skin was used because it is possible to obtain larger osmotic flows through this preparation, notably when the skin is treated with antidiuretic hormone.

It was found that very substantial drag effects could be obtained (flux ratios as high as 3). However, when the osmotic pressure was the same on both sides of the skin, the flux ratios invariably were 1. Since the two test substances are roughly the same size as the sodium ion it is safe to say that when the osmotic pressure is the same on both sides of the skin, the flow of solvent does not introduce any force upon the sodium ion. The very minute net water transport which goes on when the frog skin is exposed to Ringer solution on both sides (about one microliter per $cm^2$ per hour) thus may be a consequence of the active salt transport but is not the cause of the transport.

From the foregoing it appears that the transport of sodium in the frog skin is active and that of chloride passive. Thus it was quite reasonable to conclude [17] that the active transport of sodium was the cause of the potential. A more direct proof that the active transport of sodium is the sole reason for the electric potential difference across the frog skin can be obtained by aid of the "short-circuiting" technique of Ussing and Zerahn [19]. The principle is that the potential drop across the skin is reduced to zero by short-circuiting it through an external circuit of zero effective resistance. In practice this is done by connecting the outside and inside bathing solutions through an adjustable external EMF and a microamperemeter, using reversible electrodes. The potential drop across the skin is measured with a millivoltmeter between reversible electrodes placed in the immediate vicinity of the skin. In this setup the electromotive force of the skin has only to overcome its own internal resistance whereas the external resistance, including that of the bathing solutions is overcome by the applied EMF. If we use identical solutions on the two sides of the skin, we have now a situation where there is no concentration difference, no electric potential difference, no pressure difference and no temperature difference between the outside and inside bathing solutions. Therefore any net transport across the skin must be due to

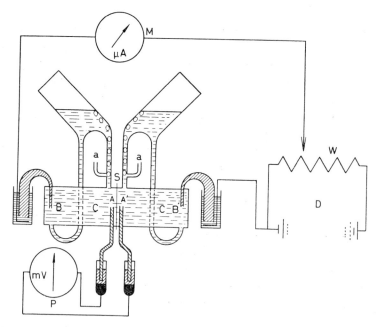

Fig. 2. Diagram of apparatus used for determining Na-flux and short-circuit current. C: Celluloid chamber, containing, on each side of the skin, 40 ml Ringer; S: Skin; a: Inlets for air; A and A': Agar-Ringer bridges, connecting outside and inside solutions, respectively, with calomel electrodes; B and B': Agar-Ringer bridges used for applying outside E.M.F.; D: Battery; W: Potential divider; M: Microammeter; P: Tube potentiometer.

TABLE II

Short-circuit current and sodium flux values for a number of short-circuited frog skins
(Rana temporaria)

(Ringer's solution on both sides)

| | $\mu$amps/cm$^2$ | | | |
| --- | --- | --- | --- | --- |
| | Na$_{in}$ | Na$_{out}$ | net Na transport | Current |
| I | 20.1 | 2.4 | 17.7 | 17.8 |
| II | 11.1 | 1.5 | 9.6 | 9.9 |
| III | 40.1 | 0.89 | 39.2 | 38.6 |
| IV | 62.5 | 2.2 | 60.3 | 56.0 |
| V | 47.9 | 2.5 | 45.4 | 44.3 |

chemical processes in the skin itself. During short-circuiting the skin will of course produce electric current, the strength of which can be read on the micro-amperemeter. The fluxes of the different ions present is too small to be measured accurately by chemical analysis, but it can easily be obtained with good accuracy as the difference between the inward flux and the outward flux as measured with suitable isotopes. Thus the fluxes of chloride can be measured with the isotopes Cl-38 and Cl-36 and the fluxes of sodium with the isotopes Na-24 and Na-22. It turned out that the chloride flux was the same in the two directions during short-circuiting, whereas the sodium influx was much larger than the sodium outflux. Actually in all experiments the sodium net flux was exactly equal to the short-circuit current. Even if, say, 30 per cent of the sodium in the two bathing solutions is replaced by potassium, the latter ion gives no contribution to the short-circuit current which still consists exclusively of sodium ions. The only ion which to some extent can replace sodium is lithium, which, however, tends to accumulate in the epithelium and ultimately inhibits the transport [21]. These results exemplify something which is very typical for active transports, namely the high degree of specificity.

The fact that there is identity between the short-circuit current of the frog skin and the active transport of sodium, makes the preparation very well suited to study the relationship between active transport and metabolism rate. Studies by Zerahn [22] and by Leaf and Renshaw [12] have shown that there exists a linear relationship between the amount of sodium transported and the amount of oxygen consumed over and above the consumption in the absence of sodium transport. Each sodium ion transported brings about an additional consumption of 1/18 molecule of oxygen. It is interesting that this relationship holds whether the sodium ions are transported between identical solutions or whether it is transported against a concentration gradient or even when it is transported "downhill" the electrochemical gradient. Thus it seems that sodium has to pass the "pump" and that there exists a sort of stoichiometrical relationship between the amount of sodium transported and the amount of oxygen consumed. It is worthy of note that about 4.5 sodium ion is transported for each electron passing through the oxydation chain. This fact puts severe restriction on the choice of possible transport mechanisms. Thus a number of hypotheses based upon the transport of one monovalent ion per electron are extremely unlikely.

The involvement of energy rich phosphate esters like ATP in the transport seems more likely. It has already been mentioned that dinitrophenol stops the active sodium transport. Furthermore, the frog skin, like most other sodium transporting cells contains a sodium activated ATP-ase [15]. This enzyme requires for activity the presence of magnesium and sodium and is further activated by potassium. In certain cell types like red cell ghosts [4] and cephalopod giant axon [6] there is direct evidence that the active sodium extrusion and potassium accumulation in the cells can take place at the expense of ATP and this enzyme is inhibitor of active sodium transport in red cells, nerve, frog skin, and many other tissues.

Active transport of sodium seems in many tissues to be coupled more or less tightly with active transport of potassium in the opposite direction (red cells, nerve, muscle) whereas in other tissues like frog skin and toad urinary bladder the use of potassium as "return freight" is less apparant. The sodium pump in many cases serves as electric charging device but one should not get the impression that all bioelectric potentials are due to active sodium transport. Thus the gastric potential is due to active transport of chloride ions [7] and also in plant cells there is evidence for the involvement of active chloride in the development of electric potentials [1]. Recently it has been demonstrated that the intestine of the silk worm (Cecropia) maintains for hours a potential difference between the lumen and the blood side of some 80 mV [5]. Short-circuit experiments have demonstrated that the electric asymmetry is maintained by a selective active transport of potassium ions from the blood to the lumen. Interestingly enough' this intestine contains an ATP-ase which does not require sodium for activation.

Finally it ought to be mentioned that sodium ions have been found to be necessary for the active transport of sugars and amino acids in the intestine. This suggests that there is some step in common for the transport of a large number of different substances.

As far as I know, no active transport mechanism has been fully explained, although the number of hypotheses is overwhelming. With our increasing knowledge the choice between the models is becoming subject to severe restrictions. This high degree of specificity of many transports has already been mentioned. To this comes the equally high energetic efficiency. Calculations show that if the oxydative energy of the frog skin is used to produce ATP which then in turn energizes the active transport of sodium, the free energy made available from the splitting of ATP may at times be exploited close to 100 per cent [compare with 20]. Furthermore the transport models are limited by the fact that the essential parts of the active transport mechanisms must be localized in or in immediate contact with the cell membranes. Thus the number of steps in the process must be relatively limited.

In conclusion I want to say that it is on purpose that I have presented experimentally well established correlations instead of a compilation of hypotheses. On an occasion like this where scientists of widely different background are gathered there is a possibility of provoking new preconceived assumptions.

## References

[1]   R. W. Blount and B. H. Levedahl, *Acta Physiol. Scand.* **49** (1960) 1–9.
[2]   E. Du Bois-Reymond, *Untersuchungen über tierische Elektrizität*, Berlin (1848).
[3]   G. Galeotti, *Z. Physik. Chem.* **49** (1904) 542–562.
[4]   G. Gárdos, *Acta Physiol. Hung.* **6** (1954) 191–199.
[5]   W. R. Harvey and S. Nedergaard, *Proc. Nat. Acad. Sci.* **51** (1964) 757–765.
[6]   A. L. Hodgkin and R. D. Keynes, *J. Physiol.* (*London*) **128** (1955) 28–60.
[7]   C. Adrian M. Hogben, *Proc. Nat. Acad. Sci.* **37** (1951) 393–395.
[8]   E. Huf, *Pflügers Arch. Ges. Physiol.* **235** (1935) 655–673.
[9]   V. Koefoed-Johnsen, H. Levi and H. H. Ussing, *Acta Physiol. Scand.* **25** (1952) 150–163.
[10]  V. Koefoed-Johnsen and H. H. Ussing, *Acta Physiol. Scand.* **28** (1953) 60–76.
[11]  A. Krogh, *Skand. Arch. Physiol.* **76** (1937) 60–74.
[12]  A. Leaf and A. Renshaw, *Nature* **178** (1956) 156–157.
[13]  K. H. Meyer and P. Bernfeld, *J. Gen. Physiol.* **29** (1946) 353–378.
[14]  T. Rosenberg, *Symp. Soc. Exp. Biol.* **8** (1954) 27–41.
[15]  J. C. Skou, *Biochim. Biophys. Acta* **23** (1957) 392–401.
[16]  T. Teorell, *Arch. Sci. Physiol.* **3** (1949) 205–219.
[17]  H. H. Ussing, *Acta Physiol. Scand.* **17** (1949) 1–37.
[18]  H. H. Ussing, *Acta Physiol. Scand.* **19** (1949b) 43–56.
[19]  H. H. Ussing and K. Zerahn, *Acta Physiol. Scand.* **23** (1951) 110–127.
[20]  H. H. Ussing, *Protoplasma* **LXIII** (1967) 292–294.
[21]  K. Zerahn, *Acta Physiol. Scand.* **33** (1955) 347–358.
[22]  K. Zerahn, *Acta Physiol. Scand.* **36** (1956) 300–318.

*Theoretical Physics and Biology,* © 1969, *North-Holland Publ. Co., Amsterdam*

# NON-EQUILIBRIUM THERMODYNAMICS OF BIO-MEMBRANE PROCESSES

A. KATCHALSKY

*Weizman Institute for Science, Rehovot, Israël*

1. The physical approach to the old standing problem "What is life?" led to the recognition that living structures are based on information-rich, self reproducing macromolecules. The biopolymers are arranged in complex patterns which permit the development of an internal cybernetic integrity and ensure a high adaptability to the external world. The functional unity of the processes within the cells and the organisms, implies a coupling between the flows, which are controled by suitable regulatory mechanisms. An adequate formal description of coupling between phenomena is provided by the thermodynamics of irreversible processes developed during the last three decades.

The fundamental equation of non-equilibrium thermodynamics is based on the local time derivative of Gibbs' equation. It gives the dissipation of free energy per unit time

$$T d_i S/dt = \sum_i J_i X_i > 0, \tag{1}$$

where $d_i S/dt$ denotes the inner production of entropy per unit time, the $J_i$ are the irreversible flows passing the system and the $X_i$ the corresponding conjugated forces.

Eq. (1) indicates that flow may be driven against the direction of their conjugate forces if another, coupled, process provides the dissipation for a contra-gradient flow. Thus for two chemical rate processes, $J_{r1}$ and $J_{r2}$, driven by the affinities $A_1$ and $A_2$,

$$\Phi = J_{r1} A_1 + J_{r2} A_2 > 0 \tag{2}$$

means that $J_{r1} A_1$ may be negative or $J_{r1}$ will flow in a direction opposite to that dictated by its own affinity, if another process $J_{r2} A_2$ will be positive enough so as to make the overall dissipation positive. Eq. (2) is the formal description of the biochemical coupling in which complex synthesis may take place on the account of metabolic dissipation.

Similarly the contra-gradient diffusional flow, known in membrane physiology as "active transport" — is based on the coupling with a metabolic dissipative process according to the equation

$$\Phi = J_d X_d + J_r A > 0. \tag{3}$$

The explicit dependence of flows on forces — all important for the application of non-equilibrium thermodynamics to real situations — may be written according to Onsager as a set of linear phenomenological equations

$$X_i = \sum R_{ik} J_k, \tag{4}$$

where the coupling coefficients $R_{ik}$ ($i \neq k$) obey the famous symmetry theorem $R_{ik} = R_{ki}$. The linear dependence of flows on forces holds only for sufficiently slow processes close to equilibrium. Investigations are carried out in various laboratories on the extension of the thermodynamic formalism also to non linear processes such as periodic oscillatory phenomena. Since most biological processes are rapid and non linear, the existing formalism should be regarded only as a first approach and useful guideline to a fuller description of living systems.

2. The present discussion is devoted to coupling in biological membranes. Cellular membranes and tissue covers are highly selective devices which regulate the interaction of cells and organisms with their surroundings. Recently they were found to be dynamic "living" organelles. Moreover there is an accumulative evidence that many of the intracellular organelles, such as the mitochondria and the chloroplasts, as well as the intracellular reticulum, are based on two-dimensional membraneous structures. There is a growing feeling that the intracellular membranes represent the two-dimensional transition step from the unidimensional macromolecules to the three-dimensional pattern of cells and organisms. Intracellular membranes play a prominent role in the organisation of enzymatic complexes and in the regulation of their coordinated function.

The operation of cellular membranes is based on the coupling between metabolic processes and diffusional transport across the membranes. As shown by P. Curie and extended to non-equilibrium thermodynamics by 1. Prigogine, such coupling cannot generally take place in an isotropic space. The biological membranes must therefore have an anisotropic structure. In the simplest case this may resemble a bilayer of membrane elements with different permeability properties.

A model system for an isotropic membrane is a series array of positive and negative elements. Bilayers made of synthetic permselective membranes exhibit non-linear dependence of electric current on applied potential in accord with the behavior of biological membranes. The bilayer exhibit rectification properties and the potential developing between two solutions does not follow the requirement of Nernst–Plank equations.

It was found that the thermodynamic formula describing adequately the experimental results is

$$\Delta \Psi = \Delta \Psi_0 + \alpha \frac{RT}{F} \ln \left( 1 + \frac{I}{I_0} \right) + (I \cdot \varrho), \tag{5}$$

where $\Delta \Psi$ is the membrane potential at current flow $I$; $\Delta \Psi_0$ is the potential at $I = 0$; $I_0$ is the limiting current of rectification; $\varrho$ the sum of resistances of the membrane elements (the term $I \cdot \varrho$ being, however, small in synthetic membranes) and $\alpha = 0 - 2$ a coefficient comprising the difference between the electric transference numbers of the membrane elements.

It is noteworthy that although the behavior of the positive and negative elements of the complex membrane is adequately described by linear phenomenological equations—the composite effect is non-linear. This could be attributed to the intermembrane space whose parameters of state are functions of the flows passing the system. In this case the intermembrane space provides a feed back mechanism which influences the permeability properties of the composite system.

When the intermembrane space becomes prominent the variation in its properties may drastically change the flow-force relations. Thus in the bilayer of Shashua composed of two monolayers of oppositely charged polyelectrolytes, the application of a constant potential releases an oscillatory flow of electric current. The oscillations which resemble those observed in nerve membranes are presumably due to a phase shift in salt and water accumulation in the intermembrane space.

3.   From the biological point of view the important case is that in which the properties of the intermembrane space are controlled by a chemical reaction. Such a system was studied experimentally by Blumenthal et al. for the case of a hydrolytic reaction. Whenever the reaction ($J_r$) changes the ionic concentration in the space between two permselective membranes, the relation of $\Delta \Psi$ versus $I$ is given by

$$\Delta \Psi = \Delta \Psi_0 = \alpha \frac{RT}{F} \ln \left( 1 + \frac{I}{2RT(\omega_1 c_1 + \omega_2 c_2)F} + \frac{J_r}{2RT(\omega_1 c_1 + \omega_2 c_2)} \right), \qquad (6)$$

where $\omega_1$ and $\omega_2$ are the salt permeability coefficients for the membrane elements 1 and 2; and $c_1$ and $c_2$ are the salt concentrations adjacent to sides 1 and 2.

Eq. (6) exhibits the regulatory function of the chemical reaction and its influence on the resting potential ($I = 0$) as well as on the limiting current ($I_0 = 2RTF(\omega_1 c_1 + \omega_2 c_2) + J_r F$). When the flows are sufficiently small, the logarithmic term may be expanded and the electrical force ($\Delta \Psi - \Delta \Psi_0$) becomes a linear function of the electrical and chemical flows. This is a special case of the general equation used by Kedem for the description of active transport

$$X_i = \sum R_{ik} J_k + R_{ir} J_r. \qquad (7)$$

The coefficient $R_{ir}$ represents the coupling between a non-chemical force $X_i$ and the flow of the reaction $J_r$, it may therefore be regarded as the coefficient of active transport. Even if all the flows $J_k$ vanish, but $R_{ir}$ and $J_r$ are non zero, a stationary force $X_i$ may be maintained across the membranes. This case corresponds to the steady distribution of ions between cell and surroundings, which disappears when the chemical process is inhibited by metabolic poisons.

4.   The transport of non-electrolytes and electrolytes across biological membranes is generally facilitated and exhibits saturation phenomena. This is generally

regarded as an expression for the participation of carriers, with a finite number of adsorption sites, in the transport process. The carrier molecules may be macro-molecules which undergo a conformational change and develop forces sufficient to carry the permeant across the membrane. A suitable model for carrier transport is a mechano-chemical engine, which transports salt through contraction-expansion cycles of polymeric filters.

Carrier models for active transport which comprise an interaction with energy rich metabolites, such as ATP, are useful means for the evaluation of phenomenological coefficients and for numerical correlations between observable data. It should, however, be clear that other models, such as the lattice model developed by Heckmann and by Hill may be used advantageously for the evaluation of coupled flows and their dependence on external forces.

5.   Recent study on ion transport in mitochondria suggest an intimate participation of the membrane matrix in the coupling between the metabolic and transport process. There is an indication of a merger between function and structure into a higher order unity, a unity in which the constituents of cellular organisation carry out an integrated system of reactions maintaining a predetermined spatial distribution.

*Theoretical Physics and Biology*, © 1969, *North-Holland Publ. Co., Amsterdam*

# CHARGE, CHARGE TRANSFER AND CELLULAR ACTIVITY

ALBERT SZENT-GYÖRGYI

*Marine Biological Laboratory, Woods Hole, Mass., U.S.A.*

I was led, lately, to the conclusion that cellular activities, like cell division or protein synthesis, are dominated, to a great extent by the electronic charges of the system. This conclusion was reached by observation on three different lines: observations made with Nitroferricyanide, Glyoxal derivatives and Thiourea. I will start with discussing these separately.

*Nitroferricyanide* (nitroprusside). W. Gordy has shown that electrons, generated on protein, tend to become localized on Sulphur atoms. In an aqueous medium negatively charged S atoms can be expected to bind a proton, forming SH. Nitroferricyanide has widely been used for the detection of SH. At a high ionic strength and pH it reduces Nitroferricyanide to a brilliant purple substance with an absorption at 510 m$\mu$. This color is due to a charge transfer in which an electron is transferred from SH to the Nitroferricyanide.

*Glyoxal derivatives.* Tissues contain a substance which retards growth. Jane McLaughlin and I called it "retine". Tissues also contain a growth promotor which we called "promine". Evidence has been obtained, in collaboration with L. Együd, that retine is a glyoxal derivative or an isomer thereof. Glyoxal derivatives strongly inhibit cell devision and protein synthesis. As ketoid electron acceptors they are oxidizing agents which may even oxidize HI to metallic I. The electron acceptor property is responsible for the inhibitory action and can be compensated by certain SH compounds. At the same time, however, the oxygen of the glyoxal is capable, with its lone pair of electrons, of acting as an electron donor in the presence of a suitable catalyst.

*Thiourea* is known to promote cell division and regeneration. If cell division depends on charge density and glyoxal derivatives suppress it as electron acceptors, then Thiourea could promote cell division by activating the lone pair of electrons of the oxygen of retine or the glyoxal, transforming thus the electron acceptor into a donor, an oxidizing agent into a reducing one, increasing hereby the charge density which, then, favours cell division.

These reactions can actually be demonstrated in a simple and striking fashion by means of retine (or a Glyoxal derivative or its isomer), Nitroferricyanide and Thiourea.

If a glyoxal derivative, say propylglyoxal (or dihydroxyacetone) is allowed to act on Nitroprusside at a high ionic strength and pH, it acts as an electron acceptor and oxidizes the Nitroprusside to some yet undefined oxide of varied and poor color. If, however, the reaction is allowed to proceed in the presence of Thiourea, the Glyoxal acts as an electron donor and reduces the Nitroferricyanide to its typical purple reduced form with an absorption at 510 m$\mu$. What, probably, happens is that the S of the Thiourea, with its available p orbitals, takes over the lone electrons of the oxygen and passes them on to the Nitroprusside. Promine may have an analogous function, passing on the electrons, *in vivo*, to the cellular structure.

All this, taken together, suggests that cellular activity may depend, to a great extent, on charge density. It was known for a long time that SH groups are essential for cell division and rapidly growing tissues are very rich in it.

Conscience and memory are cellular functions of the nervous system. I find it impossible to approach the mechanism of these functions without supposing the participation of delocalized electrons. I have shown, earlier, that indoles are good "local" electron donors. LSD, Lysergic acid, the hallucinogen, is an indole derivative. With I. Isenberg, we concluded that it may produce hallucinations by donating electrons to the nervous tissue. This suggestion was corroborated by Snyder and Merril who studied a great number of related compounds and found halucogenesis to go parallel with the electron donating ability.

I found no substance in the brain which activates ketone aldehydes as electron donors. These substances can thus act only as electron acceptors. If electron donors cause hyperactivity, halucinations, — electron acceptors, then, should reduce activity. We found that glyoxal derivatives produce a short-lived narcosis in mice.

*Theoretical Physics and Biology,* © 1969, North-Holland Publ. Co., Amsterdam

## DISCUSSIONS

I. PRIGOGINE: Au delà de l'instabilité chimique que j'ai discutée hier, la partie de la production d'entropie due à la diffusion est négative à cause de l'apparition et du maintien du gradient de concentration. Il peut ainsi y avoir un lien direct entre structure dissipative et transport actif.

A. KATCHALSKY: Dr. Prigogine's remark requires a longer answer: The first question to be answered is what are the relations between flows and forces. For sufficiently slow flow and small forces we may write that the flow is linearly proportional to the forces. Thus, we may put the flow of diffusion as equal to Conc. $C$, times the velocity $\vec{V}$. $\partial_d = C \cdot \vec{V}$; but the velocity $\vec{V}$ is assumed to be directly proportional to the diffusional force $X_d$: $\vec{V} = uX_d$ where $u$ is the mobility.

For the diffusional force we write $X_d = -d\mu/dx$ where $\mu$ is the chemical potential, which equals in an ideal case to $\mu = \mu_0 + RT \ln c$. Hence

$$X_d = - \frac{RT}{c} \frac{dc}{dx} .$$

and the diffusional flow is given by

$$J_d = c \cdot \vec{V} = cu \left( - \frac{RT}{c} \frac{dc}{dx} \right) = RTu \left( - \frac{dc}{dx} \right) = D \left( \frac{dc}{dx} \right),$$

which is Fick's equation, with the Einstein expression for the diffusional coefficient $D = RT \cdot u$.

In a similar way we may write that the rate of a chemical reaction $J_r$ is proportional to the affinity $A$ of the reaction. Moreover, the linear dependence is generalized to the case of a flow dependent on several forces. Thus, for a diffusional flow which depends on both the gradient of chemical potential and to the affinity of a chemical reaction $A$ we may write

$$J_d = L_{11} \left( - \frac{d\mu}{dx} \right) + L_{12} A,$$

i.e., we assume there is a *straight* dependence of $J_d$ on $-d\mu/dx$ and a *coupled* dependence on $A$, where $L_{12}$ is the coupling coefficient.

On the other hand, the Curie-Prigogine principle requires the coupling coefficient between scalar and vectorial flows, in an isotropic space, to be zero. Since the chemical reaction is scalar while the diffusional flow is vectorial, we might expect $L_{12}$ to be zero in isotropic space. Luckily for the phenomena of life, all biological membranes, in which active transport takes place, are anisotropic.

Hence, $L_{12} \neq 0$ and active transport is possible. Indeed, my colleague O. Kedem, suggested the nonvanishing of chemical diffusional coupling coefficients as a criterion of active transports. Thus in cells in which diffusional flow vanishes on a chemical affinity allowing the maintenance of a concentration gradient according to the relation

$$ - \left( \frac{d\mu}{dx} \right)_{J_d=0} = - \frac{L_{12}}{L_{11}} A. $$

In reality most of the biological systems are nonlinear and such a treatment as proposed by Dr. Prigogine is highly welcome. I would like, however, to point out that in many cases nonlinear relations are obtained from linear elements. This is beautifully demonstrated by a series of membranes with small intercellular spaces. The state parameters, such as concentration and pressure, in the small spaces are functions of the supply of material through the membranes which causes a nonlinearity in the behaviour. A typical example of this kind is a double membrane composed of a sheet permeable to positive ions and a sheet permeable to negative ions. Such a double membrane will either accumulate salt or lose salt according to the direction of the electrical current. If salt is accumulated in the intramembrane space it becomes a conductor of electricity. On the other hand, if it loses salt it becomes an isolator (fig. 1). Evidently the dependence of current and potential will be nonlinear and the membrane will become a typical rectifier although the elements obey linear phenomenological laws.

Dr. Prigogine's question is very important. It is really difficult to imagine life without active transport. Indeed, most students of the origins of life are not satisfied with the spontaneous formation of organic molecules similar to those found in the living organism. Neither do they regard the spontaneous formation of the biopolymers as the zero point of life formation. It is felt that at least two requirements have to be fulfilled to regard a structure as living, namely the elementary structure is to be self organizing, and it should have a capacity of selective accumulation of essential materials. To this end a membrane endowed with the capacity of transporting actively is required. There are good indications that synthetic capacity and selective permeability go together, as demonstrated, for instance, in the classical study of Jacob and Monod on the genetic regulation of permease formation. It may also be pointed out that it is not only whole cells but even intracellular organelles which are endowed with a capacity for active transport. Thus, the mitochondrian-intracellular organelle which produces the energy-rich ATP is a membrane structure composed of membrane layers and endowed with a capacity of selectively accumulating ions, and calcium, in particular. The ion pumping is always based on the coupling of the diffusional flow with a chemical reaction, and, for instance, for the red blood cell the energy source for the chemical conversion is ATP. As shown recently, however, the mitochondrion and many bacteria use the chain of oxido-reduction itself for

The composite membrane

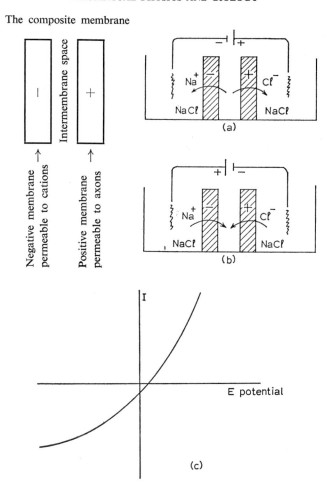

Fig. 1. (a) Ion depletion between the membranes; (b) Ion accumulation between the membranes; (c) Rectifier behaviour of double membrane.

pumping purposes, and not ATP. The discovery that an oxido-reduction chain is involved in the pumping makes the membrane an integral part of a direct membrane process, and one may speculate that the formation of the membrane is coupled with the active transport processes. From this point of view, the elementary formations of life seem to have facility of coupling, the coupling being responsible not only for the interaction of flows but also for the formation of the rudimentary living structures.

P. MAZUR: Indeed I think that Dr. Prigogine's approach and Dr. Katchalsky's may have some analogy. But I think there is also a basic difference which is the following. In Dr. Prigogine's approach the equilibrium state is the uniform state;

while in Dr. Katchalsky's treatment there is basically already a structure: one or more membranes. The coupling in Dr. Katchalsky's approach arises from the coupling coefficient in linear phenomenological equations and they exist because there is a membrane with anisotropic properties. In Dr. Prigogine's approach the "coupling" arises from the non linearity of the kinetic equations and stability conditions. So I think that Dr. Prigogine's approach would rather apply to the creation of the membrane while in Dr. Katchalsky's treatment the existence of the membrane is given. So I think those are not equivalent pictures although they may be complementary.

I. PRIGOGINE: Je suis entièrement d'accord avec vous.

P. MAZUR: I am not disputing the fact that the two pictures may be complementary. I am saying that, as it stands here, Dr. Katchalsky takes the existence of the membrane for granted. Even at equilibrium the membrane is still there.

Equilibrium simply means that on both sides of the membrane you have the same conditions. Dr. Prigogine's approach on the other hand provides the possibility to create an inhomogeneity that is perhaps of a membrane.

L. ONSAGER: I have found it difficult enough to understand even 'passive' transport through membranes. That is part of explaining the active; we have indications that passive transport facilities exist by themselves, and they are quite likely to occur as components of active systems as well. A typical hydraulic analogue would include one pump at some convenient location and a length of plumbing to guide the output.

The membrane seems eminently suited to restrict the flow of ions. It presents a hostile non-polar environment; we know that even salts of monovalent ions rarely dissolve in comparable liquids, and the soluble ones are sparingly dissociated. From a theoretical point of view these observations are quite intelligible: the electrostatic energy of a charged particle in a medium of low dielectric constant is much higher than the energy in an aqueous solvent, so that the work needed for transfer is hardly ever supplied by the molecular motion. Thus the very penetration of the membrane by ions requires some facilitating substance or structure. Anionic carriers which combine with cations to form neutral complexes may play a role; but we know that electric currents do traverse membranes, and even that specific ions may very nearly control the potential difference across a membrane. Dr. Ussing pointed out that a potential difference of the order of 100 millivolts is common.

So ions move by themselves. How do they get into the membrane? We might consider a solvating shell of dipoles, which would produce a bigger ion. That does not seem to be quite the answer. To produce a hydrated ion in a non-polar environment the water molecules must be removed from the much more congenial company of their own kind, so that the net reduction of the energy will hardly amount to much. And the idea does not fit the facts: studies of the membranes which separate two nerve cells at a synaptic junction indicate that small univalent ions pass and big ions don't.

Let us consider a lipid bilayer, as described by Dr. Ussing, which separates the interior of a cell from the exterior salt solution. We shall try to devise a reasonable facility for ions to traverse the non-polar core, some 50 A° thick, formed by the hydrocarbon tails of the lipid molecules. If the ion is to enter a "pore" of molecular dimensions, a connected sequence of polar molecules or molecular groups will be needed. It is instructive to consider first a single chain of dipoles in an environment of low dielectric constant.

To minimize the energy, all the dipoles must assume identical orientations parallel to the direction of the chain, either all inward or all outward. The dipole moment per unit length defines an effective end charge; for a chain of hydroxyl groups this might amount to about one quarter of an elementary charge, or a little less. If such a chain joins the opposite faces of a membrane then the terminal fields can spread in the adjacent aqueous phases, which entails a negligible amount of energy. The optional direction of polarization can be reversed by successive rotations of the polar groups; this process transports a "polarization charge" equal to twice the end charge. The "dielectric constant" of such a chain is extremely anisotropic. It exhibits a very large polarizability in the lengthwise direction; in fact it is more like a ferroelectric system. One such chain would compensate almost half the charge of a univalent ion so as to reduce the energy of immersion by a factor or three or four, which might or might not suffice; but a system of two or at most three chains ought to be enough. An ion may enter a polarized chain only at the end which carries an end charge of opposite sign. As the ion moves along the chain, the dipoles turn in succession; so that a compensating polarization charge travels with the ion. After a complete traverse the process can be reversed but not repeated until the passage of a separate polarization charge has restored the original condition, and a full elementary charge has not passed until this happens. The elementary charge is effectively divided into two or three parts for earies penetration and transport.

Among the types of polar groups found in the side chains of amino acids hydroxyl[a]) and carboxyl[b]) groups are particularly good candidates; amino groups [c]) and imidazole [d]) might be considered too.

Added in print: a) serine and threonine, b) aspartic and glutamic acids, c) lysine, d) histidine.

The ideas which enter into the kinetic theory of dipolar chains have evolved largely through efforts to understand the properties of protonic semiconductors like ice (Eigen and others) and $K H_2 PO_4$. These solids contain disordered networks of proton bonds, so that a reorientation of molecules or groups can occur in response to an electric field. Moreover, this property combined with a measure of intrinsic ionization enables them to transmit protonic currents; detailed studies of ice (Eigen and others) and of the phosphates have produced a wealth of interesting information. The results for ice are pertinent to chains of hydroxyl groups, and further studies of formic acid might well teach us what we should very much want to know about carboxyl chains. While the phosphates command our interest in the context of enzyme kinetics rather than passive transport, they have taught us one lesson which cannot be dismissed as irrelevant. The pertinent facts were revealed by studies of deuteron magnetic resonance. As was first suggested by Takagi, the shuttling of protons between phosphate groups occurs very much more often than the rotational transfer between bonds; thus the former gives rise to a dielectric relaxation at microwave frequences and the latter process, some $10^8$ times slower, determines the conductivity. In pure ice the rotation is about 400 times faster than the shuttling, in doped ice either may prevail. What little information we have about formic acid suggests a borderline case.

We might well be prepared to find a variety of chain systems adapted to different functions in cell membranes, and some of them could perhaps be fairly elaborate. Just how does the sodium pump manage to move 3 ions for each molecule of ATP? A measure of ferroelectric cooperation between chains might help—but it would be just as well if we wait for more facts before we speculate too much. The excitable membrane of a nerve axon exhibits a remarkable sensitivity to small changes of the trans-membrane potential. The response corresponds to the passage of several elementary charges through the imposed potential drop. This too might well depend on a elaborate structure, and again we have too much room for speculation—except, perhaps, about one detail: by far the greated part of the imposed potential drop will be found in the non-polar core between the polar faces of the membrane, because the capacitive impedance of that layer much exceeds the sum of all the others. Thus, only a charge shifted all the way through the membrane can utilize the potential drop as effectively as the observations suggest. The polarizable component may be a chain of groups or a set of chains, linked to a gate for sodium ions but probably not directly involved in the transport process.

H. HAKEN: I would like to put two questions to Dr. Katchalsky. I think his talk was so clear that even a theoretical physicist may ask a question. The first one is: Because you have oscillations one might expect a scheme of the type

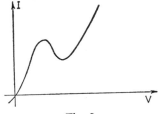

Fig. 2.

But this is also known in semiconductor physics to be used e.g. for logical switches. So the obvious question is: Is there any evidence that such things with membranes occur in neurons?

A. KATCHALSKY: The question of Dr. Haken is rather important since it is related to the intriguing problem of the molecular basis of nerve excitation.

The beautiful theory of Hodgkin and Huxley is *phenomenological* and does not provide for a physicochemical insight into the mechanism of excitation. It assumes, however, on the basis of experimental evidence, that a stimulated nerve membrane is many times more permeable than the resting membrane. While the resting membrane is almost entirely impermeable, stimulus makes it relatively open to the passage of sodium and potassium, and more recent experiments of Rojas indicate that there is also an increase of nonelectrolyte permeability.

Hodgkin & Huxley noted that the increase of permeability follows an exponential equation and there is a growing feeling that what underlies it is a cooperative phenomenon, say a local conformational change of certain macromolecules such as structural membrane proteins. It is well known that cellular membranes—and nerve membranes included—comprise a lipid bilayer which according to Dawson & Danielli carries on its outer and inner surfaces, protein layers. In our study of the viscoelasticity of red cells membranes we could demonstrate that the protein layers are made of stretched fibrillar molecules which endow the membrane with its touch mechanical properties. Now the fine investigations of Miller and Rudin and of Thompson demonstrated that while a pure lipid bilayer is highly resistant to ionic flows, the addition of suitable protein may make the membrane highly permeable and conducting.

This, however, is not the whole story: It is known since the classical work of Osterhaut that $Ca^{++}$ is involved in excitability. The removal of calcium ions by a chelating agent brings the cells into a state of constant stimulation, while the addition of $Ca^{++}$ is involved in the maintenance of the *resting* nerve membrane *structure*, while the exchange with monovalent cations releases a cooperative change, increase permeability and leads to the electric potential changes described quantitatively by the theory of Hodgkin & Huxley.

I have the feeling that some evidence for the ion exchange may be deduced from the potential changes leading to excitation. It is well known that the inner surface of the nerve membrane is negative with respect to the outer surface and has a potential of $-70$–$100$ miv. Upon making the inner surface slightly more positive, by say 20–30 miv an excitation is released. Now such change in potential is an energy investment of 0.02 eV per monovalent ion which is much too low for the opening of a chemical bond—it might however, suffice for an ion exchange process. Finally, if the lipo-protein structure is metastable the ion exchange will lead to the assumed cooperative change.

H. HAKEN: If one has one element in a system the obvious question of a physicist is: what happens if you have the same thing in a periodical structure. In other words: if you have one membrane what happens if you have them in series. Are there examples in nature that membranes are arranged in a periodic structure?

A. KATCHALSKY: The experiments of Dr. Eigen are not only beautiful but very important for the behaviour of ghosts and similar systems. But there is one peculiarity which I do not see still any experimental way of overcoming. It is known from the experiment of Miller, Rudy and Thomson and others on the bi-layers, that when you add protein the permeability changes enormously and somehow the protein does not seem to fit into the business. On the other hand all biological membranes comprise protein; there is no question. It is not only a lipid bi-layer. There is some protein. I do not know where it is. But my feeling is it is really attached somehow to the bi-layer in the bi-layer. But the feeling is that the protein which participates in the biological membrane is a network of stretched protein molecules and not spherical protein molecules. The indication I mentioned "en passant" in my talk that if you measure the rheology of the redblood cell ghost, this was measured in different ways, you find that the rheology fits a network of stretched macromolecules. From any point of view it looks like a texture of textile fibres which can move and change the porosity, whatever you like.

The real experimental problem would be to try and attach a protein which would co-agulate on the surface and give both mechanical strength and presumably elucidate of the permeability properties of the natural membranes. But whatever be the case, it does not diminish the value the lipid bi-layer structure. Allow me to mention some of the history preceding the modern theory of allosteric change: Almost twenty years ago, Gergeli and Laki proposed an explanation for the contraction of muscle based on a two state model for the conformational change of reactive macromolecules. The model, denoted as an "autone model", assumed that the segments of contractile macromolecules may exist in two forms, a long and a short—their distribution being dependent on the extent of reaction with a suitable reagent. Upon interaction the long segments converts to short, and this is how the mechanochemical transformation is released.

Several years later the theory was put on a rigorous statistical basis by Terell Hill who developed carefully the consequences of the autone model. It is rather interesting that the formation of the autone model resembles closely the allosteric equations, and there is, in reality, a deeper relation between the two approaches.

Two days ago I showed you the contractility of collagen fibers upon interaction with strong salt solutions. In an attempt to describe theoretically the contraction process, Dr. Merry Rubin in my lab applied the autone model to the protein molecules of collagen, and found that it works surprisingly well with sufficiently stretched fibers. With small stretching forces the model fails—but it was found by S. Reich of my lab, that the behaviour is adequately described by the statistical equations for random coils of finite length (The Langevenian equations for ideal rubbers).

It seems that a fuller description of the behaviour of bipolymers, subjected to both mechanical and chemical forces, requires a more general statistical treatment which would comprise both the autone-allosteric approach and the Brownian statistics of random coils. We are now working on this problem and I hope to be able to report on the conclusions at the next meeting.


L. ONSAGER: To support a chain of polar groups through a membrane we need a framework, and the most suitable material on hand would seem to be a section of protein chain in helical conformation with its axis perpendicular to the face of the membrane. Side groups attached to two or more different helices may contribute alternate links to the dipolar chains, and the entire system will be quite stably embedded in the membrane if the side chains not so involved are non-polar. Recent investigations by Johnson and Singer and independent studies by Wallach and Zahnder (Proc. Nat. Acad. Si. Nov. and Dec. 1966) reveal that membrane proteins from widely different sources assume helical conformation over 20–25 per cent of their length in liquid environments which might correspond reasonably to intact membrane.


M. DALCQ: Je crois utile de verser à ce débat une information embryologique qui établit que des enzymes déphosphorylantes (phosphohydrolases du type ATPase et ADPase) sont présentes dans certaines membranes cellulaires. C'est dans les premiers stades du développement des oeufs de Mammifères que cette constatation a été faite. Quand un oeuf de petite dimension se segmente, il édifie successivement des cloisons membranaires, prolongements de la membrane cellulaire. Si l'on plonge un oeuf de rat, de souris, de lapin ou de taupe — ce sont les espèces effectivement utilisées — dans un bain tamponné contenant un mono-nucléotide, on voit, par les moyens appropriés, sa surface tout entière s'assombrir en raison de la précipitation de phosphates libérés enzymatiquement à partir des esters phosphoriques utilisés. Lors des divisions successives, le dépôt des phosphates se produit au niveau de chacun des sillons, et cela avec une intensité bien plus forte qu'à la surface de l'oeuf ou de ses blastomères.

Les enzymes responsables de ces activités résistent à une fixation au formol, ce qui a permis une analyse cytochimique plus poussée. Le phénomène est maximal avec les triphosphates des divers nucléosides, moindre avec les diphosphates, très modéré avec les monophosphates. Il a pu être discerné que les éléments responsables de cette évolution sont de minuscules granules situés dans le cortex et qui, lors des divisions, se concentrent dans le sillon. Leur nature inframicroscopique reste à déterminer. Jusqu'à présent une spécificité stricte n'a pu être mise en évidence, et l'on ne peut donc dire qu'une ATPase est seule en cause. Il est bon d'ajouter que ces constatations ne concernent pas nécessairement d'autres oeufs que ceux des Mammifères. On ne les a notamment pas retrouvées dans les oeufs d'Invertébrés marins, mais il s'agit de recherches délicates, où un résultat négatif peut n'être pas décisif.

H. S. BENNETT: We have heard from Dr. Ussing and Dr. Katchalsky of some very interesting properties of the membranes around cells. These properties are displayed by the membrane of the red cell (*see* fig. 3). The ion pumping activities of such a membrane are usually envisioned as occurring without any gross movement of the membrane itself. Thus the process described by Dr. Ussing and mentioned by Dr. Katchalsky is usually thought of as taking place by intramolecular movements within the membrane. In this model, the whole structure of the membrane, within which are moving molecular parts, can be regarded as fixed in position in relation to the rest of the cell.

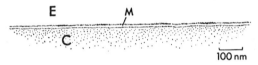

Fig. 3. A diagram of the surface of a red cell. E represents extracellular space and C the cytoplasm. The usual concepts of active transport envision the ion pumping mechanisms as located within the cell membrane, M, which is about 8 nm thick, and which displays two dense layers separated by a less dense central stratum when viewed in cross section with an electron microscope.

There are, however, additional properties of membranes which involve what I shall call "gross movements"; that is, translations in space involving large areas of membrane moving from one part of the cell to another, carrying particles bound to them and enclosed within them.

These properties may not be found in the membranes of all cells. For example, I know of no good evidence of gross translation of membranes in the red cell. But such gross movements have been observed in many cells. This phenomenon is sufficiently frequent to deserve attention.

Membrane movements of this type have been described extensively in the ameba by Heinz Holter and his colleagues, (in *Intracellular membraneous structure*, Proc. First Internat. Symposium for Cellular Chemistry, Ed. J. Seno and E. V. Cowdry, Okayama, Japan, The Chugoku Press (1965) p. 451). Some of the essential physiological changes can be observed with a light microscope. Further insight is gained by using the electron microscope. The physiological significance of the morphological changes can be followed by measurements of the movement of matter across the cell boundaries.

In the ameba, the membrane can be represented as showing the double protein-lipid sandwich structure which Professor Katchalsky mentioned as characteristic of the unit membrane (see fig. 4). In the ameba, there is, in addition, a very

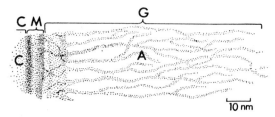

Fig. 4. A diagram of the surface of the ameba, *Chaos chaos*. The cell membrane (M) is represented as showing the trilaminate "unit membrane" structure. The cytoplasm, C, is to the left. Attached to the membrane and extending into extracellular space to the right is an extensive antennular glycocalyx (G), containing glycoprotein rich in acid mucopolysaccharides. The frond-like antennulae (A) of the glycocalyx are capable of binding large quantities of basic proteins or other cations by ion exchange mechanisms. They are capable also of entangling and embarrasing the cilia of *Paramecium* and *Tetrahymena*.

interesting structure which looks like a fringe of a lady's scarf or the nap of a carpet, representing the ameba's glycocalyx. This structure is on the outer surface of the cell membrane, and attached to it. In electron micrographs one sees this fringe as characterized by very numerous antennulae resembling the structures shown in the figure. The material constituting the glycocalyx shows chemical characteristics of acid mucopolysaccharides and displays strong ion exchange properties.

If an ameba is placed into a solution containing protein, particularly a protein rich in cationic charges, the protein is rapidly accumulated and binds, presumably by ionic forces, to the carbohydrate containing antennular glycocalyx on the outer surface of the membrane. This binding appears to serve as information for the ameba, to which the cell responds by changing the configurational pattern of its membrane in a very interesting way. Fig. 5 shows the topology of the membrane changes in essential features, without going into details.

After some of the material has been bound to the surface of the ameba, the membrane becomes indented or invaginated, forming a little recess or cave, called a caveola. The invaginated membrane carries the bound material. At a later stage, the lips of the caveola come together and a fusion and recombination of membranes takes place, so that an area of membrane which was originally on the surface of the cell now finds itself in the interior.

A similar plasticity and capacity for fusion is a familiar property of soap bubbles, which possess an arrangement of lipids rather similar to that in the membrane. Thus one supposes that this pinching off, fusion and recombination of the membrane may be associated in an important way with lipid components of the membrane. Whatever the mechanism may be, the membrane area originally on the external surface of the cell becomes labelled by virtue of particles binding to it. This labelled area can be recognized clearly in electron micrographs, and is seen at later stages in the interior of the cells. It can be moved from place to place

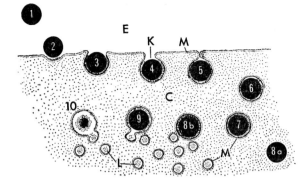

Fig. 5. A diagram of successive stages of the essential topological features of membrane movements, fusions, recombination, and translocations associated with uptake of material by encytosis, such as characterizes phagocytosis and pinocytosis. The black discs represent a particle of some sort which is to be taken into the cell. The particle might be a bacterium, a carbon particle, a protein molecule, or a group of molecules and ions bound together in some way.

At 1, the particle is represented as free in the extracellular fluid, E, prior to uptake. At 2, the particle is shown as having collided with the membrane, M, and as having been bound to it, thus labelling this particular area of membrane. The membrane has responded to the bound particle by indenting slightly. At 3 and 4, the indentation has advanced to form a caveola (K). At 5, the caveola has deepened and the margin of the orifice is converging like the opening of a purse. The membrane surfaces at the orifice have just made contact prior to fusion. At 6, one sees the particle enclosed in a vesicle formed by fusion and recombination of the cell surface membranes which formerly were in a caveola as at 5. At 7, the particle has moved further into the cytoplasm, C, still enveloped in its membrane, M, which was derived from the cell surface. The particle now lies inside the cell, where analytical methods would show it in an intracytoplasmic fraction.

From this point on several alternative sequences may be observed. 8a shows a common course followed by animal cell viruses and by proteins such as ferritin. The particle is shown free in the cytoplasmic matrix, without any surrounding membrane, following the disassembly or dissolution of the vesicle membrane M represented in 7. By this sequence, particles varying in size over a wide range can move from extracellular space (E) at 1, into the cytoplasmic matrix (C), at 8a, without going through any pore or perforation of the cell membrane. In a sense, one can say that the particle moved from one side of the cell membrane to the other without going through the membrane.

A second possible series of events is shown in 8b, which represents the vesicle containing the particle as moving into the company of many lysosomes (L). Lysosomes are membrane bound vesicles containing hydrolytic enzymes or zymogens synthesized previously in other portions of the cell. At 9, one sees these lysosomes making contact with the vesicle containing the particle. Membrane of lysosome and of the vesicle containing the particle fuse and combine in such a way that the lumen of the lysosome becomes confluent with that of the vesicle and the enzymatically active contents of the lysosome are discharged into direct contact with the particle. At 10, one sees an additional lysosome fusing and introducing its enzyme load into the vesicle, whereas the particle is represented as disintegrating under the enzymatic attack of the contents of a number of former lysosomes. The events represented in 1 to 7, 8b, 9 and 10, are characteristic of phagocytosis. They are displayed by ameba upon uptake of prey and by human blood leucocytes ingesting bacteria.

Another common consequence of encytosis not shown in the diagram, is for several vesicles similar to those represented at 6 and 7 to fuse and to combine their contents in a single membrane-bound sack.

within the cell. The membrane surrounding the vesicle can be modified at the cell's leisure in some part of the cell remote from the original surface (*see* fig. 5).

An interesting example of this phenomenon in reverse has been demonstrated by Palade and by Ichikawa (*see* J. Cell. Biol. **24** (1965) 369) in another type of cell, the exocrine cell of the pancreas (*see* fig. 6). In this organ we have a cell which is very active in protein synthesis. After a nice lunch, such as we had today, when some of us are fighting the battle between sleep and attention, our pancreas is very busy synthesizing enzymes to digest the sausages and cheese we have just enjoyed. This process is characterized by exercise of ribosomes and of messenger RNA in the cell, mediating the process of protein synthesis in a manner familiar to us all. The ribosomes in the pancreas are arranged on the surface of membranous cisternae to which Dr. Katchalsky alluded. These cisternae lie in the basal portion of the cell. Upon synthesis, the protein appears very quickly inside them. The newly formed protein then moves towards the apex of the cell, through the Golgi membranes, and thence into secretion granules, which are really membrane-bound sacks or vesicles of concentrated enzyme precursors (*see* fig. 6). After the meal is eaten, the protein in the secretion vesicles is discharged into the pancreatic ducts. This discharge is characterized by a movement of the vesicles, each containing many millions of molecules, to the surface membrane of the cell facing the ducts. There the secretion granule membrane fuses with the cell membrane, and the vesicle comes to open to the exterior, forming a secretory caveola.

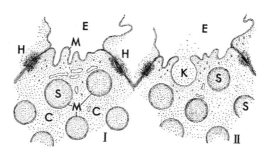

Fig. 6. A diagram of the discharge of secretion products by pancreatic acinar cells. I shows the apical portion of a cell prior to receiving a stimulus to secrete. In the cytoplasm (C), one sees several secretion granules (S), which are really spheroidal membrane-bounded vesicles or sacks containing many millions of enzyme or zymogen molecules. The cell membrane, M, separates the cytoplasm from the extracellular space (E), which here represents the lumen of an acinus, which is continous with the pancreatic ducts. Two desmosomes (H) bind a portion of the lateral cell membrane to that of adjacent cells.

II shows the same cell after responding to the hormone, pancreozymin, which stimulates the pancreatic acinar cell to release secretory products. One of the secretion granules (S) has moved to the surface of the cell. Its membrane has made contact with the cell membrane and has fused and combined with it, causing the lumen of the former vesicle to open to the extracellular space, forming a transient secretory caveola (K). Zymogen molecules are shown in the process of passing into the acinar lumen (E).

The contents are then discharged into the duct. The vesicle membrane, which was inside the cell, is now part of the surface membrane of the cell.

These events tell us that protein has been carried, millions of molecules at a time, from one part of the cell to another, and has finally been discharged — not molecule by molecule, but in bulk, many molecules at a time. One thinks of the analogy of a large bulk carrier, such as a big oil tanker. Just as it is cheaper to carry oil in large bulk than it is to convey it drop by drop, so is it cheaper to carry protein in big sacks, millions of molecules at time, rather than molecule by molecule. As Palade has pointed out, transport of protein in bulk is economical. Thus the cell uses moving membranes to transport millions of protein molecules together, to carry them to the surface, and to discharge them under control. Gross movements of large areas of membrane are involved in this transport process.

Dr. Ussing focused attention on the capacity of the cells to transport small ions. Cells are able to transport specifically other substances also. May I remind you of the classic experiment of Avery, McLeod and McCarty (see J. Exper. Med. **79** (1944) 137), in which they demonstrated that the transforming principle of the pneumococcus, which is DNA, can be taken up through the cell membrane of bacterial cells and enter the genome of the cell. Implicit in this experiment was the demonstration that DNA from outside the bacterial cell found its way to the inside. This is an example of a large molecule traversing the cell membrane.

There are many other similar examples known of proteins or other large molecules moving from the outside of the cell to the inside, or in the reverse direction. A familiar example of the transport of chemical material from the out-side of the cell to the inside is the process of phagocytosis. Certain cells can take up a whole red cell, or one or more bacteria, and translate them from the external surface of the cell to the interior. In effecting these movements, the cell utilizes membranes movement of the kind I have explained. From a formal point of view, phagocytosis meets all the definitions of active transport. That is, phagocytosis involves a specific translation of chemical material from one side of the membrane to the other against a concentration gradient. For example, if one examines the cell after it has moved bacteria into the interior by phagocytosis and counts the number of bacterial inside, and compares it to the concentration outside, one arrives at a result which is formally similar to the transport process that Dr. Ussing has described.

The mechanism first involves a bonding of the particle to be taken up, whatever its size, to the exterior surface of the cell (see fig. 5). A bacterium, an ion, or a protein which will not bind to the surface is unlikely to be taken up by this process. If the sum of all the bonding forces is sufficiently strong to counteract the kinetic forces which may drive the particle away from the cell surface, the cell may respond to this binding by invagination its membrane and then pinching it off by membrane fusion and recombination, thus taking the particle into the interior of the cell. In many cases the cell, having taken up the material, can manipulate it and maneuver it about in the cytoplasm.

Hirsch (*see* Ann. Rev. Microbiol. **19** (1965) 339), at the Rockefeller University has shown that leucocytes contain many membrane-bound packets of enzymes similar to those which appear in the pancreatic acinar cell. After phagocytosis of bacteria, these enzyme-containing vesicles are moved to the vesicle containing the bacteria. There the two vesicles fuse and enzyme is discharged into the vicinity of the bacteria (*see* fig. 5). This process offers a simple variation of the features of secretion by the pancreas.

In some instances, however, particles originally outside the cell, having been taken into a membrane lined pocket, can be detected later outside membranes in the portion of the cell which we call the "cytoplasmic matrix". For example, some virus particles or ferritin molecules are often taken up by cells through this process and, after uptake, can be seen, first in membrane-like pockets and later in the cytoplasmic matrix. Apparently, in these cases, the cell disassembles the membrane which was around the particle so that it comes to lie directly in the cytoplasm. The particle then faces the cell membrane from below, whereas before, it looked at it from above. Yet, at no moment was it going through the membrane itself, nor did it traverse any membrane pore, nor was the integrity of the membrane compromised at any moment.

These comments may bring to your attention some of the interesting properties of cell surfaces, which are capable of gross movements, of fusion, of recombination, of pinching off, and of specific binding at the surface.

*Journée du 29 juin 1967*

---

# L'INFORMATION EN BIOLOGIE

*1ère séance*
PRÉSIDENT H. C. LONGUET-HIGGINS F.R.S.

---

### S. OCHOA
Genetic Coding

### F. GROS
Remarques sur le Code Génétique

### A. FESSARD
Les Problèmes du Code Nerveux

### M. A. BOUMAN
Quantum Noise and Vision

Discussions

*Theoretical Physics and Biology,* © 1969, *North-Holland Publ. Co., Amsterdam*

# GENETIC CODING

## S. OCHOA

*University School of Medicine, New York, U.S.A.*

## Background

### Genetic Expression

The genetic information of living organisms and DNA viruses is contained in one of the two DNA strands. It is transcribed by transfer to a special messenger RNA through a DNA-directed synthesis of messenger. This RNA is an exact replica of the DNA strand that bears the genetic information and it programs the synthesis of proteins with features specified by the original DNA blueprint. Genetic and other experiments show that a linear sequence of deoxyribonucleotides in DNA specifies a corresponding sequence of ribonucleotides in messenger RNA and this in turn directs the synthesis of polypeptide chains with a unique sequence of amino acids. Thus, the four character (the four nucleotide bases) language of nucleic acids is translated into the twenty character (the twenty amino acids) language of the proteins.

### Genetic Code

Clearly a linear sequence of several nucleotide bases must specify each of the twenty amino acids. A doublet code (two bases for one amino acid) would be insufficient to specify twenty amino acids for it would have only $4^2 = 16$ doublets, but a triplet code, with $4^3 = 64$ triplets, would contain enough information. There is evidence that the genetic code is a triplet code. Moreover, the code is non-overlapping and commaless. This means that in a sequence ABCDEFGHI...XYZ, ABC would specify one amino acid, DEF another one, and so forth.

### Molecular Mechanism of Translation

Assembly of the polypeptide chains of proteins takes place on the ribosomes as they move along the messenger. The amino acids are taken to the site of synthesis in an activated form linked to special transfer RNA molecules each of which is specific for one of the twenty amino acids. Their alignment in a sequence prescribed by the nucleotide sequence of the messenger depends on the recognition of the various base triplets of the messenger (codons) by triplets of complementary base sequence (anticodons) of the amino acid-carrying RNA's. Codon-anticodon recognition and interaction are believed to occur through a Watson-Crick base pairing mechanism.

## Deciphering Of Genetic Code

Cell-free systems of protein synthesis can be obtained from bacteria, reticulo-cytes, and other cells. They consist of ribosomes and supernatant fluid. The latter contains, among other things, various soluble enzymes required for the process. When supplemented with transfer RNA's, ATP, GTP, and messenger RNA these systems synthesize proteins characteristic of a given messenger, e.g., viral coat proteins, when viral RNA messengers are used. Natural messengers can be replaced by synthetic polyribonucleotides of known base composition. Poly U directs the synthesis of polyphenylalanine, poly A that of polylysine. Random polynucleotides such as poly UG, direct the synthesis of peptides containing phenylalanine, cysteine, valine, lysine and trytophan among other amino acids. These observations opened the way for deciphering the genetic code for they showed that UUU and AAA are phenylalanine and lysine codons, respectively, whereas triplets containing 2 U and 1 G or 2 G and 1 U are codons for cysteine, valine, glycine and tryptophan.

With use of a variety of synthetic polynucleotides, the base composition of some fifty codons was established three years ago. Clearly there is more than one codon for each amino acid indicating redundancy of the genetic code. The base sequence of the individual codons has been recently established by studies of (a) the specific binding of aminoacyl-transfer RNA's to ribosomes in the presence of trinucleotides of known base sequence, and (b) the synthesis of polypeptides with artificial messenger polynucleotides of alternating base sequence. For example, poly $(UG)_n$ with UGU and GUG codons, promotes the synthesis of polypeptides containing strictly alternating cysteine and valine residues.

## Polarity of Translation

Since polynucleotide chains have a polarity, it was of interest to know the direction in which the messenger RNA chain is read during translation. This question has been answered with use of short polyadenylic acid messengers with a unique triplet of known base sequence at either end of the chain. For example, polynucleotides such as AAAAAA......AAAAAC direct the synthesis of poly-peptides of the structure lysine-lysine . . . . . lysine-asparagine with $NH_2$-terminal lysine and COOH-terminal asparagine (lysine codon, AAA; asparagine codon, AAC). Since polypeptides are assembled from the $NH_2$- through the COOH-terminal end, the above results unequivocally establish the direction of reading of the message. The same conclusion has been reached from experiments on hybridization of insertion-deletion mutants of T2 bacteriophage affecting the synthesis of the phage-induced enzyme lysozyme.

Other experiments indicate that the ribosomes start the reading of synthetic polynucleotide messengers at one end of the chain (the so-called 5'-end) and that a special codon (AUG) at this end sets the reading frame.

*Beginning and End of Translation*

Synthesis of natural polypeptide chains appears to require signals for initiating and terminating translation of individual cistrons. Initiation involves codons (e.g., AUG) that direct the introduction of N-formylmethionine as the first ($NH_2$-terminal) amino acid of the polypeptide chain. Two protein factors, normally associated with the ribosomes, are required for initiation of translation. Termination involves release of the completed peptide from the ribosomes preceded or followed by elimination of the transfer RNA attached to the peptide chain. The fact that certain mutations give rise to premature release of unfinished polypeptide chains suggests the existence of special codons for chain termination. Recent experiments with oligonucleotide messengers of specified base sequence have confirmed the prediction that UAA is a chain termination codon.

## Translation of polycistronic messengers

The main topic for discussion will center on some aspects of the translation of the RNA of RNA-containing bacteriophages. Its small size would seem to be advantageous for these studies.

*Translation of MS2 RNA*

MS2 RNA has a molecular weight of about $1 \times 10^6$ (about 3000 nucleotides) with potential information for the specification of 1000 amino acids if the full length of the RNA is translated.

Thus, this RNA could program the synthesis of from three (average mol. wt. $\simeq 30000$) to five (average mol. wt. $\simeq 20000$) virus-specific polypeptides. To date, genetic studies and studies of the viral proteins synthesized in *E. coli* cells infected with MS2 (or the related phages f2 or R17) both show the production of three viral polypeptides designated, in order of their electrophoretic mobility, as polypeptides I, II, and III. They have been characterized as viral RNA synthetase (s), a protein (maturation factor) required for the formation of viable phage particles, and viral coat protein, respectively. The possibility that one or more, as yet undetected, polypeptides are formed cannot be excluded at present.

*Control of Translation*

Initiation of translation may be controlled by the already mentioned ribosomal factors. They function, at least in part, by enhancing the binding of fromyl-methionyl-transfer RNA to ribosomes mediated by the AUG codon. Particularly intriguing are the mechanisms concerned with the relative rates of translation of the various genes of the polycitronic messenger. Considerably more coat than synthetase or maturation polypeptide is produced either *in vivo* or *in vitro*. In the latter case it has been found that the ratio of coat to RNA synthetase polypeptide formed is at least 15 : 1. This appears to be due to repression of translation of the synthetase cistron by the coat polypeptide. Moreover the coat also appears to repress translation of the maturation cistron. Other observations suggest that the maturation factor may enhance the translation of the coat cistron.

*Theoretical Physics and Biology,* © 1969, *North-Holland Publ. Co., Amsterdam*

## REMARQUES SUR LE CODE GENETIQUE

FRANÇOIS GROS

*Institut de Biologie Physico-Chimique,* 13 *Rue Pierre Curie, Paris 5ème*

On distingue aujourd'hui deux phases distinctes dans les mécanismes mis en oeuvre pour convertir l'information génétique présente dans le DNA en des structures protéiques définies. La première représente une transposition de la séquence constituant le texte génétique en une séquence chimique nouvelle, celle que représente l'alignement des nucléotides dans le RNA. Cette étape est généralement dénommée "transcription". La seconde est essentiellement la traduction en protéines spécifiques du texte représenté par les groupes de nucléotides qui forment la charpente du RNA. On appelle cette seconde étape la *traduction informative.*

### Transcription

Cette étape repose sur un mécanisme enzymatique relativement simple. Elle est "catalysée" en effet par une protéine oligomérique qui a pu être obtenue sous un très grand état de pureté chimique, l'enzyme RNA polymérase. Le système issu des cellules d'E. coli a été particulièrement bien étudié. Il s'agit d'une protéine dont le poids moléculaire est voisin de 500 000 daltons. Elle comporte probablement plusieurs protomères, lesquels sont séparemment inactifs. Les conditions qui régissent l'équilibre entre le polymère et ses sous-unités ne sont que partiellement connues. De fortes concentrations en cations monovalents, particulièrement en ions $K^+$ ou $Na^+$ favorisent la dissociation en sous-unités. Il y a des raisons de penser que l'enzyme se dissocie en ses sous-unités avant de se fixer sur les sites d'initiation.

Puisque le DNA est composé de deux chaînes complémentaires, la possibilité s'offrait a priori que chacune d'elle soit transcrite simultanément. Cette situation eut abouti à la production de deux chaines de RNA de polarités opposées. Il est cependant établi que seule l'une des deux chaînes est transcrite (transcription asymétrique). Le RNA messager est donc composé de molécules monofilaires. Ceci ne veut pas dire, toutefois, que des *portions distinctes* de chacune des deux chaînes de DNA ne puissent servir de modèle dans la synthèse des messagers formés. C'est ainsi que les RNA messagers formés au cours du développement d'un virus bactérien, le bactériophage lambda, sont transcrits à partir de territoires distincts présents sur chacune des deux chaînes complémentaires du DNA viral.

215

Ceci démontre aussi incidemment que le système transcripteur (la RNA polymérase) peut se "mouvoir" dans des directions opposées selon les phases considérées
de l'expression génétique.

Bien qu'il soit possible de synthétiser des RNA messagers *in vitro* en mélangeant
une préparation de DNA, de la polymérase et les substrats de la réaction (les
nucléosides triphosphates), on ne possède pas encore la certitude que les molécules
de RNA ainsi formées soient tout à fait conformes à celles des messagers naturels
correspondants. La seule preuve tangible sera la démonstration que du RNA
ainsi fabriqué *in vitro* peut induire à son tour la synthèse d'une protéine spécifique.
Les expériences récentes du biochimiste allemand Zillig paraissent indiquer que
le RNA produit par copiage "acellulaire" du DNA provenant du bactériophage
$T_4$ est capable d'orienter la synthèse d'un enzyme fonctionnel, la cytosine désami-
nase, lorsqu'il est incubé avec les extraits adéquats.

Une autre inconnue qui paraît subsister a trait à la nature chimique de la
punctuation qui, sur le DNA, démarque les signaux de départ et d'arrêt dans la
transcription d'un gène, lequel est, comme on sait, un segment de DNA de
longueur définie. L'idée qui prédomine actuellement est que la ponctuation
"départ" —ce que les généticiens dénomment à la suite des travaux de Jacob,
un *promoteur*—serait constituée par une séquence "réitérative" de pyrimidines,
telle que par exemple une succession de 3, 4, 5 résidus des bases thymine ou
désoxycytosine, ou plus (voir fig. 1, tirée d'une publication de Szybalski).

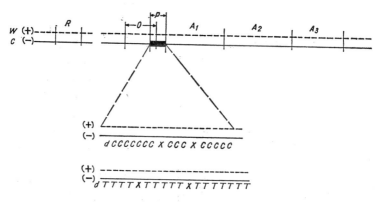

Fig. 1.

Deux arguments plaident en faveur de cette hypothèse: on sait en premier
lieu que, si la RNA polymérase opère la transcription chimique d'un DNA
préalablement dénaturé, en présence d'ATP comme seul substrat, de grandes
quantités de polyribo A (poly A) sont formées. Une telle réaction s'observe
également si l'on utilise comme matrice synthétique des oligonucléotides à chaînes
courtes (tels que (oligo dT) 5–7). Le poly A formé peut comporter plus de 50

résidus d'adénine car l'enzyme "dérape" au contact des courts segments formés par l'union des nucléotides à thymine et "incorpore ainsi un très grand nombre de résidus adényliques en liaisons covalentes.

Un autre argument en faveur du rôle des pyrimidines dans la ponctuation de "départ" réside dans l'observation aujourd'hui bien étayée selon laquelle les *premiers* résidus nuclétidiques incorporés *in vitro* dans une chaîne de RNA messager sont toujours des dérivés de *purines*. Selon l'origine du DNA servant de matrice, les purines incorporées en position proximale sont soit l'*adénine*, soit la *guanine*. Ceci est conforme à l'idée que les points d'attachement au niveau desquels l'enzyme commence (initie) sa réaction de copiage comportent des séquences de thymine ou de désoxycytosine (Hurwitz, Bremer et Konrad, etc.).

Egalement à l'appui de ces hypothèses viennent les expériences de Sheldrick et Szybalski. Ces auteurs ont observé que, si l'on dénature du DNA par la chaleur ou l'alcali, les chaînes monocaténaires peuvent facilement former des complexes avec le poly A ou le poly G, résultat suggérant l'existence de faisceaux localisés (clusters) de thymine ou de désoxycytosine à l'intérieur de l'une ou l'autre de ces chaînes.

Le fait que la polymérase manifeste beaucoup plus d'affinité pour les chaînes de DNA (ou de RNA) monofilaires que pour les structures bihélicoïdales (DNA natif, complexe poly A, poly U, etc.) suggère que les sites au niveau desquels débute la transcription (les promoteurs de Jacob) pourraient fort bien être des régions partiellement dénaturées de la double hélice (voir fig. 1).

Certains biochimistes vont jusqu'à penser que les résidus de purines présents dans la "boucle" ainsi créée existeraient peut-être à l'état "méthylé" afin d'empêcher un réappariement local (Stent).

Depuis les travaux désormais classiques de Hurwitz, Bremer et Konrad, et de Hayashi, on a tout lieu d'admettre qu'au fur et à mesure que la polymérase s'éloigne de son point d'origine (le "start signal" des auteurs anglais), la double hélice de DNA se déroule sur son passage mais se reforme immédiatement après. Le DNA ayant servi de matrice à la formation de RNA messager conserve en effet intactes ses propriétés de double hélice; il demeure "natif" tant en se référant aux critères d'hyperchromicité qu'aux tests biologiques (transformation). La chaîne de messager "naissant" est probablement expulsée par reformation progressive de la double hélice. Elle demeure toutefois rattachée au DNA par la molécule même de l'enzyme en déplacement.

On ignore la nature des signaux chimiques qui, sur le DNA, déterminent l'*arrêt* de la transcription, signaux présents à l'extrémité distale des gènes. Certains auteurs contestent même l'existence d'une ponctuation d'arrêt en tirant argument du fait que, si l'on change les conditions ioniques au cours de la transcription *in vitro*, on peut synthétiser des chaînes de RNA messagers démesurément longues. Il se pourrait donc que les chaînes de RNA soient normalement "découpées" par des éléments du système traducteur.

**Traduction informative:** *Synthèse des protéines*

La seconde étape dans le mécanisme chimique qui permet de transférer l'information du DNA aux protéines est la traduction du RNA messager. C'est ici qu'il convient de parler de décodage ou décryptage puisque nous voyons qu'une macromolécule (le RNA) dont la structure est constituée par l'assemblage de 4 types de lettres (les quatre bases du RNA) peut orienter l'assemblage colinéaire des protéines, lesquelles sont formées par l'union selon un enchaînement spécifique de 20 types distincts d'acides aminés.

Pour comprendre, même dans ses grandes lignes, le mécanisme de la traduction, il importe, d'une part, de saisir la nature même du code génétique (c'est-à-dire quels types de combinaisons de bases codent pour tel ou tel acide aminé) et de connaître les éléments de la machinerie cellulaire qui participent à la traduction.

*Considérations générales sur le code:* Puisqu'il existe 20 types distincts d'acides aminés et seulement 4 types principaux de bases, l'idée s'est de bonne heure imposée (Gamow) que le code ne pouvait être biunivoque. En d'autres termes, seules des combinaisons (des enchaînements) de bases selon une séquence définie, devaient exister en nombre égal ou supérieur à celui des acides aminés présents dans les protéines. Ce type de raisonnement devait conduire Crick et Brenner à rejeter un code de type binaire comme fort peu vraisemblable puisqu'il n'existe que 16 combinaisons de 4 lettres sous la forme de doublets. Un code à triplet semblait pouvoir convenir puisque le nombre de triplets possibles à partir de 4 types de lettres est de $4^3 = 64$, soit plus qu'il n'apparaissait nécessaire pour la représentation codée de chaque amino-acide. Aussi l'idée selon laquelle la plus petite unité d'information devait être un triplet composé de nucléotides a-t-elle été avancée sur des bases purement théoriques.

La démonstration de la nature ternaire du code a été fournie ultérieurement en étudiant, les protéines formées par recombinaison entre des mutants du bactériophage $T_4$, lesquels avaient été obtenus séparément sous l'influence d'un agent particulier, la proflavine, un composé acridinique. Mais, avant de décrire le principe de cette très importante expérience, il importe de faire quelques remarques générales sur les modalités qui peuvent a priori se présenter dans le déchiffrement d'un code à 3 lettres.

Si les gènes sont définis, en première approximation, comme les segments délimités de la double hélice d'ADN, on doit pouvoir se représenter les ARN messagers qui en dérivent comme une succession de nucléotides au sein de laquelle des groupes de 3 nucléotides définissent un code. Il existe cependant de nombreuses façons d'entrevoir la nature d'un tel code ou, si l'on préfère, plusieurs cadres de lecture.

Considérons une séquence composée de 4 types de lettres, soit:

$$B \ C \ D \ A \ A \ B \ C \ D \ B \ A \ A \ C \ldots$$

Plusieurs dispositions des lettres peuvent être imaginées. Celles-ci peuvent être *adjacentes* comme dans l'exemple ci-dessous où la lecture débute par la première lettre:

(1)          B C D     A A B     C D B     etc. . . .

Les triplets peuvent également être séparés par des lettres ne correspondant à aucun acide aminé précis mais servant par exemple de virgules:

(2)          B C D , A, A B C , D, B A A     etc. . . .

Dans les deux types de codes décrits ci-dessus, chaque groupe de 3 lettres ne code que pour un seul aminoacide *à la fois*. On peut cependant imaginer un groupage tel que chaque lettre d'un triplet donné puisse être considéré comme faisant également partie du triplet adjacent. C'est le cas par exemple des codes dits "chevauchants":

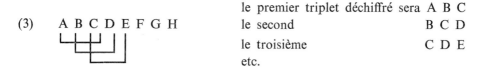

(3)     A B C D E F G H

le premier triplet déchiffré sera A B C
le second                              B C D
le troisième                          C D E
etc.

L'expérience démontre en fait que le code de lecture est bien un système à base 3 et précise en outre qu'il est *non* "chevauchant" et ne comporte pas de virgule. Il correspond donc à l'exemple N° 1.

Les évidences en faveur d'un code ternaire sont nombreuses et ne peuvent toutes être analysées ici.

Une des approches consiste à déterminer le nombre de nucléotides entrant dans la composition d'un ARN messager spécifique ainsi que le nombre de résidus "amino-acides" présents dans la protéine dont la synthèse est codée par ce messager. Il est malheureusement très difficile d'isoler d'une cellule l'ARN messager particulier à une protéine. En outre, ainsi que nous le verrons, bon nombre d'ARN messagers cellulaires contiennent de l'information pour plus d'une protéine spécifique (messagers polycistroniques). Les ARN messagers naturels qui se prêtent le mieux à l'isolement et dont le contenu informatif correspond le plus souvent à une seule protéine (ou à un petit nombre d'entre elles) sont les acides ribonucléiques des virus animaux ou végétaux ou encore ceux des bactériophages. C'est ainsi que les acides nucléiques d'un grand nombre de petits virus ont des poids moléculaires proches de $1,5 \times 10^6$ (ce qui correspond à environ 4 300 nucléotides) et ils contiennent fréquemment 3 types distincts de protéines dont les PM sont compris entre 30 000 et 40 000, c'est-à-dire qui renferment 400 résidus. Le rapport spécifique de codage doit donc être voisin de $4\,300/400 \times 3 = 3,5$. L'ARN du virus nécrotique des feuilles de tabac renferme 1200 nucléotides et code pour une seule protéine de capside, laquelle comporte 400 résidus. D'où un rapport de 3 pour la plus petite unité de codage, etc.

Des données beaucoup plus précises sur les valeurs du rapport de codage (coding ratio) proviennent cependant des expériences de Crick et Brenner sur le brouillage du texte génétique par les acridines. Nous reviendrons sur leur principe dans un instant.

Sans doute non moins directe est la démonstration apportée par Nirenberg et Leder que, lorsque l'on mélange *in vitro* des trinucléotides (généralement obtenus par hydrolyse enzymatique d'acides nucléiques suivie de fractionnements appropriés), des ribosomes et des ARN de transfert estérifiés par leurs aminoacides correspondants, on forme des complexes trinucléotides-ribosomes-ARN de transfert spécifiques. Les mono et dinucléotides ne permettent pas la "rétention spécifique" des ARN de transfert.

Ainsi, il s'avère que le cadre de lecture est bien élaboré par des groupes de 3 nucléotides.

Les expériences de Crick, Brenner et Watt-Tobins auxquelles nous nous sommes déjà référés plus haut démontrent en outre que la traduction des ARN messagers débute en un point précis, l'extrémité proximale de la chaîne, et chemine d'une manière unidirectionnelle. Le déchiffrement est réalisé par groupes de triplets et ne fait pas intervenir de virgules.

Ces expériences tirent parti de ce que les acridines causent facilement des mutations dans l'ADN des bactériophages en provoquant soit des micro-insertions (de bases nouvelles) dans la séquence d'un gène, soit au contraire en causant des micro-délétions.

Supposons un texte génétique dont la lecture par groupe de 3 lettres ait une signification précise. Pour simplifier, nous utiliserons ici des combinaisons de mots:

T O N A M I E S T I C I

Lu correctement, en commençant par le début, ce texte signifie "TON AMI EST ICI". Si, à la suite d'une mutation, nous "insérons" par exemple une lettre supplémentaire dans ce texte et que nous continuions à le déchiffrer sur le mode précédent, il est évident que nous faussons le cadre de lecture:

↓<br>T O R N A M I E S T I C I

donne un message brouillé

TOR NAM IES TIC I ...

Si nous affectons d'un signe "+" une telle mutation qui allonge d'une lettre le texte génétique, il devient clair qu'il existe seulement deux moyens pour rétablir la signification du moins partielle du message:

– l'un est d'introduire une mutation de signe—c'est-à-dire déléter l'une des lettres dans le texte faussé:

$$\overset{+}{\downarrow} \qquad \overset{-}{\downarrow}$$

$$\text{T O R N A M E S T I C I}$$

Dans un tel message, la fin du texte redevient compréhensible.

– l'autre consiste à introduire deux mutations supplémentaires de même polarité (+) que la précédente:

$$\overset{+}{\downarrow} \qquad \overset{+}{\downarrow} \qquad \overset{+}{\downarrow}$$

$$\text{T O R N A U M O I E S T I C I}$$

Il est clair que, si une telle situation peut être génétiquement vérifiée, elle apporte la preuve que l'information est déchiffrée par groupe de 3 nucléotides, non séparés par des virgules, la traduction étant unidirectionnelle.

Ce but a été atteint par les auteurs anglais en étudiant les effets mutationnels, simples ou combinés, de la proflavine sur la synthèse d'un enzyme spécifique, le lysozyme, par le gène correspondant présent dans l'ADN du bactériophage T₄. Par le jeu des recombinaisons génétiques, on peut en effet introduire diverses mutations de signes + (insertion d'une base) ou — (délétion d'une base) dans le gène considéré et définir les situations qui *restaurent* l'activité de l'enzyme correspondant.

Les conclusions de Crick et ses collaborateurs ont d'ailleurs été étayées peu après grâce à l'analyse de la *séquence* peptidique du lysozyme normal, et des pseudo-lysozymes dont l'activité avait ainsi été rétablie. Il a été vérifié que, dans le cas d'un pseudo-enzyme de type (+ —), l'un des peptides de la chaîne renfermait un amino acide *de plus* que le peptide correspondant dans l'enzyme sauvage, tandis qu'un peptide plus distal était déficient en un amino acide par rapport à l'homologue normal.

### Nature du code génétique

Il n'a pas fallu beaucoup plus de 5 années pour établir la nature chimique du code génétique. Notre propos n'est pas de décrire ici les diverses phases des recherches qui ont permis cette remarquable réalisation de la Biologie Moléculaire.

Il est toutefois intéressant de rappeler quelles ont été les grandes voies d'approche.

Celle qui devait à coup sûr ouvrir le champ à ces investigations est sans nul doute l'observation due à Nirenberg et Matthaei, selon laquelle il est possible d'orienter *in vitro* la synthèse de polypeptides artificiels en substituant aux ARN messagers naturels des hauts polymères dont les biochimistes savent, depuis les travaux d'Ochoa et Grunberg-Manago, réaliser la synthèse. De tels polymères peuvent en effet être obtenus sous l'influence de la polynucléotide phosphorylase, un enzyme capable de condenser selon un enchaînement statistique n'importe quel nucléotide, que celui-ci soit d'un type donné ou qu'il s'agisse d'un mélange de plusieurs types. On obtient ainsi toute une gamme d'homopolymères et de copolymères contenant 1, 2 et jusqu'à 4 types de bases.

Lorsque ces polymères synthétiques sont mis en présence de ribosomes purifiés extraits de la bactérie E. coli, et que le mélange réactionnel comporte en outre les aminoacides appropriés ainsi que divers autres effecteurs, on assiste à une synthèse d'homo ou de copolypeptides dont la composition en aminoacides dépend, fait remarquable, de celle des polynucléotides ajoutés.

Cette observation et son exploitation systématique par les laboratoires de Nirenberg et de Ochoa devait rapidement conduire à élucider la *composition* de presque tous les codons.

Restait à établir leur séquence. Il existe en effet 9 combinaisons de 3 lettres. Ici, deux méthodes ont principalement été utilisées. L'une déjà évoquée ci-dessus a consisté à synthétiser chimiquement ou à préparer par digestion enzymatique la quasi-totalité des trinucléotides de *séquences* connues et à identifier avec chacun d'entre eux la nature des amino-acyles-t-ARN dont ils permettent l'attachement aux ribosomes. Ceci peut être réalisé très commodément grâce à l'emploi de membranes millipores qui se laissent normalement traverser par les ARN adaptateurs, estérifiés ou non, mais qui retiennent ces derniers si un appariement correct est réalisé à la surface du ribosome entre le trinucléotide et l'anticodon spécifique du t-ARN (voir ci-après). Il est dès lors facile de préciser parmi les différents types de trinucléotides comportant une séquence de 3 bases données quel est celui, ou ceux, dont l'interaction avec un aminoacyl-t-ARN en permet (ou permettent) la fixation sur les ribosomes.

Une autre méthode a consisté à utiliser des copolynucléotides dont la séquence n'est pas distribuée au hasard (comme c'est le cas dans les copolymères formés par la polynucléotide phosphorylase) mais répond au contraire à une alternance définie telle que: A U A U A U A U etc.... ou A A C A A C A A C. Il est clair en effet que, si de tels polymères orientent la synthèse de copolypeptides *in vitro*, on peut, en déterminant l'alternance des aminoacides, définir du même coup la *séquence* des codons correspondants.

Toutes ces études ont en définitive abouti à préciser la signification des 64 trinucléotides pouvant normalement exister. Ceci a conduit à un certain nombre de conclusions générales.

1. – La première—et sans doute la plus importante—est qu'il existe toujours plus d'un triplet (encore appelé codon) capable de spécifier un aminoacide donné. En d'autres termes, à part trois exceptions que nous envisagerons dans un instant, *tous* les trinucléotides ont un sens dans l'alphabet nucléique. Il n'y a pas de correspondance biunivoque entre codons et aminoacides. Pour employer une terminologie consacrée, on dit encore que le code est dégénéré. L'importance de ce fait dans le cadre des mécanismes moléculaires de l'évolution est considérable:

qu'il existe plusieurs systèmes de codage pour un même aminoacide rend possible l'existence, au sein d'espèces dont les matériels génétiques (ADN) sont de compositions très différentes, de protéines-enzymes possédant des structures primaires suffisamment voisines pour y jouer le *même rôle* biologique. La dégénérescence du code permet donc à la cellule d'exercer une sorte d'*effet tampon* contre les variations que l'évolution physiologique et les mutations ne manqueraient pas d'apporter à la structure du matériel génétique.

Bien qu'il existe souvent deux ou trois codons distincts, voire davantage, pour un même aminoacide, les codons "homonymes" ont une particularité commune très importante: les deux premiers des nucléotides qui les constituent sont les mêmes. Seuls diffèrent les nucléotides situés en *troisième* position dans la séquence.

On était en droit de se demander s'il existe un ARN de transfert possédant une structure appropriée à *chacun des* 64 codons possibles ou si l'ARN de transfert correspondant à un aminoacide défini pouvait servir d'adaptateur sinon pour tous, du moins pour plusieurs, les codons homonymiques. Cette dernière alternative a été suggérée par Crick dans sa théorie dite du "Wobble", d'après laquelle la specificité de l'appariement du codon et de l'anticodon du t-ARN repose surtout sur la séquence des *deux premières lettres du triplet* (ou des deux dernières dans l'anticodon, puisque l'appariement messager-t-ARN se fait de manière antiparallèle). En d'autres termes, un certain degré de "jeu" (Wobble) est permis dans les interactions entre le troisième nucléotide présent dans le triplet du messager et le premier appartenant à la séquence de l'anticodon. Le code a donc une certaine *flexibilité* qui, jointe à son caractère de dégénérescence, le prémunit contre des changements mutationnels qui à tout coup risqueraient d'être léthaux.

2. – Trois codons n'ont pas de signification définie dans le langage protéique. Il s'agit des triplets UAG, UAA et UGA. Ces triplets jouent le rôle de signes de ponctuation dans la traduction. Ce rôle apparaît des plus essentiels si l'on réalise que bien souvent, du moins chez les microorganismes ou leurs virus, plusieurs gènes *contigus* sont transcrits en un ARN messager *unique* (transcription polygénique). Il est donc très fréquent qu'une même chaîne d'ARN contienne l'information pour plusieurs protéines distinctes. Cette organisation particulière du matériel génétique est liée en fait au mécanisme de la régulation (Jacob et Monod). Il est en effet plus facile de contrôler l'activité des gènes impliqués dans une même fonction physiologique si ces gènes sont exprimés d'une manière coordonnée, ce que rend possible leur "cotranscription".

Pour que le système de lecture (voir plus loin) puisse traduire un messager unique en plusieurs protéines distinctes, il faut nécessairement une ponctuation qui délimite le début et la fin de la lecture. Nous dirons quelques mots ci-après des mécanismes d'initiation.

En ce qui concerne les signaux de terminaison de lecture, notons que les 3 codons UAA, UAG et UGA ont précisément ce rôle. Il semble cependant que dans les conditions normales l'arrêt de lecture ne soit signalé que par le seul triplet *UAA;* UAG et UGA peuvent apparaître au sein d'un ARN messager *sous l'effet d'une mutation*, ces deux triplets *interrompant* alors artificiellement la propagation des ribosomes et leur présence se manifestant par la synthèse de "fragments" protéiques. Les codons UAA et UAG sont souvent dénommés "ochre" et "amber" par les auteurs anglo-saxons.

3. – Le code est universel. Un trinucléotide donné codera pour le même amino acide quelque soit l'organisme envisagé. Ceci peut être aisément démontré par l'emploi de systèmes acellulaires.

La signification du code peut cependant être perturbée de deux façons, soit par *mutations*, soit sous l'effet d'agents chimiques ou physiques qui modifient la "reconnaissance" d'un codon donné par les ribosomes et peuvent ainsi en changer le sens.

Une mutation est un changement dans la séquence chimique du texte génétique ADN. Certaines de ces mutations entraînent une délétion d'une région plus ou moins importante de l'ADN. Le plus souvent, il s'agit simplement du *remplacement* de l'une des bases d'un triplet par une base différente, perturbation apparaissant lors de la réplication. Lorsque l'on examine le tableau de code génétique, il est clair que, par le simple changement d'un nucléotide au sein d'un triplet, on peut modifier complètement le sens de ce triplet.

C'est un fait remarquable, lequel apporte à l'établissement du code une confirmation essentielle que toutes les substitutions d'aminoacides qui s'observent dans une chaîne à la suite *d'une seule mutation*, (mutation ponctuelle) s'expliquent parfaitement par le *remplacement d'une seule des lettres dans un codon*.

Cette conclusion est étayée par les remarquables études effectuées par l'école de Yanofsky sur les structures chimiques des tryptophane synthétases d'Escherichia coli, altérées par des mutations distinctes ainsi que par celles de Whitman sur les propriétés des protéines génétiquement modifiées que l'on isole des mutants du virus de la mosaïque du tabac.

Un codon de séquence définie peut cependant, nous l'avons souligné, changer de signification selon le contexte biochimique. C'est le phénomène d'*ambiguité* ou d'*infidélité* dans la traduction. De telles ambiguités peuvent être créées par des changements thermiques, des variations de pH, la présence de certains agents dont l'activité repose sur une modification des propriétés des ribosomes. C'est le cas par exemple de certains antibiotiques, telles la streptomycine ou la néomycine, qui, ajoutés à faibles concentrations dans des extraits renfermant des ARN messagers artificiels, peuvent modifier la composition des polypeptides formés. L'ambiguité "spontanée", c'est-à-dire le fait qu'un codon défini puisse être traduit d'une manière aberrante au sein d'une cellule, par ailleurs normale, est un phénomène très rare, encore que décelable (Von Ehrenstein).

*Comment s'effectue la traduction des codons du messager en protéines?:* Telle est la question que nous envisagerons en dernier lieu. L'hypothèse a priori la plus simple voudrait qu'un aminoacide soit *directement* attaché en regard de chaque codon présent dans la séquence d'un ARN messager.

Une telle possibilité, pour économique qu'elle puisse apparaître à première vue, se heurte d'emblée à maintes objections théoriques. En effet, ce qui distingue chaque aminoacide de son voisin, c'est la nature du groupe latéral ou radical associé au $(R) - CH \overset{\displaystyle NH_2}{\underset{\displaystyle CO_2H}{<}}$. Ces radicaux ont des structures chimiques extrêmement différentes (noyaux aromatiques, aliphatiques ramifiés ou non, etc....). Par ailleurs, rien dans la structure linéaire de l'ARN messager ne constitue un élément de stéréospécificité suffisant pour "reconnaître" chacun de ces radicaux. Même si tel était bien le cas, le positionnement des amino acides au voisinage *direct* des codons disposerait ces aminoacides à des *distances* variables de la matrice et rendrait impossible la mise en jeu des enzymes nécessaires au cimentage de la liaison peptidique.

Ces considérations ont depuis plusieurs années conduit F. Crick à prédire l'existence d'adapteurs spécifiques, eux-mêmes de nature nucléique, capables de "placer" chaque aminoacide en regard du codon correspondant. On sait que, peu de temps après, Hoagland a découvert l'existence de ces ARN adaptateurs encore appelés ARN de transfert ou en abrégé "t-ARN".

Ces t-ARN ont en quelque sorte une vertu double qui en fait des éléments clés dans le déchiffrement du code. Ce sont de petits polynucléotides (ils renferment environ 80 nucléotides et la structure complète de maints d'entre eux est connue). Ils abritent à l'intérieur de leur chaîne une séquence trinucléotidique, l'*anticodon*,

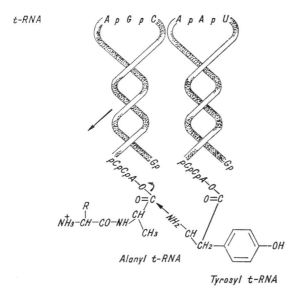

Fig. 2.

laquelle est capable de s'apparier directement à un triplet spécifique du messager et ils sont porteurs à leur extrémité d'une séquence particulière (le résidu CCA) à laquelle se trouve rattaché l'aminoacide (fig. 2).

Les premiers stades dans la synthèse des protéines n'ont donc d'autre but, on s'en doute, que d'activer thermodynamiquement les aminoacides pour en permettre l'attachement à l'extrémité CCA. Cet attachement est, bien entendu, catalysé par un enzyme spécifique (il en existe un pour chaque aminoacide). Une fois fixés à leurs t-ARN adaptateurs, on pourrait penser que les aminoacides sont disposés le long de la chaîne du messager et peuvent y être condensés en polypeptide. A peu de chose près, c'est en gros ce qui se passe mais il faut apporter quelques touches supplémentaires à ce tableau. En effet, bien que l'interaction primaire des amino-acyl-t-ARN et du messager soit assurée par un appariement entre les deux triplets complémentaires, les liens hydrogènes qui en sont cause sont si fragiles qu'un tel complexe fondrait très rapidement s'il ne se trouvait renforcé par d'autres facteurs. La cellule doit donc mettre en oeuvre un dispositif spécial qui assure le renforcement de ces liaisons. C'est ici qu'interviennent les ribosomes, ces granules cytoplasmiques qui renferment la quasi totalité de l'ARN stable de la cellule.

Les ribosomes sont en effet des dimères, en ce sens qu'ils comportent une grande sous-unité (50 S) et une plus petite (30 S), l'ensemble sédimentant à 70 S. Leur rôle essentiel est de renforcer transitoirement l'association du t-ARN et du messager et de contenir les enzymes cimentant le lien peptidique entre deux aminoacides voisins. D'une part, en effet, ils se fixent par la sous-unité 30 S à l'ARN messager (Takanami). D'autre part, leur sous-unité 50 S contient deux

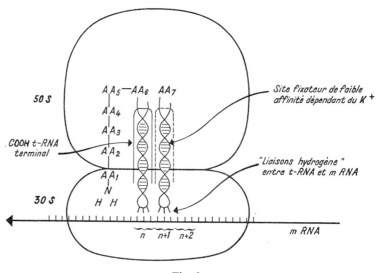

Fig. 3.

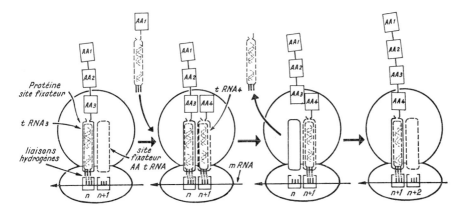

*Croissance graduelle de la chaîne polypeptidique*

Fig. 4.

logettes spécifiques et contigües dans lesquelles peuvent s'insérer les amino-acyl-t-ARN correspondant à deux codons adjacents (voir fig. 3).

Une peptide synthétase (Traut et Monroë) qui fait partie intégrante de l'unité 50 S scelle les deux amino-acyles voisins par un lien peptidique (fig. 4). Un dipeptidyl-t-ARN est alors formé. L'étape suivante peut être définie comme une "contraction" du ribosome qui déplace le dipeptidyl-t-ARN de la logette où il se trouvait dans celle renfermant le t-ARN libre. Ce dernier est expulsé et un nouvel amino-acyl-t-ARN vient occuper la place laissée libre par le dipeptidyl-t-ARN. A chaque contraction (laquelle est catalysée par un enzyme) la translocase, et par un cofacteur allostérique, le GTP, le ribosome se meut d'une distance équivalente à un triplet du messager: ainsi se trouve "tissé" le long ruban polypeptidique qui doit constituer la protéine finale. Au fur et à mesure que les premiers ribosomes attachés à la chaîne du messager se sont suffisamment déplacés vers les régions "distales", d'autres ribosomes s'attachent en début de chaîne et commencent un nouveau cycle de synthèse. A l'état de régime, une même chaîne de messager peut donc être "parcourue" par un train de ribosomes se déplaçant dans une même direction et tissant, chacun pour son propre compte, une chaîne polypeptidique. On donne le nom de polyribosomes (ou polysomes) à ces formations, lesquelles sont facilement décelables au microscope électronique. Une fois arrivé au niveau d'un codon UAA, signal de fin de lecture, le ribosome "relargue" le polypeptidyb-t-ARN. Cette opération est préalablement catalysée par un facteur protéique qui vient d'être identifié par Capecchi. Une hydrolase spécifique (Cuzin et Chapeville) clive alors le lien entre le t-ARN et le polypeptide tandis que celui-ci revêt sa configuration secondaire et tertiaire.

Dans le processus d'élongation du polypeptide, on sait que le premier amino acide incorporé dans la chaîne est celui qui correspond au résidu α-aminé présent dans la chaîne protéique terminale, le dernier correspondant au résidu carboxyle libre de la protéine (Dintzis-Goldstein).

Il semble que chaque messager naturel débute par une séquence particulière qui servirait de point d'attache au premier ribosome, lequel serait d'abord fixé à l'état de sous-unité 30 S (Nomura). Cet attachement paraît catalysé par un facteur protéique spécial, dénommé facteur C, dont Revel et Gros ont récemment démontré l'existence. La séquence de fixation ne serait pas à proprement parler "traduite" en protéines. La traduction ne commencerait qu'un peu plus loin vers l'intérieur de la chaîne. Des précisions ont été apportées sur ce dernier point:

a – le premier codon traduit (chez les microorganismes) est le codon AUG (Capecchi; Khorana);

b – le premier aminoacyl-t-ARN incorporé est un dérivé découvert par Marcker et Sanger: le formyl-méthionyl-t-ARN. Ainsi, *toutes* les chaînes protéiques chez les bactéries (et chez de nombreux virus) débutent par le résidu formyl-méthionine. Ce résidu est ensuite clivé par une peptidase spécifique (Waller-Capecchi). Ainsi, le mécanisme qui assure l'initiation des synthèses protéiques revêt-il un caractère de généralité tout à fait remarquable. C'est au niveau de la sous-unité 30 S qu'a lieu la fixation du formyl-méthionyl-t-ARN, et elle est catalysée par deux facteurs protéiques particuliers (appelés $F_1$ et $F_2$) (Wabba, Ochoa, Brawerman-Revel et Gros).

Il existe en somme une polarité stricte dans le processus traducteur qui aboutit à la synthèse des protéines. De remarquables travaux dus à Ochoa, d'une part, ainsi qu'au groupe de Streisinger, d'autre part, ont clairement établi que l'extrémité du messager codant pour le résidu $\alpha$-aminé de la chaîne polypeptidique était la terminaison 5′-phosphate de l'ARN. Ainsi la traduction progresse selon une polarité $5'p \rightarrow 3'OH$.

Si l'on songe que la transcription (synthèse du messager) se déroule également dans la direction $5'P \rightarrow 3'OH$, on voit du même coup que les deux phases de l'expression génétique (transcription et traduction) évoluent selon la même polarité. Tout porte d'ailleurs à croire que ces deux étapes sont, dans la cellule, intimement couplées (Stent-Gros). Ce sont les ribosomes qui, en s'attachant aux premières séquences transcrites du messager, détacheraient cet ARN de sa matrice génétique. Ce point est certainement d'une grande importance dans les phénomènes cellulaires qui ajustent le taux de formation des messagers au taux de synthèse des protéines.

## Bibliographie

Les références se rapportant à ce problème sont trop nombreuses pour pouvoir figurer *in extenso*.

Le lecteur pourra cependant se reporter aux revues suivantes:

[1]  J. Monod, F. Jacob et F. Gros, Structural and rate determining factors in the biosynthesis of adaptive enzymes, *Biochem. Soc. Symp.* **21** (1961) 104.
[2]  J. D. Watson, *Molecular biology of the gene* (Benjamin, 1965).
[3]  V. Ingram, *The biosynthesis of macromolecules* (Benjamin, 1965).
[4]  M. Grunberg-Manago, Le code génétique, *Ann. Pharmac. Franc.* **25** (1966) 147.
[5]  J. Tavlitski, Le code génétique, *Sciences* **34** (1964) 19 et **35** (1965) 14.
[6]  *The genetic code*, Symposia on quantitative biology, Vol. **31** (1966), Cold Spring Harbor Laboratory of Quantitative Biology.

*Theoretical Physics and Biology,* © 1969, *North-Holland Publ. Co., Amsterdam*

# LES PROBLEMES DU CODE NERVEUX

ALFRED FESSARD

*Laboratoire de Neurophysiologie Générale, Institut Marey, Paris, France*

Les rapports qui s'établissent entre la physique – théorique ou expérimentale – et la physiologie du systeme nerveux, ou neurophysiologie, sont de deux ordres. On peut distinguer ces deux ordres de la même façon que l'on ne confond pas, si l'on y prête attention, les termes *fonctionnement* et *fonction*. Je m'explique:

Le système nerveux, en particulier le cerveau des vertébrés supérieurs – qui nous intéresse au premier chef en ce qu'il nous aide à comprendre le cerveau de l'homme – est sans doute le produit organisé le plus complexe de la nature. Nous admettons qu'il obéit comme tout autre organe dans son fonctionnement, aux strictes lois de la physique et de la chimie, et le neurophysiologiste cherche tout d'abord à expliciter ces lois de fonctionnement. Lorsqu'il attaque les problèmes en physicien – on le dit alors biophysicien – les concepts et les grandeurs qu'il manie sont ceux des propriétés physiques de la matière et des divers aspects de l'énergie, avec leurs deux composantes – de tension et de quantité – qu'il s'agisse de chaleur, d'électricité ou d'énergie mécanique. On les rencontre à différentes échelles, des molécules constitutives au neurone, et de celui-ci au système nerveux tout entier. Un des mécanismes nerveux fondamentaux est celui que révèle et qu'explique assez bien aujourd'hui la biophysique des membranes cellulaires, ou membranes "plasmiques", avec leurs propriétés électriques, passives et actives, parmi lesquelles celle qui se manifeste sous la forme d'une brève impulsion, le "potentiel d'action" (en anglais "nerve impulse"). Celui-ci, selon nous, pourrait être appelé en français neuroquantum ou neuropulsion. Il est cependant toujours désigné entre spécialistes par le terme anglais imagé de "spike".

Bien entendu, la seconde tâche du neurophysiologiste – c'est même la plus proprement "physiologique" – consiste à établir quelles sont les *fonctions* de ce système nerveux, c'est-à-dire les opérations auxquelles il participe majoritairement avec d'autres organes (les organes des sens, les muscles, les glandes endocrines, etc...) pour accomplir certaines catégories d'actions auxquelles nous reconnaissons (conventionnellement, mais avec beaucoup de bonnes raisons!) une certaine utilité biologique, utilité (préférons ce mot à "finalité") pour la survie de l'organisme total, pour son adaptation aux conditions changeantes du milieu, pour sa croissance et, éventuellement sa reproduction. Les fonctions accomplies par le système nerveux sont des performances qui, lorsque tout va bien, sont remarquablement ordonnées. En langage moderne, nous dirions qu'elles exécutent un *programme*; les lignes directrices de ce programme doivent être inscrites quelque part dans les centres nerveux; elles doivent y être *codées*. Nous savons d'autre part que les opérations correspondantes, qu'elles soient appliquées à l'environnement ou cachées dans l'organisme, sont souvent déclenchées, et généralement guidées, corrigées, ajustées, grâce à des *informations* captées au dehors ou au dedans, au cours même de l'action, puis *communiquées* aux centres de commande, après certains "traitements"; ce qui suppose également un code. Nous reconnaissons-là le langage de l'informatique et c'est en cela que l'on peut parler de la deuxième forme sous laquelle se présentent aujourd'hui les relations entre physiciens ou ingénieurs d'une part, biologistes ou plus spécialement neurophysiologistes de l'autre. Le neurophysiologiste est rarement capable de suivre le physicien ou l'ingénieur dans les développements mathématiques qui découlent de la théorie de l'information — selon Shannon, selon Hartley, selon Fischer, mais il comprend parfaitement que dans les opérations *fonctionnelles* dont il s'occupe, ce sont les structures formelles que l'on peut dégager des séquences et des ensembles de signaux, et celles des réseaux nerveux où ces derniers se propagent, qui sont importantes à élucider, et non leur support physico-chimique. C'est là du moins une façon claire et efficace de diviser la difficulté afin d'essayer de la mieux résoudre, comme aurait dit Descartes.

Il reste à savoir si la direction prise par les mathématiciens physiciens ou ingénieurs dans leurs travaux, qui furent à l'origine inspirés par les problèmes pratiques posés par la communication télégraphique des messages, le calcul automatique, et l'automatisation dans les machines, est bien ce qui convient pour faire progresser la neurophysiologie fonctionnelle. De part et d'autre on entend bien dire que le cerveau est analogue à un ordinateur. Grey Walter et A. M. Uttley [1] prétendent qu'il doit essentiellement se comporter comme "un calculateur de probabilités conditionnelles", cela sans trop nous dire comment il le fait; tandis que d'autres auteurs sont plus circonspects, tels G. P. Moore, D. H. Perkel et J. P. Segundo [2] pour lesquels: "At best, in fact, the concepts and constructs of information theory are metaphors when applied to the nervous system." Nous n'entreprendrons pas ici d'approfondir ce sujet; nous nous contenterons d'examiner une question préalable et fondamentale, celle de la possibilité de définir un (éventuellement plusieurs) *code nerveux*, sans la connaissance duquel il serait vraiment inutile de parler d'information.

Les Neurophysiologistes doivent donc commencer par chercher quels peuvent être les symboles de ce code, comment ils se trouvent incarnés dans ce bel édifice de connexions que constitue le Système nerveux central (SNC), comment ils traduisent les phénomènes avec lesquels ce SNC se trouve confronté, comment les messages résultants sont transmis et décodés. Disons tout de suite que ces problèmes sont loin d'être résolus et que dans ce secteur du code neurophysiologique on est nettement moins avancé que dans celui du code génétique. Il y a à cela plusieurs raisons. L'une d'elles est que le code génétique sert à donner des "ordres", ceux impliqués dans le programme du génome, mais que le génome n'en reçoit pas; tandis que le code nerveux, s'il est utilisé aussi à transcrire des messages porteurs d'ordres (information *efférente*) doit d'autre part refléter la structure formelle des atteintes excitatrices qui ébranlent le SNC (information *afférente*). Autrement dit, pour employer la terminologie en cours, tout le SNC a une *entrée* et une *sortie*, ou plutôt plusieurs entrées et plusieurs sorties, et fait ainsi figure de *canal de communication*, du moins en première approximation (fig. 1); d'un côté un champ de stimulations (S), de l'autre un champ de réponses (R) dont l'ensemble constitue le comportement. Des effects rétroactifs (ligne pointillée) compliquent généralement le schéma. Nous les négligerons pour l'instant, bien que, finalement, ils soient à la base de phénomènes caractéristiques du fonctionnement nerveux. Nous ne pouvons cependant tout expliquer ici.

Fig. 1. SNC représente par une "boîte noire" le système nerveux central, S un ensemble de stimulations, R un ensemble de réponses. En pointillé, une boucle de rétroaction.

Une autre raison de notre retard à bien comprendre le code nerveux vient de la diversité des éléments possibles du code — il y en a trop — sans que nous sachions bien reconnaître lesquels jouent réellement un rôle ni quand ils jouent ce rôle, même si tous les états discernables identifiés peuvent le cas échéant être comptés comme *bits* dans l'accomplissement d'une opération déterminée (hypothèse optimiste selon laquelle le SNC serait capable d'utiliser tout indice informationnel disponible, du moins dans la mesure où celui-ci émerge du "bruit"). Ces éléments ou états discernables — et éventuellement discernés — nous les distinguons soit dans l'Univers spatial des structures (neurones et ensembles de neurones, garnitures de synapses, et peut-être macromolécules spécifiques), soit dans l'univers temporel des signaux (modulation des niveaux du potentiel membranaire local et trains de spikes propagés); mais nous ne savons pas toujours si ce que nous distinguons en tant qu'"observateurs" correspond à ce qu'un SNC en activité distingue effectivement, en tant qu'"acteur", pour accomplir ses fonctions. Devons-nous faire le compte, un par un, des neurones activés, ou des "spikes" émis, ou considérer plutôt leurs ensembles spatiaux, temporels ou spatio-temporels? Devons-nous distinguer ces ensembles en tenant compte de leurs effectifs, de leurs densités spatiales, ou temporelles, ou encore de leurs configurations globales d'éléments (*patterns*)? Une certaine "épaisseur" d'espace-temps, plutôt qu'une séparation absolue des deux cadres — spatial et temporel — n'est-elle pas à considérer comme un meilleur découpage de la réalité nerveuse, du moins en matière d'information?

On voit quelle est l'ampleur de nos incertitudes. En outre, les symboles une fois définis, le niveau de "bruit" dans chaque cas ayant été évalué, nous aurions encore à déterminer les probabilités de mise en jeu de chaque symbole si nous voulions estimer la valeur moyenne du flux d'information mobilisé dans une opération donnée. En réalité, selon les opérations, nous avons de bonnes raisons de penser que c'est tantôt un mode de codage, tantôt un autre, qui entre en jeu, et, dans une opération donnée, que plusieurs codes peuvent être simultanément ou successivement utilisés.

Si l'on considère ce qui se passe dans une seule fibre de nerf de Mammifère, fibre sensitive tactile par exemple, qui fait synapse au niveau spinal, et si l'on suppose que chaque neuropulsion est isolément capable de produire une opération sélective (soit un bit par pulsion), le débit maximum possible de la ligne est de l'ordre de 1000 bits par seconde. En fait, il dépasse rarement deux ou trois centaines par seconde. L'inverse de la période 1/T est probablement un symbole *élémentaire* plus significatif, car nous savons que le SNC se sert souvent de modulations d'intervalles entre "spikes" pour contrôler ses opérations synaptiques. D. MacKay et W. Culloch [3] ont calculé dans ce cas, et pour un faible bruit temporel moyen de 0.05 ms, le débit maximum de la fibre, soit 2900 bits/s. La valeur moyenne la plus commune des fréquences de pulsion étant de l'ordre de 100/s, on tire de la formule un débit *moyen* d'environ 1500 bits/s, soit 15 bits par influx, pour le nombre d'opérations sélectives possibles, mais non nécessairement réalisées au cours des décodages centraux.

Dans la réalité, un grand nombre de fibres sont presque toujours simultanément actives, qu'il s'agisse d'un transport d'informations afférentes d'origine sensorielle, de communications inter-centrales à l'intérieur du système, ou finalement d'une émission de signaux efférents, engendrant les réponses. La connaissance du plan de connexions est donc indispensable. A cet égard, beaucoup de faits nouveaux ont été constatés au cours des vingt dernières années, grâce à l'association de l'électrophysiologie à la neuroanatomie, et, sur le plan technique, grâce à l'emploi de microélectrodes ultra-fines, qui ont permis de dresser un bilan approximatif des types de messages reçus par *un seul neurone cérébral*, au sein de populations qui peuvent atteindre une densité de 40.000 cellules par millimètre cube. Cette prouesse technique est aujourd'hui devenue routine dans les laboratoires spécialisés. Un autre progrès a consisté à ne plus utiliser d'anesthésiques : ils dénaturent toujours le fonctionnement. On opère alors en deux temps : implantation d'électrodes sur l'animal endormi, puis recueil des signaux, élémentaires ou non, sur l'animal éveillé, en l'absence de contraintes douloureuses.

Un des résultats les plus importants de ce type d'expériences, faites en général sur des chats ou des singes, est que beaucoup de neurones, la majorité certainement, reçoivent des informations de plusieurs origines, selon les assortiments les plus variés. Notre laboratoire a largement contribué à établir ce fait fondamental. Cette *convergence d'informations* sur un même neurone va de pair avec un système de structure hiérarchique des réseaux nerveux afférents et est vraisemblablement à la base d'un code selon lequel chaque neurone apparaît comme un "indicateur spécifique" de l'ensemble de ses afférences, ou tout au moins de certains sous-ensembles de celles-ci. L'ingénieur informaticien anglais A. M. Uttley a fait d'intéressantes spéculations théoriques à ce sujet (fig. 2, flèche de gauche).

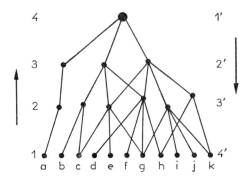

Fig. 2.  Schéma symbolisant la structure hiérarchique du système nerveux central, a, b, . . . , k, . . . représentent ou bien des *entrées* (organes récepteurs) si l'on considère les voies afférentes (flèche ascendante), ou bien des *sorties* (organes d'action), si l'on considère les voies efférentes (flèche descendante).

Une structure hiérarchique analogue, mais fonctionnant en sens inverse et distribuant par conséquent de façon *divergente* l'information, caractérise les réseaux efférents, ceux qui portent les ordres aux organes de réaction — principalement aux muscles et aux glandes (fig. 2, flèche de droite). En fait, que l'on suive le cheminement de l'information dans un sens ou dans l'autre, de la périphérie sensorielle aux centres les plus élevés ou de ceux-ci à la périphérie effectrice,

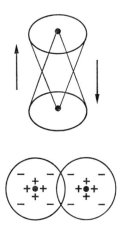

Fig. 3.   En haut, cônes emboîtés symbolisant l'organisation divergente-convergente des communications nerveuses afférentes aussi bien qu'efférentes.
En bas, cercles figurant deux ensembles neuroniques ayant une intersection, avec deux foyers d'activation (signes +) entourés de zones d'inhibition (signes —).

c'est presque toujours à une étroite imbrication d'architectures neuronales convergentes et divergentes que l'on a affaire, d'un côté rassemblant et concentrant les signaux, de l'autre les diffusant avec plus ou moins d'extension (fig. 3, en haut). En direction latérale, cette "diffusion" mérite rarement son nom, en ce sens qu'il ne s'agit pas en général d'un envahissement inorganisé de signaux excitateurs, mais d'une répartition spatialement ordonnée de signaux excitateurs et inhibiteurs, qui découpent en quelque sorte dans les champs neuroniques des plages d'activité distinctes, parfois chevauchantes, mais souvent isolées aussi les unes des autres par des franges on zones latérales d'inhibition (fig. 3, en bas); le tout en perpétuel changement selon les modalités de l'excitation d'origine externe (ou interne) et la configuration des programmes d'action déclenchés.

Cette vue d'ensemble, très schématique, de l'organisation fonctionnelle des activités nerveuses centrales, n'est pas une simple construction de l'esprit. Elle est étayée par de nombreuses expériences microphysiologiques qui visent avant tout à démêler dans leurs grandes lignes les principaux rapports de connectivité, à distinguer les effects inhibiteurs des effets excitateurs, à mettre en évidence les contrôles hiérarchiques et ceux qui prennent la forme de boucles de rétroaction, à établir, quand c'est possible, la loi qui relie le contenu informatif des signaux sortant d'un réseau nerveux bien délimité au contenu informatif des signaux qui y sont entrés (fonction de transfert).

Nous mentionnerons seulement ici trois modalités d'organisation fonctionnelle des transferts d'information qui semblent jouer un rôle décisif dans la façon selon laquelle l'information contenue dans un ensemble de signaux propagés se trouve traitée par le système nerveux central.

La fig. 4 montre une courbe en S qui est la forme la plus courante des fonctions de transfert lorsque des messages élémentaires multiples abordent en parallèle

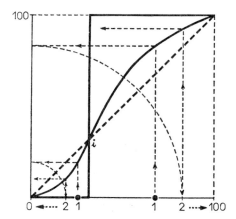

Fig. 4. Graphique destiné à faire comprendre comment une "fonction de transfert" peut, dans un réseau de neurones, passer d'une forme sigmoïde à une forme en marche d'escalier (*step function*).

un ensemble de barrières mono-synaptiques. Si plusieurs barrières analogues se présentent successivement, en série, au flux de signaux, ces derniers tendront soit à s'atténuer jusqu'à disparaître, soit à s'accroître jusqu'à atteindre leur régime maximum. La courbe en S tendra vers une marche d'escalier (et sa dérivée vers une fonction $\delta$ de Dirac), la réponse du réseau vers un régime de "tout ou rien", favorable à l'établissement de ségrégations à l'intérieur des champs neuroniques.

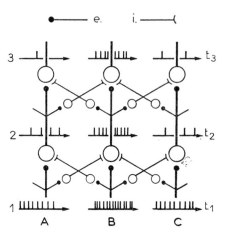

Fig. 5.   Schéma d'un réseau de neurones supposé parcouru, dans la direction 1–3, par des trains d'influx représentés sur les échelles de temps $t_1$, $t_2$, $t_3$.
Les interneurones croisés (petits cercles blancs) sont inhibiteurs. Il en résulte que les densités d'influx au niveau 3 sont atténuées, mais beaucoup plus en A et C qu'en B, d'où un effet de "focalisation" dû au jeu réciproque de cette inhibition latérale.

La fig. 5 indique comment le jeu de l'inhibition latérale réciproque permet de concevoir comment se réalisent dans le système nerveux — et tout d'abord au niveau même de la rétine — les effets d'accentuation de contrastes, ici de focalisation, constatés si fréquemment par le psychologues à propos des phénomènes perceptifs, quelle que soit la modalité sensorielle. Ces effets résultent simplement de ce que les chaînes neuronales fortement excitées inhibent de ce fait leurs voisines plus qu'elles ne sont inhibées par elles. Une distribution A, B, C peu

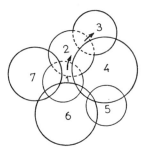

Fig. 6.

contrastée d'intensités au niveau 1, codée en fréquences moyennes de trains de neuropulsions, devient bien davantage contrastée en 3. Il se fait une focalisation qui serait encore plus marquée si à cette influence de l'inhibition latérale s'ajoutait celle du phénomène précédent (fonction de transfert de forme sigmoide).

La fig. 6 est une interprétation schématique vraisemblable de ce qui se passe dans des champs neuroniques voisins présentant des intersections, et dont les activités se trouvent associées simultanément ou successivement. Lorsqu'un ensemble de neurones, 1 par exemple, est excité fortement, il peut entraîner une suractivité des champs neuroniques voisins avec lesquels il chevauche, et une décharge généralisée peut s'ensuivre par entraînement de proche en proche; mais c'est là un processus anormal, qui se produit dans la crise épileptique. Normalement, apparaissent des zones d'inhibition qui limitent et protègent le champ actif. Cependant, si deux ensembles présentant une intersection ont été activés plusieurs fois l'un après l'autre, régulièrement (comme dans les expériences classiques de conditionnement) ou même assez irrégulièrement [5], il se produit des modifications persistantes dans les caractéristiques de transmission, de telle sorte que les barrières d'inhibition semblent céder. Au bout d'un certain nombre de ces "associations", entre 1 et 2 par exemple (fig. 6), la seule stimulation qui, entraînant l'activité de l'ensemble 1, entraîne évidemment celle de la partie de 2 commune aux deux ensembles amènera 2 à s'activer tout entier; de même pour 2 avec 3, etc. . . . Ainsi se forment entre les champs de neurones impliqués au cours d'un conditionnement ou d'un apprentissage, ou même au hasard de certaines circonstances, des chaînes d'associations fonctionnelles qui peuvent éventuellement se refermer sur elles-mêmes (comme si 3 entraînait l'ensemble 4 et celui-ci 2 ou 1).

Notons qu'il s'agit ici de représentations purement symboliques qui rappellent les "cercles d'Euler". Elles sont destinées à figurer dans chaque cercle des neurones particulièrement bien interconnectées et, pour ainsi dire, "habitués à souvent fonctionner ensemble". Ils ne sont pas nécessairement rassemblés dans un même petit espace anatomique; ils peuvent devoir leur parenté fonctionnelle à des voies associatives à longue distance, le corps calleux par exemple.

Dans la mesure où les enchaînements associant nouvellement des ensembles neuronaux survivent longtemps à leurs causes, nous pouvons parler de mémoire, ajoutant ainsi à tout ce qui précède un élément évidemment capital, qui fait apparaître bien insuffisante la représentation d'une opération nerveuse centrale comme simulable par un simple canal de communication (fig. 1) qui transmettrait simplement dûment codé, un flux informationnel continu.

Un schéma formel certainement plus adéquat est celui de la fig. 7. M y représente la fonction "mémoire", dans sa relation bidirectionnelle avec le reste du système nerveux, lequel figure le simple niveau réflexe. Cela signifie que tout message afférent parti de S, s'il nous intéresse par les réactions immédiates qu'il peut déclencher, nous intéresse aussi par les modifications durables qu'il laisse dans certains "compartiments" du système nerveux (flèche ascendante), modifications que l'on nomme souvent du terme général d'"engrammes". Conservées à l'état latent, ces modifications sont de deux sortes: celles qui changent banalement pendant un certain temps les conditions générales de fonctionnement du Système (par exemple en accélérant l'irrigation sanguine d'un territoire) celles qui, conservant une structure de code, représentent une information potentielle emmagasinée, prête à servir dans des opérations ultérieures, en se combinant avec les informations actuelles (flèche descendante). La lettre P, initiale du mot "programme", jointe à la liaison M → P, symbolise une conception d'ensemble à

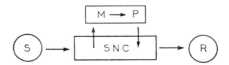

Fig. 7.  Organigramme simple destiné à détacher la fonction "mémoire" (M), et les programmes (P) qui peuvent en résulter, de l'ensemble fonctionnel restant du système nerveux central (SNC). S le stimulus, R la réponse. Cette dissociation est purement symbolique et ne correspond à aucune subdivision anatomique.

laquelle nous attachons du prix, à savoir que les engrammes évoluent, ne restent pas figés, et se transforment assez souvent en programmes d'action ou d'évocation (nous admettons qu'il n'y a pas de différence essentielle, dans le mécanisme nerveux de base, entre une action extériorisée et une activité mentale consciente, qui reste d'évocation intérieure: R, la réaction, représente dans notre schéma l'une comme l'autre). Il n'y a mémoire que s'il y a conservation d'une trace *codée*; et il n'y a de contenu réellement *informatif* que si l'engramme peut être décodé, c'est à dire utilisé dans une opération fonctionnelle. Tout cela est cohérent, indiscutable, et c'est bien ce que représente notre schéma. La difficulté commence au moment où nous voulons l'incorporer dans les structures réelles du système nerveux et préciser la nature et la forme du ou des codes mis en jeu.

Nous n'entreprendrons pas ici de dire ce que les travaux les plus récents nous suggèrent sur la localisation du "compartiment mémoire". C'est une question qui, à elle seule, mériterait d'être longuement traitée, sans qu'il y ait à espérer autre chose qu'une multiplicité d'hypothèses attendant le verdict d'expérimentations laborieuses.

Sur la vraisemblance de l'évolution des engrammes vers des programmes d'évocation ou d'action, nous nous en tiendrons à ce que chacun de nous a pu constater sur lui-même en matière de souvenirs et d'habitudes. A ces évidences psychologiques et subjectives, pourraient s'ajouter de nombreuses preuves tirées de l'expérimentation animale (dressages, conditionnements). Nous ne pouvons en faire état ici. Par contre, nous pouvons dire quelques mots des preuves neurophysiologiques de la présence de véritables programmes d'action au sein des structures cérébrales.

L'existence de programmes d'action dans le système nerveux central est bien démontrée par les expériences de stimulation électrique ou par des opérations pratiquées chez l'animal, et parfois chez l'homme. On a établi ainsi de nombreuses correspondances topographiques entre différentes régions du cerveau (cortex moteur, hypothalamus, rhinencéphale et lobe limbique, etc. . . .) et des manifestations bien définies du comportement. La stimulation focalisée d'une de ces régions déclenche des actes stéréotypés accomplissant une *fonction* (prise de nourriture, comportement sexuel, peur et fuite, colère et agression, sommeil, etc. . . .). Chez l'Homme, les observations de W. Penfield [6] sont bien connues. Au cours d'opérations cérébrales destinées au traitement des épilepsies temporales, un même souvenir organisé peut être évoqué de façon répétée par la stimulation du même point du lobe temporal. Le cerveau apparait alors, pour ainsi dire, comme une collection de programmes "presse-bouton", certains imposés génétiquement, d'autres acquis par l'expérience, et ceci est vrai non seulement pour le comportement visible mais aussi pour les activités mentales; image certainement plus exacte que celle qui limiterait le cerveau, spécialement le cortex cérébral, à n'être qu'une surface de projection point par point des excitations du monde extérieur. Ces surfaces de projection spécifique ne sont guère que des stations-relais pour des messages qui trouvent plus loin leur véritable destination fonctionnelle: plus loin, c'est-à-dire vers les ensembles neuroniques organisés en programmes d'action (ou d'évocation) latente. Même lorsqu'il s'agit de stimulations sensorielles simples, la structure de la perception n'est jamais un reflet passif de la structure du stimulus, mais elle dépend largement des propriétés dynamiques des réseaux nerveux qui reçoivent les messages. Un véritable programme de "traitement" leur est imposé. Les psychologues modernes ne parlent plus "d'écran de la conscience" comme si nous assistions passifs à un spectacle de cinéma. Ils disent que les structures perceptives, pour ne prendre que les moins complexes, portent la marque d'une certaine adaptation à l'action. Elles sont donc, en un sens, des programmes. Les expériences mentales qu'elles évoquent dépendent autant, parfois plus, du passé que de la situation présente.

\* \* \*

Nous sommes maintenant mieux armés pour revenir au problème du codage. Laissant de côté les phénomènes de mémoire pour ne retenir que leur aspect "programme", nous pouvons dès le début décomposer le problème en un minimum de trois étapes, parcourues dans le sens où progresse l'information dans le système nerveux central (SNC), c'est-à-dire du pôle récepteur (1) au pôle effecteur (3). La fig. 8 schématise le processus, et y ajoute un observateur O, l'expérimentateur qui reçoit pour les comparer les informations captées aux différentes étapes, après passage à travers les instruments appropriés (a). Ce sont le plus souvent des amplificateurs suivis d'oscilloscopes, puisque c'est généralement par le captage de signaux électriques que se fait le recueil des informations caractéristiques de

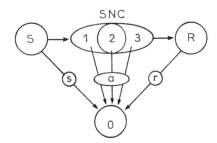

Fig. 8. Schéma méthodologique destiné à représenter comment l'expérimentateur en neuro-physiologie (O) reçoit les informations qui lui arrivent des différents secteurs soumis à une observation, assistée d'appareils variés (s, a, r, voir texte).
Dans une situation expérimentale où un S bien défini donne lieu à un R bien identifié, le travail du neurophysiologiste consiste à essayer de reconstituer avec ce qu'il reçoit en O la succession et les liens de causalité des phénomènes qui se passent en 1, 2 et 3 et leurs intermédiaires.

l'activité nerveuse. Les trois tâches principales qui incombent à l'observateur-expérimentateur sont alors les suivantes:

1° – Disposant des moyens modernes pour spécifier le (ou les) stimulus S, à l'aide des instruments de mesure appropriés (s) et selon les procédés classiques de la physique (codage physique de la réalité extérieure), 0 va comparer cette information à celle qu'il croit pouvoir reconnaître dans la structure spatio-temporelle des signaux nerveux captés en 1. Il essayera en somme de comprendre comment les structures réceptrices du SNC codent (en 1) les caractères du stimulus. En cela il a assez bien réussi, notamment dans les domaines de l'Audition et de la Vision. Diverses modalités du codage nerveux sensoriel ont été mises en évidence, ainsi qu'il a été rappelé plus haut. Les récepteurs sensoriels et leurs prolongements nerveux jouent le même rôle que les instruments de mesures, de telle sorte qu'un observateur neurophysiologiste auquel le stimulus S serait masqué pourrait acquérir une certaine information sur la *nature*, la *localisation*, la *durée* et l'*intensité* de ce stimulus en interrogeant la première étape de l'activité nerveuse. Il y gagnerait parfois en sensibilité, mais il y perdrait en précision et en stabilité. Revenant à la comparaison des deux codes, le physique et le neurophysiologique, il peut se rendre compte de ce qui est préservé et de ce qui est irrémédiablement perdu en information au cours de ce transcodage.

2° – Passant alors à l'autre extrêmité, nous trouverons une situation compa-rable, mais en sens opposé, à savoir la confrontation d'un code nerveux en 3, relatif à l'exécution d'un programme, avec l'ensemble des symboles avec lesquels on peut caractériser la réponse R. La situation est moins claire que précédemment, du fait tout d'abord que les procédés de codification (r) de la réponse R ne sont pas toujours bien au point. Le cas le plus simple est celui où la réponse peut se noter par *oui* ou par *non*, comme dans une recherche de seuil. Le cas sans doute le plus complexe est ce code "d'expression" extrêmement raffiné que représente notre langage. Du côté nerveux, il n'est pas encore question de reconnaître les nuances du langage à partir des variations de l'état électrique des zones cérébrales mises en jeu. Peut-être obtiendrons-nous quelques lumières en étudiant chez des Singes supérieurs comme travaille leur cerveau pendant qu'ils émettent les sons articulés de leur langage primitif. En attendant, de nombreuses recherches sur des parties isolées du SNC, segment de moelle par exemple, ont bien montré l'existence d'une correspondance étroite entre l'organisation anatomo-fonctionnelle des centres réflexes et la distribution spatio-temporelle des activités musculaires et des inhibitions qui sont à la base d'une réponse motrice coordonnée. La polygraphie électromyographique est un procédé de choix pour coder les réponses motrices au niveau des muscles impliqués.

Quoi qu'il en soit de l'insuffisance des données actuelles, il n'y a pas de difficulté de principe à reconnaître à cette étape une transmission d'information, que nous appellerons "ordre", ou "commande" plutôt que "message", une sorte de transcodage entre les programmes du niveau 3 et la notation r qui informe l'observateur des caractères de la réponse, active ou expressive. A cet observateur de juger si cette transmission d'information, qui est équivalente à la réalisation d'un programme, est bien conforme au schéma nerveux de la commande lorsque ce schéma a été localisé et exploré.

3° – La véritable difficulté, qui est de principe, est celle qui attend l'observateur lorsqu'il essaye de continuer à appliquer un schéma de communication d'information à l'analyse des transferts 1–2 et 2–3. Tout se passe comme si le système nerveux parlait un certain langage en 1, un autre en 3, mais comme si ces deux langages ne pouvaient être traduits l'un dans l'autre, bien que le stock de symboles utilisés ne puisse être réellement différent dans tous les cas. Le passage de 1 à 3 ne peut être conçu comme une simple transmission d'information, même en tenant compte de sa dégradation par le bruit et de son enrichissement par l'information en réserve dans la mémoire. Un processus beaucoup plus compliqué a lieu au sein de ces relais centraux, un processus que l'on peut assimiler à des "reconnaissances de formes", grâce auxquelles, par une certaine "congruence" entre la structure des messages afférents et celle des "récepteurs" intra-centraux qu'ils abordent, les programmes d'action (ou d'évocation) se trouvent déclenchés. La vieille analogie de la clef dans la serrure se présente à l'esprit: le message, ou tel ensemble de messages convergents, ouvre une ou plusieurs "portes". Il est clair que le nouveau spectacle qui s'affiche (image mentale ou manifestation comportementale, souvent associées à des réactions végétatives engendrant la coenesthésie) n'a rien à voir, structuralement, avec le profil de la clef. Cela revient à dire que l'ensemble des programmes latents constitue le "répertoire" à partir duquel chaque message est déchiffré, décodé, prend signification. Ce n'est d'ailleurs que dans ce sens qu'on peut lui attribuer un contenu informationnel; car il n'y a pas d'information sans „lecteur" capable de déchiffrer le code, et de se comporter en conséquence.

C'est peut-être le moment de rappeler le risque de confusion qui guette le neurophysiologiste: à savoir que se trouvent ici en présence deux lecteurs bien différents. L'un est le cerveau soumis à l'expérimentation; l'autre est l'observateur qui échantillonne un petit nombre de ses activités. Les procédés de décodage qu'ils emploient ne sont pas les mêmes: le premier dispose de la totalité des informations élémentaires, souvent redondantes, qui l'abordent; mais il doit aller vite pour y répondre et est astreint à l'ordre chronologique; le second ne dispose que d'un échantillonnage très pauvre, mais il a tout son temps et surtout, il n'est pas contraint de suivre l'ordre séquentiel; et il possède un jeu varié de programmes de traitement que ses machines calculatrices se chargent de mettre en oeuvre: histogrammes d'intervalles, chronohistogrammes de post-stimulation, distributions d'intervalles associés, corrélogrammes sériés, autocorrélogrammes et inter-corrélogrammes, analyses factorielles, etc. . . . Il y a évidemment peu de chances pour que le traitement par les "structures d'accueil" du tissu nerveux (réseaux de neurones, champs synaptiques, sites récepteurs membranaires, macromolécules) ressemble à celui que l'observateur met en jeu, sauf si celui-ci pouvait étudier le fonctionnement d'un cerveau de physiologiste en train de traiter les données de ses enregistrements!

Pourtant, dans un cas simple, celui d'un Coléoptère placé dans des conditions où se déclenche une réaction optomotrice, W. Reichardt [7] a pu montrer que tout se passe comme si les ganglions nerveux de l'insecte étaient capables de faire un calcul d'autocorrélation. Un modèle mathématique, qui pourrait être concrétisé par un montage électronique, schématise ce fonctionnement.

L'exemple précédent m'amène, au moment de conclure, à dire un mot de la méthode globale du "Modèle". Nous avons vu en effet que la méthode classique, analytique, qui vise à décomposer le système nerveux en ses éléments fonctionnels et qui est amenée en conséquence à s'interroger sur les processus de communication et sur le code nerveux qu'ils impliquent, ne peut prétendre aller très loin dans la connaissance des mécanismes essentiels. Ceux-ci mettent en jeu des relations qui ne sont jamais linéaires; ils se manifestent dans des systèmes de liaisons dont on commence seulement à savoir représenter, à l'échelle macroscopique, les organigrammes; mais la complexité et la diversité des matrices nerveuses qui reçoivent et traitent les signaux, aux diverses échelles où l'on trouve de l'ordre structural, est véritablement déconcertante. Sauf dans les cas les plus simples, réalisés dans les conditions artificielles du laboratoire, ou bien lorsqu'on peut supposer que domine l'aléatoire, il y a peu de chances que l'on arrive, par une synthèse des fonctionnements étudiés partie par partie, à donner une explication satisfaisante d'un fonctionnement bien intégré, à l'oeuvre dans l'accomplissement d'une fonction.

On peut alors se demander si la méthode du "Modèle", sans être exactement dans la ligne de l'esprit cartésien, ne serait pas capable d'apporter au moins une certaine satisfaction aux exigences de la compréhension là où la méthode analytique aurait échoué. Elle n'est pas opposée à la recherche des meilleures "pièces détachées"; mais, ayant bien souvent constaté l'impossibilité de prévoir ration- nellement le comportement d'ensemble à partir des connaissances sur le fonc- tionnement des parties, elle assemble celles-ci en systèmes qu'elle soumet ensuite à des essais globaux. C'est de ces essais qu'elle attend un verdict de réussite ou d'échec à imiter la nature. Elle estime avoir atteint son but explicatif lorsque la réaction finale du modèle à une stimulation d'entrée dûment codée imite suffisam- ment bien la réaction d'un système nerveux soumis à la même stimulation. C'est ce que représente schématiquement la fig. 9. Notons que le degré de ressemblance

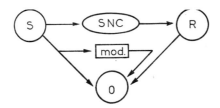

Fig. 9.   Organigramme correspondant à la méthode du Modèle.

entre les deux "sorties" pourrait faire l'objet d'une évaluation automatique par ordinateur. L'évaluation du résidu de dissemblance pourrait même servir à informer un dispositif de contre-réaction correctrice de telle sorte que certains paramètres du modèle s'en trouveraient modifiés dans le sens d'une meilleure imitation de la réalité: le modèle s'optimaliserait tout seul. Le rôle de l'observateur serait alors bien diminué, mais il lui resterait la tâche somme toute essentielle de proposer, plus ou moins intuitivement, un nouveau choix de pièces détachées et de nouvelles structures d'assemblage, dans l'espoir d'aboutir à de meilleures solutions, à l'"optimum optimorum". Il est d'ailleurs à prévoir que pour un certain degré de tolérance, plusieurs solutions également acceptables se présen- teraient, sans que l'on puisse décider quelle est celle qui correspond le mieux à la réalité. L'étude des effets de certaines altérations faites parallèlement sur la préparation vivante et sur le modèle pourrait aider à lever l'indétermination.

Instrumentaux ou formels, mis à l'essai concrètement ou traités à l'ordinateur, les modèles, dès qu'ils atteignent une certaine complexité, peuvent avoir des comportements inattendus que l'on doit étudier systématiquement pour en con- naître les "habitudes". Au même titre que le système nerveux, ils illustrent à leur tour le principe que le tout est plus que la somme des parties et ils mettent le plus souvent en échec notre capacité d'en rendre compte analytiquement, par la connaissance préalable des interactions (un physicien évoquera ici le fameux "problème des trois corps"). Ils "expliquent" donc mal, non seulement parce qu'ils ne peuvent être que des imitations extrêmement simplifiées de la réalité vivante, mais surtout parce qu'ils ne répondent pas à une exigence profonde de notre esprit analytique. Ils nous apportent cependant une certaine satisfaction dans la mesure où ils nous montrent la possibilité d'obtenir, avec des procédés ne mettant en jeu que des lois physiques, des comportements de système analogues à ceux que nous observons lorsque travaille tout ou partie d'un système nerveux, d'un cerveau notamment. Ce sont, en outre, des outils de travail qui permettent de mettre à l'épreuve les hypothèses sur les propriétés et sur l'arrangement des parties constituantes, et sur les principes généraux que l'on est conduit à invoquer pour le fonctionnement global.

Finalement, que deviennent dans tout ceci les problèmes du code nerveux? Il est facile de comprendre qu'ils ne représentent qu'un des aspects du problème général de l'activité nerveuse supérieure, c'est-à-dire celle qui est responsable des fonctions les plus complexes et à propos desquelles se posent particulièrement les questions fondamentales de l'intégration des parties, de l'émergence des propriétés de l'ensemble, de la stabilité du système, de la conservation et du rappel ordonné des séquences d'activité (problèmes de la mémoire), de la capacité de réagir à des formes, indépendamment du support, etc. . . . En face de ces problèmes majeurs, ceux du code nerveux semblent s'estomper et avec eux la réelle utilité de tout ce mouvement créé autour de la Théorie de l'Information dans l'espoir d'y trouver un nouvel outil pour l'analyse des mécanismes fondamentaux. A notre avis, si effectivement on peut douter de l'utilité de la théorie "télégraphique" de l'Information en tant que "modèle" de certaines activités nerveuses, la notion même d'*information*—dans son sens étymologique d'"action de mettre dans une forme", disons autrement "le fait de considérer les phénomènes sous l'angle de leur forme"—a rénové chez les neurophysiologistes leur manière d'*aborder*— mais non certes de résoudre entièrement—les grands problèmes du système nerveux. Les notions d'éléments ou d'états discernables, de symboles et de code, de bruit, de redondance, de probabilité conditionnelle, etc. . . . sont venues enrichir non seulement le vocabulaire, mais surtout les modes de penser du neuro-physiologiste moderne, et parallèlement ceux de son confrère en "cérébrologie", le psychologue. C'est en effet au moyen des concepts et du langage informaticiens que s'articulent au mieux ces deux champs de connaissance, que l'on pourrait de façon imagée dénommer, le premier de "cérébrologie intérieure", le second de "cérébrologie extérieure". La recherche d'une incarnation nerveuse d'un ou de plusieurs codes n'est qu'une première étape, mais une étape après laquelle on voit tout naturellement s'annoncer les autres problèmes.

Pour nous résumer en terminant, nous répéterons que les problèmes spécifiques du code nerveux se placent essentiellement à trois niveaux: aux entrées réceptrices du système nerveux (codage des stimulations externes ou internes), aux sorties effectrices (programmes d'action ou d'expression), et dans une articulation en-gramme-programme dont le schéma opérationnel n'est autre que celui de la *reconnaissance des formes*. Il appartiendra aux expérimentateurs et aux théoriciens, dès maintenant à l'oeuvre, de nous faire de mieux en mieux comprendre les modalités de ce dernier mécanisme, en le répartissant aux différentes échelles structurales et temporelles auxquelles il opère. C'est là croyons-nous, exprimé dans toute sa généralité, le problème le plus important parmi tous ceux que soulève l'introduction de la notion de "code" dans les sciences du système nerveux.

## Références

[1] A. M. Uttley, The Engineering approach to the problem of neural Organization, *Progr. Biophys.* **11** (1961) 26–52.

[2] G. P. Moore, D. H. Perkel and J. P. Segundo, Statistical Analysis and functional inter-pretation of neuronal spike data, *Ann. Rev. Physiol.* **28** (1966) 493–522.

[3] D. MacKay and W. MacCulloch, The limiting information capacity of a neuronal link, *Bull. Math. Biophys.* **14** (1952) 127–135.

[4] A. Fessard, Réflexions en marge d'un exposé sur le codage cérébral. *Actualités Neuro-physiol.*, 6° Série, (Masson, 1965) 269–276.

[5] G. Lelord, *L'acquisition libre. Aspects neurophysiologiques chez l'animal et chez l'Homme*, Thèse, Faculté des Sciences de Paris, (1967) (en cours de publication).

[6] W. Penfield and P. Perot, The Brain's record of auditory and visual experience, *Brain* **86** (1964) 595–696.

[7] W. Reichardt, Autocorrelation, a principle for the evaluation of Sensory information by the Central nervous System; in: *Sensory Communication*, (MIT Press, 1961) 303–317.

*Theoretical Physics and Biology*, © 1969, *North-Holland Publ. Co., Amsterdam*

# QUANTUM NOISE AND VISION

MAARTEN A. BOUMAN

*Utrecht University, The Netherlands.*

Because of the short time available I will restrict my contribution to a very limited presentation of a few simple points.

**1.** There is general agreement upon the fact that *visual receptors* are activated by single quantum absorptions and thus are *single quantum* counters. I will not go into experimental details of the studies which were the bases for this first point. However, it has been ascertained for the human eye's *rod as well as cones* and for quite a collection of other species' eyes. It holds almost for any illumination level that occurs in nature. At night time a receptor might catch the quanta at a rate of about once an hour and when playing on a sunny midsummer day at the beach the rate might reach hundreds of counts per second.

**2.** The capability of the visual system for discrimination between intensity differences represents under simple stimulus conditions the performance of an ideal physical detector: this fact has for the human eye been found in psychophysical experiments in which test subjects were asked to respond by yes or no dependent on whether they had or had not perceived a particular test flash. The background illuminance upon which these flashes were presented was varied over a large range of intensities. In case area and duration of the test flashes are small, it proved that over quite a number of log units of the background illumination $\bar{n}$, the test flash energy $\Delta N$ is proportional to $\bar{n}^{\frac{1}{2}}$. In case sample-time and sample-area for the background $\bar{N}$ are taken equal to duration and size of the test flash, $\Delta \bar{N} \simeq 3\,\bar{N}^{\frac{1}{2}}$ which indeed represents very closely the way how the sensitivity of an ideal physical detector for corpuscular radiation behaves: the limit for detection of changes in flux is proportional to the square root of the flux. For the eye it says: *the threshold for discrimination is proportional to the square root of luminance level and is determined by the statistical fluctuation in actual numbers of quanta of flash and background.*

**3.** Besides intensity discrimination, another essential capability of the visual system is the discrimination between different spectral energy distributions: colour-discrimination. This capability is mediated by three different types of receptors that have their main sensitivities in the red, in the green and in the blue part of the spectrum. These three types of cones have recently definitively been identified by microspectrophotometric measurements. From psychophysical

246

studies on discrimination between equally bright stimuli but with different spectral energy distributions it could be shown that human capability for detection of colour differences behaves under simple stimulus conditions similar to the performance of a group of three independently acting ideal physical detectors. It means that *colour discrimination thresholds $\Delta K$ are determined by the combination of the statistical fluctuations $\bar{\Delta}\bar{R}$, $\bar{\Delta}\bar{G}$, $\bar{\Delta}\bar{B}$ in quantumnumber $\bar{R}$, $\bar{G}$ and $\bar{B}$ in each of the three cone-systems for test flash and reference stimulus:*

$$\frac{(\bar{\Delta}\bar{R})^2}{R} + \frac{(\bar{\Delta}\bar{G})^2}{G} + \frac{(\bar{\Delta}\bar{B})^2}{B} = (\bar{\Delta}\bar{K})^2.$$

For the participants of this symposium it is interesting to note that the theoretical physicist Schrödinger originally postulated a similar equation for colour discrimination on purely empirical arguments and not referring to ideal physical detectors for corpuscular radiation.

**4.** The fourth point refers to a property of the transmission of action potentials through the nervous system. Although each absorbed quantum deposes a signal in the retina, not all these signals are individually transmitted to the brain. The retina applies a data processing action upon these quantum signals by a scaling procedure. Even a cascade of such procedures, one in each of the successive neuron layers, might be present in the visual pathways. This finally makes that we can speak of a certain quantum to spike ratio $k$ under any particular illuminating conditions. Each spike is then connected individually with the last quantum of a package of $k$ quanta. Along different ways has this quantum-spike ratio been studied. There have psychophysical experiments as well as objective experiments been made around this topic.

In the latter category Hartline's work is especially worth mentioning. He demonstrated that the chance for occurrence of action-potentials were in agreement with a $k$-coincidence mechanism applied to a Poisson-type of input events. He moreover showed in his studies that the number $k$ increased by increase of the illumination level to which the eye of the Limulus was adapted. The loss of sensitivity under the higher illumination levels proved only to be due to changes in $k$. This conclusion was also made by us on various arguments. One of them was that in case like for an ideal physical detector, the detection limit varies proportional to the square root of the average flux, there is no room left for decrease in quantum efficiency: quantum absorption must be proportional to quantum input. So significant changes in photopigment concentration are in the range of luminances in which these facts on the detection threshold were found, excluded. In more recent experiments Weale and Rushton indeed found here no significant changes in photopigment concentration via their measurements of spectral remittance of the retina. Inherent to the quantum coincidence mechanism and the changes of its order $k$ by changes in illumination level, is the peculiar fact that in case only a part of the retinal area over which the packages of $k$ quanta are collected, is illuminated by an adapting light, the whole area including the part that is kept in the dark becomes by the increase of the $k$-values less sensitive. Recently Rushton made some relevant experiments in this respect. His results were another confirmation of our coincidence theory.

As a statistical mathematical analysis can show, there is a critical limit to the quantum to spike ratio $k$ beyond which the performance of the ideal physical detector can no longer be maintained. This leads us directly to the conclusion that in case for any particular stimulus condition the performance is like that of the ideal detector the acting k-value is directly related to this limit: square root of the average flux in the retinal area of the coincidence units: $k \leqslant \bar{N}^{\frac{1}{2}}$. So this point reads: *the retinal tissues apply a scaling procedure upon the output of the quantumcounters (receptors). The resulting quantum to spike ratio k varies proportional and is near to the square root in the average flux of quanta upon the scaler's retinal areas.*

5. Scaling by $k \sim \bar{N}^{\frac{1}{2}}$ makes that with the smallest allowable number of action potentials and also in the smallest allowable frequency bandwidth, still maximum performance is maintained. How knows the eye so readily as it does without a time consuming statistical analyses, which $k$-value it must install in order to reach this optimal situation?

Statistical analysis of this problem results in the conclusion that the necessary and sufficient condition is "*arithmetical adaptation*". In its simplest form this means that for each next outgoing action potential a fixed number ($\varrho$) more input impulses are needed than were required for the last output pulse. This indeed means a square root relation between input and output: In case $k$ output pulses have been registered, they were produced by $\frac{1}{2}\varrho k(k+1)$ input pulses. I here will not go into problems related to the time over which the unit counts the input and how the scale factor $k$ that is reached at the end of such a sample time follows changes in intensity in the next sample periods. Here I only mention the fact that in case the intensities in two such periods differ just significantly the scale factors $k$ and thus the number of output pulses differ by one, no matter at what absolute energy level this significant difference occurs.

## 6. Moral of the story

Any living system's vital duty is to detect changes in environmental conditions and react "adequately" upon them. The only crucial criterion for this adequacy seems to be that this reaction must not impeed the organism's capability to detect and handle another future change in environmental conditions.

In the "arithmetical adaptation" mechanism the retina possesses an optimalized system to eliminate only the unevitable and essential statistical variations in radiation from a constant environment.

It is this *arithmetical adaptation mechanism that is most valuable and perhaps the only one able to survive through evolution from random structures into biological substances.* The merits of the arithmetical adaptation mechanism are not restricted to poisonean inputs but for other probability distributions, like the normal and the gamma distributions, the system is equally valuable. This means that *arithmetical adaptation might represent a general and essential property of the living system* in various other instances as well, like in signal transmission by nerve networks more centrally located in the nervous system.

## References

[1] H. de Vries, The quantum character of light and its bearing upon threshold of vision, the differential sensitivity and visual acuity of the eye, *Physica* 10 (1943) 553.

[2] H. A. van der Velden, The number of quanta necessary for the perception or light of the human eye, *Ophthalmologica* 111 (1946) 321.

[3] M. A. Bouman and H. A. van der Velden, The two quanta hypothesis as a general explanation for the behavior of threshold values and visual acuity for the several receptors of human eye, *J. Opt. Soc. Am.* 38 (1948) 570.

[4] M. A. Bouman, Peripheral contrast thresholds of the human eye, *J. Opt. Soc. Am.* 40 (1950) 832.

[5] M. A. Bouman, Peripheral contrast thresholds for various and different wavelengths for adapting field and test stimulus, *J. Opt. Soc. Am.* 42 (1952) 820.

[6] M. A. Bouman, Mechanisms in peripheral dark adaptation, *J. Opt. Soc. Am.* 42 (1952) 941.

[7] M. A. Bouman and J. ten Doesschate, Nervous and photochemical components in visual adaptation, *Ophthalmologica* 126 (1953) 222.

[8] M. A. Bouman and J. ten Doesschate, The mechanism of dark adaptation, *Vision Research* 1 (1962) 386.

[9] M. A. Bouman, "Sensory phenomena", in: *Physico mathematical aspects of Biology*, (Academic Press, 1961).

[10] M. A. Bouman, "History and present status of quantum theory in vision", in: *On principles of sensory communication*, (Wiley, 1961).

[11] M. A. Bouman and P. L. Walraven, Quantum theory for color discrimination of dichromates, *J. Opt. Soc. Am.* 51 (1961) 474 (abstract).

[12] P. L. Walraven and M. A. Bouman, Fluctuation theory of color discrimination of normal trichromates, *Vision Research* 6 (1966) 567.

[13] G. van den Brink and M. A. Bouman, Quantum coincidence requirements during dark adaptation, *Vision Research* 3 (1963) 479.

[14] M. A. Bouman, Efficiency and economy in impulse transmission in the visual system, *Acta Psychologica* 23 (1964) 239.

[15] M. A. Bouman and C. G. F. Ampt, Fluctuation theory in vision and its mechanistic model, in: *Performance of the eye at low luminances*, Excerpta Medica Foundation's International Congress Series (1965).

[16] W. A. van de Grind and M. A. Bouman, A model of retinal sampling-unit based on fluctuation theory, *Kybernetik* 4 (1968) 136.

[17] W. A. van de Grind, T. van Schalm and M. A. Bouman, A coincidence model of the processing of quantum signals by the human retina, *Kybernetik* 4 (1968) 141.

[18] M. A. Bouman, Quanta explanation of vision, *Documenta Ophthalmologica* 4 (1950) 23.

[19] H. K. Hartline, Light quanta and the excitation of single receptors in the eye of limulus, *Proc. 2nd Int. Congress Photobiol.*, Turin (1959).

[20] F. Ratliff, Some interrelations among physics, physiology and psychology in the study of vision, in *Psychology: Study of a Science* IV, (McGraw–Hill, 1962) 444.

[21] W. A. H. Rushton, Visual adaptation, *Proc. Roy. Soc. B.* 162 (1965) 20.

*Theoretical Physics and Biology*, © 1969, North-Holland Publ. Co., Amsterdam

# DISCUSSIONS

L. ROSENFELD: Une question à M. Gros, sur les ordres de grandeur entrant en jeu dans les longueurs de chaînes, et les temps, etc. . . . N'a-t-on pas d'exemples, dans la nature, de systèmes plus élémentaires, qui ne fassent intervenir que deux lettres?

F. GROS: Ce qui vous intéresse, ce sont les ordres de grandeur en termes d'angströms?

L. ROSENFELD: Oui.

F. GROS: La chaîne de messager responsable de la synthèse du polypeptide alpha de l'hémoglobine, a une dimension de l'ordre de 1.500 angströms. Les ribosomes eux-mêmes ont un diamètre d'environ 150 angströms. Si l'on considère un état de régime au cours duquel le messager est pleinement fonctionnel, c'est-à-dire où il dirige avec une capacité maximum la synthèse des protéines dans un réticulocyte, la chaîne polynucléotidique qui le constitue est parcourue par un train de 5 à 6 ribosomes (il y a donc un certain espacement entre chaque ribosome). C'est un processus qui se déroule, dans l'ensemble, très rapidement. On a pu calculer que dans une cellule d'Escherichia coli la vitesse requise pour synthétiser une chaîne de protéines de poids moléculaire voisin de 100.000, c'est-à-dire contenant environ 1.000 résidus, était de l'ordre de la minute, ou un peu moins.

En ce qui concerne les codes à deux lettres, on n'en connaît pas formellement dans la nature. En effet, les copolynucléotides à séquences répétitives de deux lettres telles que UGUGUGU admettent le même cadre de lecture que des messagers ordinaires en ce sens que le codon fonctionnel n'est pas UG, mais soit UGU, soit GUG. Toutefois, votre question est très pertinente si l'on se rappelle que, d'après l'hypothèse de Wobble, seules deux des trois lettres du triplet doivent être strictement définies, la troisième pouvant varier sans affecter le sens. En bref, le fait que le code génétique repose sur une succession de triplets impose un *cadre de lecture* par groupe de trois lettres, mais la *signification véritable* de chaque codon découle seulement de deux lettres sur trois. Il n'est pas interdit de penser qu'à l'origine deux lettres suffisaient à coder un acide aminé spécifique.

H. FRÖHLICH: Je voudrais poser une question sur la transcription du DNA à la protéine. Il est possible parfois que des erreurs se fassent, qui auraient des conséquences importantes. Y a-t-il des moyens de corriger les erreurs?

Deuxième question: si vous avez ces deux éléments qui se font face et qui ne sont pas corrects, si la correction se trouve être en eux-mêmes, elle pourrait être en deux sens (à double sens). La protéine pourrait réagir sur le DNA, au lieu que ce soit le DNA qui réagisse sur la protéine. Alors, y a-t-il un double sens de correction?

F. GROS: En effet, c'est une question fort importante que vous posez là, à savoir la modification d'une lettre particulière dans un triplet donné. Si vous considérez un triplet quelconque (A-B-C): puisqu'il n'existe que quatre types de lettres ABCD et que toutes les lettres de ce triplet *peuvent* être modifiées, il existe donc 9 combinaisons possibles en plus de la combinaison originelle.

Quand un changement de lettre intervient au cours de la replication du DNA, on engendre ainsi un nouveau triplet, en un endroit donné du texte génétique. Il lui correspondra, par le jeu des transcriptions, un nouveau codon dans la séquence du messager. Ceci est à l'origine des mutations. Certaines de ces mutations *altèrent* la signification de l'unité de codage; elles ont pour effet ultime le changement dans la chaîne protéique d'un résidu d'aminoacide par un autre. Les codons produits par de telles mutations sont appelés "mis-sense" par les auteurs anglo-saxons.

Parfois, le nouveau codon produit par mutation ne présente *aucune* signification en termes d'aminoacides. C'est le cas des triplets UAA, UAG ou UGA. L'apparition de tels codons à l'intérieur d'un messager aura pour effet d'interrompre la synthèse de la chaîne protéique à l'endroit correspondant (terminaison anticipée). De tels codons sont appelés non-sens. La cellule dispose de mécanismes aptes à *corriger* ces mutations. Elle peut en effet (à la suite d'une mutation supplémentaire) modifier certains ARN de transfert de telle sorte qu'ils deviennent capables soit de traduire un codon mis-sense comme le codon d'origine, soit de traduire en un aminoacide quelconque le codon "non sens", ce qui restaure la *propagation* du système traducteur.

La réparation de telles mutations confère-t-elle un avantage évolutif à la cellule? C'est difficile à dire. En tout cas, c'est peu vraisemblable.

J. DUCHESNE: Je voudrais me permettre de poser à M. Gros une question, qui ne peut être et n'est probablement que la mesure de mon ignorance dans cette chimie remarquable et subtile du code génétique.

Sait-on quelque chose sur la nature de la liaison des codons avec les différents acides aminés? Sait-on s'il s'agit de liaisons d'énergie faible, comme le seraient, par exemple, les liaisons hydrogène, ou des transferts de charge? Ou bien s'agirait-il de liaisons plus fortes, comme les liaisons de nature covalente? Et, enfin, qu'est-ce qui fait cette spécificité remarquable qui oriente les acides aminés sur des codons tout à fait spécifiques?

F. GROS: Sans aucun doute, l'appariement entre codon et anticodon dépend en premier lieu de l'existence d'une complémentarité entre leurs séquences et les interactions qui maintiennent celles-ci en contact reposent pour une part sur l'échange de liens hydrogènes. Néanmoins cette interaction est en soi insuffisante pour assurer la stabilité d'un tel complexe. Il est indispensable que d'autres éléments du système traducteur la renforcent. C'est le rôle des ribosomes, lesquels, ainsi que nous l'avons expliqué, peuvent également contracter des liaisons indépendantes avec le messager, d'une part, et avec les ARN de transfert, d'autre part.

En ce qui concerne l'appariement du premier amino-acyl-tARN à être mis en place dans la séquence protéique en formation et du codon "initiateur" AUG, il fait intervenir en outre des facteurs (protéiques) spéciaux, dits facteurs d'initiation, dont la nature et le rôle ont fait l'objet d'études approfondies, tant dans le laboratoire d'Ochoa qu'à l'Université de Yale et dans notre propre laboratoire (Revel).

P. AUGER: Une question, qui n'est pas d'un physicien théoricien, mais d'un physicien expérimental. Autrefois au laboratoire, j'ai entendu M. Langevin dire d'un air demi-sérieux que la physique expérimentale était une question de support. C'est à dire que le controle de la place et du mouvement, dans l'espace des appareils, est une chose essentielle pour réussir une expérience. Or, dans cette description, si remarquable, de l'usage du code génétique par une cellule, une bactérie par exemple, il est évident que des choses se déplacent, qui viennent à l'endroit où il faut. On peut comparer cela au système de fabrication industrielle par les machines-transfert, celles-ci apportant l'élément nécessaire à la reconstitution d'un ensemble avec des rubans qui relient les différentes parties. Elles amènent à l'endroit voulu, au temps désiré, ce qu'il faut pour constituer le tout. Pourrait on admettre qu'il y ait des liaisons permanentes entre différentes parties, qui permettraient, grossièrement parlant, de les attirer comme avec des rubans ou des fils, à l'endroit voulu pour qu'elles se trouvent en bonne place? Le dispositif qui a été décrit ici est évidemment analogue à celui des machines à écrire automatiques, recevant une bande perforée, et produisant automatiquement un texte en lettres d'imprimerie, c'est-à-dire en termes lisibles—alors qu'elles ont reçu une bande perforée? C'est quelque chose dans ce genre que fait le ribosome. Mais, pour cela, il faut qu'on amène la bande et le texte à l'endroit voulu, qu'il y ait donc des déplacements coordonnés à l'intérieur de la cellule—déplacements assez complexes certainement—et qui se trouvent produits au bon moment. Probablement, les mécanismes qui existent sont-ils analogues à ceux qui gouvernent le déplacement total des chromosomes dans la mitose ou dans la méiose, ou l'on observe effectivement des déplacemenss dans l'espace d'objets qui viennent se placer à un endroit déterminé, au moment voulu.

Est-il possible qu'il existe normalement, dans la cellule, des liaisons permanentes que nous ne voyons pas, ou que nous ne savons pas distinguer, et yui voot stocker, préparer à l'avance, pour pouvoir les amener, à l'endroit voulu, les objets visés? Ou bien y a-t-il attente de la cellule, pendant le temps nécessaire pour que les mouvements browniens intérieurs aient apporté les objets voulus, à l'endroit voulu? Cette seconde hypothèse ne correspond vraisemblablement pas avec la très grande rapidité du fonctionnement de ces dispositifs, et, en tout cas, leur sûreté de fonctionnement. S'il y a, effectivement, des liaisons préexistantes, on obtient une image de la cellule beaucoup plus complexe, disons anatomiquement, que celle que l'on a en ne considérant que des organelles placés ici et là et dont nous ne voyons pas clairement en quoi ils sont reliés les uns aux autres.

F. Gros: Evidemment, il s'agit d'une question fort complexe. En fait, les biologistes ont de plus en plus le sentiment que les structures responsables de la transcription génétique, et même probablement les ribosomes eux-mêmes, sont, le plus souvent, "empaquetés" dans un territoire bien défini de la cellule. Dans le cas des bactéries, c'est assez difficile à dire, comme vous pouvez aisément l'imaginer. Car il n'y a rien qui équivaille, à proprement parler, à un réticulum; on décèle naturellement des structures qui lui ressemblent, des invaginations, qui ont été appelées "méso-somes", lesquelles sont liées d'une manière très étroite à la membrane. Il semble en effet bien établi que si on lyse une cellule bactérienne, la majorité du DNA et des systèmes de transcription ainsi qu'une partie importante des ribosomes se distribuent après centrifugation dans des vésicules membranaires. Il est cependant difficile de tirer, de ces observations, une conclusion définitive. Vous savez à quel point les interprétations se heurtent à des difficultés d'artefact. On peut toujours emprisonner n'importe quoi dans un conglomérat.

Il y a aussi des données de la microscopie électronique, allant dans le même sens. Il semblerait que les ribosomes ne sont pas distribués vraiment au hasard, mais sont au contraire reliés d'une manière assez étroite avec le réticulum endo-plasmique. Bref, les éléments qui constituent la machinerie cellulaire forment des associations spécifiques régies par des mécanismes de "reconnaissance stéréo-spécifique". Toutefois, leur assemblage ne résulte pas de simple collisions statis-tiques. Il semble bien que l'intérieur d'une cellule même aussi simple que celui d'une bactérie soit plus ou moins compartimenté en régions dans lesquelles sont groupés les particules, les enzymes et les macromolécules destinés à interagir d'une manière fonctionnelle.

Th. Vogel: J'ai été frappé par le fait que la correspondance entre les triplets et les produits ne soit pas bi-univoque. Ne pourrait-on pas expliquer cela par la théorie des treillis, en particulier en prévoyant des sous-treillis bipolaires?

F. Gros: Je n'ai pour ma part aucune explication raisonnable, n'étant pas à proprement parler un spécialiste du code, mais je serait surpris si, dans cette salle, ne se trouvaient pas des personnes susceptibles d'apporter une réponse à cette question. On peut envisager une raison, "a posteriori", pour laquelle le code n'est pas bi-univoque. C'est qu'il peut y avoir dans cette situation un intérêt évolutif—je ne sais si le M. Monod est d'accord sur ce point—pour conserver à l'abri des modifications mutationnelles certains "sites" essentiels, certaines positions clés des protéines. Ainsi, il est très intéressant de voir que les codons, pour la leucine, sont extrêmement nombreux: il y en a cinq. Il y a probablement intérêt à ce que la leucine change le moins possible dans les protéines.

TH. VOGEL: C'est finaliste!

F. GROS: Oui, mais en fait c'est une finalité qui fait appel à la conservation d'un mécanisme précis.

M. DALCQ: Je m'excuse de m'écarter de cette brûlante question du code, pour revenir à un point que le M. Gros a posé initialement, à savoir que, des deux fibres de DNA, il n'y en a qu'une qui intervient dans les phénomènes de synthèse, et l'autre seulement dans la réduplication. J'aimerais apprendre quelles sont les bases logiques ou expérimentales de cette distinction.

F. GROS: Les bases expérimentales sont en fait, très solides. Elles ont été établies d'une manière certaine, par des expériences d'hybridation moléculaire. Les ARN-messagers qu'on isole des cellules peuvent tous être considérés comme des poly-nucléotides monofilaires. Et on a pu montrer, en les marquant par des isotopes, qu'ils ne formaient d'hybrides complémentaires qu'avec l'une des deux chaînes de l'ADN. Ce point a été établi pour d'assez nombreux systèmes. En réalité, la transcription, si elle est toujours *asymétrique* et s'effectue *préférentiellement* sur l'une des deux chaînes de la double hélice, peut pour un petit nombre de gènes, être réalisée sur l'autre chaîne (cas des phages). Pourquoi l'une des deux chaînes est-elle transcrite *de préférence* à l'autre, on l'ignore.

Quant à la raison logique de l'asymétrie de transcription, cela revient à se demander s'il y aurait un intérêt, pour le RNA-messager, à être une structure bihélicoïdale. On pourrait le concevoir, à la rigueur, s'il fallait par exemple faire intervenir des considérations de stabilité relative des messagers, puisque l'on sait que les structures à deux brins sont plus résistantes aux enzymes nucléolytiques. Mais il y aurait un inconvénient très grand à ce que le messager soit bihélicoïdal: son appariement avec les ribosomes serait beaucoup plus difficile, sinon impossible. On a pu montrer en effet sur des modèles concrets que les doubles hélices poly-U–poly-A n'étaient pas capables d'être traduites en protéines, par suite de leur inaptitude à s'associer aux ribosomes alors que chacune des chaînes qui les constituent peut diriger pour sa propre part la synthèse d'homopolypeptides.

J. MONOD: Quelle est la fraction du génome d'un homme qui est employée du point de vue informatif à construire son système nerveux central?

A. Fessard: Naturellement, je ne peux pas répondre à cette question. Un généticien le pourrait-il? J'en doute. Ce que je peux répondre, c'est que je doute, a priori, de ce que toute structure, jusqu'à la distribution des synapses, soit prédéterminée dans le génome. Le contenu informationnel devrait être absolument fantastique!

On a fait des expériences récemment: de jeunes animaux, dont le cerveau n'est pas mûr encore, sont, les uns mis depuis la naissance à l'abri de la lumière, les autres vivant normalement. On compare leurs lobes occipitaux (partie consacrée à la vision), histologiquement. Les neurones des animaux qui ont été à l'obscurité sont beaucoup moins développés que les autres. Il y a donc des facteurs de croissance qui dépendent des conditions externes et dont la forme même de l'information n'est pas contenue dans le patrimoine génétique.

J. Monod: Vous croyez que, dans la constitution du système nerveux central, une part importante, significative de l'information, pourrait venir d'ailleurs que de l'œuf. Votre réponse, M. Fessard signifie, qu'une part importante de la structure d'un système nerveux central pourrait être déterminée par autre chose que par l'œuf initial.

A. Fessard: Elle est déterminée en partie par les conditions extérieures dans lesquelles l'œuf se développe. Dans la construction du système nerveux central, interviennent des facteurs hormonaux, par exemple . . .

J. Monod: Ils sont génétiquement déterminés.

A. Fessard: Oui. En fait, je suis très embarrassé pour exprimer correctement, et sur le champ, toute ma pensée.

J. D. Cowan: M. Monod's question is a very important and relevant one for the determination of neuronal structure. One has to be very careful, whenever one tries to assign information measures. In information theory the function which is used to enumerate the number of possibilities is entirely dependent on the statistical structure of the source of information and on the nature of the code.

If there is any hierarchical structure leading to complicated statistics, the decoder may not use any of the first order or second order statistics, it may use only higher order, statistics, so that unless one knows exactly what is the code, these numbers mean absolutely nothing. It's meaningless to ask for the number of bits specified by the code. It would be very difficult to enumerate and it wouldn't tell one anything useful.

H. C. Longuet-Higgins: It seems to me that any estimate of the number of bits needed to specify a nervous system, even if this quantity could be properly defined, would be likely to be out by at least a factor of ten, and possibly by several powers of ten. In these circumstances perhaps the simplest and best way of obtaining an idea of the (anatomical) complexity of the human brain would be to compare the amount of genetic material needed to specify a human being with the amount needed to specify a related animal with a much smaller brain—perhaps a rat.

I. PRIGOGINE: Une simple petite remarque, toute proche de celle que M. Cowan a faite. Je ne pense pas que cette notion d'information, ici, soit très intéressante, et je ne pense pas que des calculs de ce genre puissent conduire très loin. Je pense à des analogies disons physiques—. . . j'ai cité un exemple dans mon exposé: le problème de Benard. C'est l'exemple le plus simple, que je connaisse, dans lequel vous voyez apparaître des structures extrêmement régulières, une organisation très complexe sur le plan moléculaire mais peut-être encore insignifiante par rapport à la biologie.

Vous pourriez aussi vous poser la même question, vous demander quelle information il faut fournir au fluide pour construire cette structure. Vous arrivez à un nombre fantastique! . . . qui ne signifie vraiment rien. Il suffit de dépasser, le nombre critique de Raighley, pour que, spontanément, avec une probabilité 1, on observe cette structure.

L. TISZA: I am much impressed by this beautiful application of information theory. I may be allowed to make a short remark that is not directly related to this talk, but that is connected with the concept of information. I am alluding to the well known claim that information is essentially identical to entropy, or neg-entropy. This conclusion is based on the fact that both concepts can be defined in terms of a formula $-\sum p_i \log p_i$, where $p_i$ is a probability distribution function. Without questioning the use of this formula, we may still point out that the nature of the averaging over the distribution function has a different character in the two instances. Let us consider a complicated molecule, say, DNA. The variables describing this systems can be classified into free and fixed parameters. The former specify, say the vibrational states of the molecule. These parameters are rapidly changing over all the possible values compatible with the nature of the system and the temperature of the environment in which the molecule is embedded. Therefore we call them *free variables*. The summation over these states in the above expression yields the entropy of the system. Consider now a class of molecules that are isomers or other well defined variants of each other. We assume also that these isomers are stable, and, apart from occasional "mutations" do not transform into each other. That is what we express by saying that the isomeric states are specified by *fixed parameters*. By selecting a particular isomer, we can code an amount of information that is obtained by summing the above expression over the class from which the selection was made. It is apparent that there is a significant difference between the two cases.

J. POLONSKY: Ma remarque concerne le lien entre l'information et l'entropie dans un système organisé à l'échelle moléculaire. La première difficulté que l'on rencontre ici provient du fait que l'entropie est un concept thermodynamique tandis que le concept de l'information y est ignoré. La deuxième difficulté est encore plus grave, car la thermodynamique ne traite que des paramètres opérationnels à l'échelle macroscopique, les paramètres microscopiques n'y interviennent qu'à titre statistique. Pour tourner ces difficultés, rien ne nous interdit de passer par l'intermédiaire des concepts utilisés en thermodynamique et dans la théorie de l'information; notamment: les degrés de liberté et les contraintes dans un système. Les degrés de liberté dans un système moléculaire concernent les états électroniques, les états de vibration, de rotation et de translation. Les contraintes sont déterminées par les barrières quantiques qui protègent les états fondamentaux et par l'ensemble des restrictions quantiques que fixent les probabilités et les corrélations dans les transitions. On peut distinguer ainsi, dans un système moléculaire, trois catégories de contraintes. Leur classement dépend de la nature quantique de l'énergie d'excitation:

1. *Les contraintes fixes* (non modifiées par l'excitation):

   Elles contribuent d'une façon plus ou moins sélective:
   a) à la réduction de la température statistique ou du "bruit" du système (c'est le cas du squelette structural) et
   b) aux corrélations fixes dans un organe cybernétique.
   En théorie de l'information, ces contraintes apparaissent sous forme de redondance c.a.d. d'une information comme à priori.

2. *Les contraintes réversibles* (régies par des matrices des probabilités):

   concernent l'information potentielle fournie par les transitions non aléatoires dans le système.

3. *Les contraintes non réversibles*

   forment le "bruit" et concernent les transitions aléatoires d'états (dont les barrières sont inférieures à kT).
   La troisième catégorie des contraintes s'apparente à l'entropie thermodynamique tandis que les deux premières peuvent être rattachées aux paramètres extensifs de travail, c'est-à-dire à une entropie négative.

H. C. LONGUET-HIGGINS: May I impose just one more remark? There are no fewer than $10^{18}$ bits in one calorie per degree Centigrade—this being the practical unit of entropy in ordinary thermodynamic experiments. So the entropy change when I take in a lecture by Dr. Polonsky is quite negligible compared with the entropy change when I digest a piece of bread.

P. Auger: Un mot sur ce qui a été dit tout à l'heure par M. Fessard, se demandant si l'information pouvait être contenue, effectivement, dans le génome. Il n'y a aucune espèce de doute: cette information est contenue dans le génome. Le fait que l'on ait pu constater, sur des animaux élevés dans l'obscurité, un cerveau un peu différent de celui des mêmes animaux élevés dans la lumière, veut dire simplement que cette information est bien utilisée ou mal utilisée. Bien utilisée dans le cas où l'animal est dans des conditions normales, c'est-à-dire éclairé. Mal utilisée, quand il est dans des conditions d'obscurité. Et, alors, il y a une baisse (si je puis dire) d'information. Des choses perdues. C'est pourquoi la structure est différente. Je ne pense pas qu'on puisse tirer, de cette expérience, une indication en faveur d'une action informative du milieu. L'information vient seulement du génome, mais elle est plus ou moins bien utilisée suivant les conditions dans lesquelles l'animal a été élevé. Cependant, dans le cas de l'homme, il y a des expériences probantes, montrant que l'anatomie du cerveau n'est pas tout à fait la même lorsqu'il s'agit d'un enfant élevé dans des conditions familiales normales, auquel on a parlé, par exemple, très tôt, ou d'un enfant qui n'a pas, pour une raison quelconque, été traité de cette manière. Il y a là, peut-être, une injection d'information dans la structure anatomique des synapses du cerveau. Mais dans ce cas il est venu du milieu extérieur une information véritable et pas seulement des conditions générales. Il peut y avoir une injection d'information nouvelle, qui serait parvenue au cerveau pendant la période de formation de la jeunesse. Mais il faut une *information* supplémentaire, et des conditions générales de milieu ne suffiraient pas.

O. Klein: In fact many theoretical physicists are rather sceptical as to the possibility of explaining the extremely complicated organisations brought about through biological evolution entirely on the basis of random mutations and natural selection. This attitude does by no means question the overall importance of natural selection in the evolution process, but what seems doubtful is whether purely random mutations can produce the necessary material for selection to act on, the number of generations available—while extremely large from an ordinary viewpoint—being still quite limited.

I have just heard from Dr. Löwdin that he has made a theoretical estimate of the probability for the occurrence of spontaneous mutations by means of quantum mechanics, his result being in fair agreement with experimental estimates. It would be extremely interesting if on this background a theoretical evaluation could be obtained of the number of generations needed for the evolution of a given new character, depending on a series of small spontaneous mutations, under selection pressure due to slowly changing environment. This is certainly no easy problem, the main difficulty being probably a realistic estimate of the selection pressure under conditions, near to those which have produced important features of adaptation.

It is not astonishing that biologists are on their guard against premature belief in the inheritance of acquired characters—like that of Lysenko. So were chemists —warned by alchemists—towards the belief that elements may be transmuted into one another. Still this, so to say, normal attitude proved to be exaggerated. In fact, already at the early time, when atomic weights began to be known, a hypothesis on the lines of that proposed by Prout—that atoms are not indivisible but built of still smaller constituents of the atomic weight of hydrogen—was quite reasonable. It was easily refuted, however, when it became clear, that atomic weights deviate quite considerably from integer multiples of that of hydrogen— isotopes being then unknown. Now we know, that our principal energy source is the transmutation of elements in the suns interior by means of a process which is much too slow to be observed under ordinary laboratory conditions.

Now, biological evolution is very slow as compared with laboratory conditions. This would seem to demand some caution against generalizing short-time expe- rience regarding non-inheritance of acquired characters. Moreover, the objects used for such studies may not have been the best for the purpose. Insects with their rigidly fixed instinctive behaviour would probably acquire very little during life. On the other hand, human beings learn so easily, that small mutations directed by an acquired state of the parent—although such mutations may have played a decisive role during the many generations preceding the appearance of the present human species—would probably be very hard to detect.

Apart from these reasons for not believing too firmly in the non-inheritance of acquired characters there seems to be one reason to believe that the probability of fixing an acquired character by mutations is not negligible. This comes from the existence of two molecular codes, both directing the behaviour of individuals and often in a strikingly similar way. One is the genetical code, which does only rarely change but has been slowly built up during the long history of the evolution of living beings. The other, the memory—the word used in a wide sense—is built up during the life of the individual. It would seem strange, if there were no such interaction between these two codes—probably having a similar chemical background—which could give rise to directed mutations, implying a certain, although small, probability for the fixation of an acquired character. Also it seems that under conditions of strong selection pressure a strain possessing this property would have a decisive advantage over other strains in the same population lacking it.

My aim in mentioning these still rather undeveloped ideas is just to warn against too dogmatic an attitude to the question whether acquired characters can be inherited which may become a hindrance to further investigation.

L. ONSAGER: In Ochoa's table I looked for some correlation between the physical properties of the amino acids and the coding system. There is quite a strong correlation. In the left half of the table you find not a single amino acid with a polar side chain [a]). Among the entries in the right half, the code words for aspartic and glutamic acids differ in one base.

This might have a bearing on the prebiological evolution. What came first, the chicken or the egg? One is encouraged to think that somehow proteins and nucleic acids might have evolved concurrently.

a) Added in print: This was an overstatement. Most of the code words whose middle letter is U or C (pyrimidines) code for amino acids with hydrocarbon side chains (Phe, Leu, Ileu, Val, Ala, Pro) and one or two for methionine, but two sets of four (UCX, ACX) code for serine and threonine, which carry hydroxyl groups.

R. P. DOU: I would like to make a very brief comment. I think that the question of Monod is very interesting and it makes sense if we understand brain as not too much determinate, that is brain as qualifying for the species. Then it would be useful to try to know how many bits of information or what amount of information would be required. This leads to another question. If one tries to define more precisely the required amount of information for one very determinate brain, say of one mammel, then I rather suspect that not only it is impossible to determine the amount of information, but that is not determined at all. I do not see any contradiction in the fact that the number of bits or amount of information, depending not only on the subject but also on the environment, determine the brain in such a way, that, in the present state of science and perhaps forever, the brain itself cannot be determined. Not even theoretically. For instance, because required differential equations may have plenty of indeterminacies.

H. C. LONGUET-HIGGINS: As Chairman may I just say that M. Monod's question seems to have been forgotten. He asked not about the number of bits of information needed to describe the human brain but the amount of genetic material which was primarily responsible for the making of a brain—at least I *think* this was the question he asked. Questions about the "amount of information" are unreal questions, in my own view.

R. WURMSER: Un mot au sujet de l'analogie hydrodynamique présentée par M. Prigogine. Lorsque sont réunies dans une cellule les éléments nécessaires pour constituer les substances protéiques et autres, propres au système nerveux par exemple, il faut comprendre que, compte tenu des vitesses des réactions enzymatiques, de la diffusion et des divers gradients, la formation des structures s'en suit nécessairement. Le problème de la différentiation sous son aspect thermodynamique très général serait donc ainsi résolu. C'est bien, je pense, l'idée de M. Prigogine.

I. PRIGOGINE: Tout à fait!

R. WURMSER: Cela pourrait répondre à la question de M. Monod. En conséquence de la création spontanée de structures par des réactions chimiques, elles-mêmes dépendant des structures déterminées par le génome, la quantité d'information nécessaire pour construire le système est moindre que si tous les détails structuraux étaient commandés indépendamment par le génome. Par suite de la création spontanée de structures, il y aura moins de types de protéines à synthétiser.

J. MONOD: Combien en faut-il? A cela revient la question.

R. WURMSER: En effet, si l'on possédait cette donnée, il serait alors facile de faire un calcul puisqu'il y a une relation entre la néguentropie de chaque proto-type de protéine, et la néguentropie, donc la structure, de la fraction du génome qui porte les directives nécessaires à la synthèse de ce prototype.

M. DALCQ: Permettez-moi de considérer une fois de plus l'objet de la présente discussion du point de vue de l'embryologiste. A mon avis, il y a certes une part de vérité dans les diverses interventions que nous venons d'entendre. Il me semble qu'il faudrait intégrer ce que nous a appris l'étude du développement du système nerveux central.

De ce point de vue, la question ne semble pas être de supputer d'après la situation observée chez l'être adulte combien de "bits" d'information sont néces-saires pour expliquer l'apparition de tel groupement fonctionnel de neurones, mais de considérer dans leur ordre la série des étapes ontogénétiques qui aboutissent à l'existence des diverses localisations qui nous préoccupent au niveau du cortex cérébral.

Ces événements successifs sont complexes. Faisant abstraction ici du jeu des facteurs qui localisent les organes axiaux dans un certain territoire, je rappellerai que la première ébauche de l'organe neural est induite par le chordo-mésoblaste qui vient se glisser sous cette future ébauche. De la part de celle-ci, la seule "in-formation" qui intervienne alors est celle qui confère à l'ectoblaste d'un certain âge la capacité de réagir globalement au flux inducteur qu'il reçoit. La modifi-cation suivante est due à l'activité d'un petit groupe de cellules posté en avant du chordo-mésoblaste. Il s'agit de la plaque préchordale, laquelle émet vers la partie antérieure de la plaque neurale un flux inducteur plus actif que le précédent. Ce flux confère au territoire ainsi imprégné la capacité d'édifier le prosencéphale, futur cerveau. Il s'agit donc d'une nouvelle "information" d'origine exogène, et qui est reçue, apparemment, de façon passive. La situation va alors se compliquer pour diverses raisons, qui sont loin d'être élucidées. On aperçoit que dans la nappe inductrice de la plaque préchordale des foyers spéciaux vont s'individualiser. Ils intensifient—ou modifient—le flux inducteur de façon à faire apparaître les ébauches destinées à l'olfaction et à la vision. C'est-là, semble-t-il, une troisième "information" exogène, de nouveau plus localisée que la précédente. Toutefois, à la même période, l'appareil circulatoire s'est constitué et entre en fonction. Il assure vers le futur cerveau des apports trophiques qui ne seront pas uniformément répartis. Par ailleurs, les parties du tube neural situées en arrière du futur prosencéphale subissent aussi une induction, mais quelque peu différente. Il s'y forme des structures en continuité avec les précédentes, et des interactions mutuelles vont s'établir. A ce moment, on est tenté de croire que des déterminants intrinsèques entrent davantage en jeu, mais c'est probablement par des signaux informateurs concernant d'assez vastes territoires et relativement peu nombreux.

Il serait hasardeux de pousser davantage cette esquisse analytique. Elle suffit à faire voir que, dans ce cortège d'événements, les "instructions" fournies par le génome ne sont pas nécessairement prépondérantes, sans cependant être négligeables. En tout cas, l'information toute préparée qu'elles représentent n'est pas seule à agir. Elle se combine constamment avec de l'épigenèse introduite par des relations de voisinage, et qui d'ailleurs comporte elle-même l'éclosion progressive de signaux informateurs, dus soit au génome de l'inducteur, soit à d'autres facteurs propres à celui-ci.

Cette vision des événements ne me permet pas de me déclarer d'accord avec la conception que M. Auger a brièvement formulée, et je me sens au contraire en accord avec M. Fessard quant à l'importance de l'épigenèse au cours des étapes de la vie foetale. Ces remarques valent, à mon avis, pour tous les aspects du développement, et non moins pour le cerveau que pour les autres organes.

B. B. LLOYD: I have a very simple question, probably very naive, but it would certainly help me to get my ideas into some sort of proportion if one of the mathematicians or engineers here could tell us how much DNA would be needed to code the largest computers so far built, that is the construction program for building that computer. Is this much more than the genetic material in a man or is it a thousandth part of it?

H. C. LONGUET-HIGGINS: As a representative—though a novice—of computer science I think it's a very difficult question to ask. The answer would inevitably depend on what kind of knowledge of mechanics or electricity one had before one started.

J. D. COWAN: There is no answer; there is no question.

B. B. LLOYD: There must be an answer to this because I am sure that the computer constructors to a great extent do use a program which must be codable on a tape.

H. C. LONGUET-HIGGINS: If you ask about a computer, how much paper tape with how many holes in it is needed to convert it into a particular special purpose machine, then the answer *can* be given, and it comes out in tens of thousands of holes.

P. AUGER: Il me paraît qu'il existe une réponse simple. Vous écrivez, dans un livre, ce qu'il faut pour construire votre machine à calculer. Vous écrivez cela en lettres, et vous pouvez donc parfaitement mesurer le nombre de "bits" qu'il y a dans le livre, sans son ensemble.

P. O. LÖWDIN: I would like to remark that the information contained in the genetic code is enormous. We know that the bacterium Escherichia Coli has a circular chromosome which contains at least three million nucleotides, and if one would print the genetic information and put a letter for each nucleotide, one would need about one volume of Encyclopedia Brittanica. So, if one would like to print the content of the human genetic code, he would probably need one thousand volumes of the Encyclopedia size. With the same space available, one could give extensive instructions for building a rather big electronic computer!

H. C. LONGUET-HIGGINS: If I may make another remark—the code only has a meaning in a very special context; its interpretation depends on the environment into which it is placed. The genetic code will not mean anything except in an environment in which protein can be synthetized.

F. GROS: Au sujet de la structure tri-dimensionnelle d'une protéine, déjà pré-déterminée, en quelque sorte, dans la colinéarité d'assemblage de ses acides aminés, la réponse est très certainement oui. Une grande partie de cette séquence, tout au moins, a une influence déterminante sinon décisive sur le mode de repliement de la protéine. Ceci est étayé par une série de faits expérimentaux.

# L'INFORMATION EN BIOLOGIE

*2ème séance*

PRÉSIDENT F. LYNEN

J. MONOD, J. WYMAN AND J. P. CHANGEUX

On the Nature of Allosteric Transitions: a Plausible Model


E. WOLFF

L'Origine de la Symétrie Bilatérale chez l'Embryon


Discussions

*Theoretical Physics and Biology*, © 1969, North-Holland Publ. Co., Amsterdam

# ON THE NATURE OF ALLOSTERIC TRANSITIONS: *)
# A PLAUSIBLE MODEL

JACQUES MONOD, JEFFRIES WYMAN and JEAN-PIERRE CHANGEUX

*Service de Biochimie Cellulaire, Institut Pasteur, Paris, France
and Istituto Regina Elena per lo Studio e la Cura dei Tumori, Rome, Italy*

> "*It is certain that all bodies whatsoever, though they have no sense, yet they have perception; for when one body is applied to another, there is a kind of election to embrace that which is agreeable, and to exclude or expel that which is ingrate; and whether the body be alterant or altered, evermore a perception precedeth operation; for else all bodies would be like one to another.*"
>
> *Francis Bacon*
> (*about* 1620)

## 1. Introduction

Ever since the haem-haem interactions of haemoglobin were first observed (Bohr, 1903), this remarkable phenomenon has excited much interest, both because of its physiological significance and because of the challenge which its physical interpretation offered (cf. Wyman, 1948, 1963). The elucidation of the structure of haemoglobin (Perutz *et al.*, 1960) has, if anything, made this problem more challenging, since it has revealed that the haems lie far apart from one another in the molecule.

Until fairly recently, haemoglobin appeared as an almost unique example of a protein endowed with the property of mediating such indirect interactions between distinct, specific, binding-sites. Following the pioneer work of Cori and his school on muscle phosphorylase (see Helmreich & Cori, 1964), it has become clear, especially during the past few years, that, in bacteria as well as in higher organisms, many enzymes are electively endowed with specific functions of metabolic regulation. A systematic, comparative, analysis of the properties of these proteins has led to the conclusion that in most, if not all, of them, *indirect* interactions between *distinct* specific binding-sites (allosteric effects) are responsible for the performance of their regulatory function (Monod, Changeux & Jacob, 1963).

By their very nature, allosteric effects cannot be interpreted in terms of the classical theories of enzyme action. It must be assumed that these interactions

*) Jacques Monod's report "Some considerations on the functional importance of molecular symmetry of some biological macromolecules" was based on this paper (published in the Journal of Molecular Biology **12** (1965) 88–118). We cordially thank Messrs. Monod, Wyman and Changeux, the editor in chief, J. C. Kendrew, and the publisher of the Journal of Molecular Biology (Academic Press) for their permission to reproduce that paper.

are mediated by some kind of molecular transition (allosteric transition) which is induced or stabilized in the protein when it binds an "allosteric ligand". In the present paper, we wish to submit and discuss a general interpretation of allosteric effects in terms of certain features of protein structure. Such an attempt is justified, we believe, by the fact that, even though they perform widely different functions, the dozen or so allosteric systems which have been studied in some detail do appear to possess in common certain remarkable properties.

Before summarizing these properties, it will be useful to define two classes of allosteric effects (cf. Wyman, 1963):

(a) *"homotropic"* effects, i.e. interactions between *identical* ligands;
(b) *"heterotropic"* effects, i.e. interactions between *different* ligands.

The general properties of allosteric systems may then be stated as follows:

(1) Most allosteric proteins are polymers, or rather oligomers, involving several identical units.
(2) Allosteric interactions frequently appear to be correlated with alterations of the *quaternary* structure of the proteins (i.e. alterations of the bonding between subunits).
(3) While heterotropic effects may be either positive or negative (i.e. co-operative or antagonistic), homotropic effects appear to be always co-operative.
(4) Few, if any, allosteric systems exhibiting *only* heterotropic effects are known. In other words, co-operative homotropic effects are almost invariably observed with at least one of the two (or more) ligands of the system.
(5) Conditions, or treatments, or mutations, which alter the heterotropic interactions also simultaneously alter the homotropic interactions.

By far the most striking and, physically if not physiologically, the most interesting property of allosteric proteins is their capacity to mediate homotropic co-operative interactions between stereospecific ligands. Although there may be some exceptions to this rule, we shall consider that this property characterizes allosteric proteins. Furthermore, given the close correlations between homotropic and heterotropic effects, we shall assume that the same, or closely similar, molecular transitions are involved in both classes of interactions. The model which we will discuss is based upon considerations of molecular symmetry and offers primarily an interpretation of co-operative homotropic effects. To the extent that the assumptions made above are adequate, the model should also account for heterotropic interactions and for the observed correlations between the two classes of effects.

We shall first describe the model and derive its properties, which will then be compared with the properties of real systems. In conclusion, we shall discuss at some length the plausibility and implications of the model with respect to the quaternary structures of proteins.

## 2. The Model

Before describing the model, since we shall have to discuss the relationships between subunits in polymeric proteins, we first define the terminology to be used as follows:

(a) A polymeric protein containing a *finite*, relatively small, number of *identical* subunits, is said to be an *oligomer*.

(b) The *identical* subunits associated within an oligomeric protein are designated as *protomers*.

(c) The term *monomer* describes the fully dissociated protomer, or of course any protein which is not made up of *identical* subunits.

(d) The term "subunit" is purposely undefined, and may be used to refer to any chemically or physically identifiable sub-molecular entity within a protein, whether identical to, or different from other components.

Attention must be directed to the fact that these definitions are based exclusively upon considerations of identity of subunits and do not refer to the number of different peptide chains which may be present in the protein. For example, a protein made up of two different peptide chains, each represented only once in the molecule, is a monomer according to the definition. If such a protein were to associate into a molecule which would then contain two chains of each type, the resulting protein would be a dimer (i.e. the lowest class of oligomer) containing two protomers, each protomer in turn being composed of two different peptide chains. Only in the case where an oligomeric protein contains a single type of peptide chain would the definition of a protomer coincide with the chemically definable subunit. An oligomer the protomers of which all occupy exactly equivalent positions in the molecule may be considered as a "closed crystal" involving a fixed number of asymmetric units each containing one protomer.

The model is described by the following statements:

(1) Allosteric proteins are oligomers the protomers of which are associated in such a way that they all occupy equivalent positions. This implies that the molecule possesses at least one axis of symmetry.

(2) To each ligand able to form a *stereospecific* complex with the protein there corresponds one, and only one, site on each protomer. In other words, the symmetry of each set of stereospecific receptors is the same as the symmetry of the molecule.

(3) The conformation of each protomer is constrained by its association with the other protomers.

(4) Two (at least two) states are reversibly accessible to allosteric oligomers. These states differ by the distribution and/or energy of inter-protomer bonds, and therefore also by the conformational constraints imposed upon the protomers.

(5) As a result, the affinity of one (or several) of the stereospecific sites towards the corresponding ligand is altered when a transition occurs from one to the other state.

(6) When the protein goes from one state to another state, its molecular symmetry (including the symmetry of the conformational constraints imposed upon each protomer) is conserved.

Let us first analyse the interactions of such a model protein with a single ligand (F) endowed with differential affinity towards the two accessible states. In the absence of ligand, the two states, symbolized as $R_0$ and $T_0$, are assumed to be in equilibrium. Let $L$ be the equilibrium constant for the $R_0 \rightleftharpoons T_0$ transition. In order to distinguish this constant from the dissociation constants of the ligand, we shall call it the "allosteric constant". Let $K_R$ and $K_T$ be the microscopic dissociation constants of a ligand F bound to a stereospecific site, in the R and T states, respectively. *Note that by reason of symmetry and because the binding of any one ligand molecule is assumed to be intrinsically independent of the binding of any other, these microscopic dissociation constants are the same for all homologous sites in each of the two states.* Assuming $n$ protomers (and therefore $n$ homologous sites) and using the notation $R_0, R_1, R_2, \ldots R_n$; $T_0, T_1, T_2, \ldots T_n$, to designate the complexes involving $0, 1, 2, \ldots n$ molecules of ligand, we may write the successive equilibria as follows:

$$R_0 \rightleftharpoons T_0$$

$$R_0 + F \rightleftharpoons R_1 \qquad T_0 + F \rightleftharpoons T_1$$
$$R_1 + F \rightleftharpoons R_2 \qquad T_1 + F \rightleftharpoons T_2$$
$$\cdots \cdots \cdots \qquad \cdots \cdots \cdots$$
$$R_{n-1} + F \rightleftharpoons R_n \qquad T_{n-1} + F \rightleftharpoons T_n$$

Taking into account the probability factors for the dissociations of the $R_1, R_2, \ldots R_n$ and $T_1, T_2 \ldots T_n$ complexes, we may write the following equilibrium equations:

$$T_0 = LR_0$$

$$R_1 = R_0\, n\, \frac{F}{K_R} \qquad T_1 = T_0\, n\, \frac{F}{K_T}$$

$$R_2 = R_1\, \frac{n-1}{2}\, \frac{F}{K_R} \qquad T_2 = T_1\, \frac{n-1}{2}\, \frac{F}{K_T}$$

$$\cdots \cdots \qquad \cdots \cdots$$

$$R_n = R_{n-1}\, \frac{1}{n}\, \frac{F}{K_R} \qquad T_n = T_{n-1}\, \frac{1}{n}\, \frac{F}{K_T}$$

Let us now define two functions corresponding respectively to:
(a) the fraction of protein in the R state:

$$\bar{R} = \frac{R_0 + R_1 + R_2 + \ldots + R_n}{(R_0 + R_1 + R_2 + \ldots + R_n) + (T_0 + T_1 + T_2 + \ldots + T_n)}$$

(b) the fraction of sites actually bound by the ligand:

$$\bar{Y}_F = \frac{(R_1 + 2R_2 + \ldots + nR_n) \quad + \quad (T_1 + 2T_2 + \ldots + nT_n)}{n[(R_0 + R_1 + R_2 + \ldots + R_n) + (T_0 + T_1 + T_2 + \ldots + T_n)]}.$$

Using the equilibrium equations, and setting

$$\frac{F}{K_R} = \alpha \text{ and } \frac{K_R}{K_T} = c$$

we have, for the "function of state" $\bar{R}$:

$$\bar{R} = \frac{(1+\alpha)^n}{L(1+c\alpha)^n + (1+\alpha)^n} \tag{1}$$

and for the "saturation function" $\bar{Y}_F$:

$$\bar{Y}_F = \frac{Lc\alpha(1+c\alpha)^{n-1} + \alpha(1+\alpha)^{n-1}}{L(1+c\alpha)^n + (1+\alpha)^n}. \tag{2}$$

In figs. 1(a) and (b), theoretical curves of the $Y_F$ function have been drawn, corresponding to various values of the constants $L$ and $c$. In such graphs the co-operative homotropic effect of the ligand, predicted by the symmetry properties of the model, is expressed by the curvature of the lower part of the curves. The graphs illustrate the fact that the "co-operativity" of the ligand depends upon the values of $L$ and $c$. The co-operativity is more marked when the allosteric constant $L$ is large (i.e. when the $R_0 \rightleftharpoons T_0$ equilibrium is strongly in favour of $T_0$) and when the ratio of the microscopic dissociation constants ($c = K_R/K_T$) is small. *) It should be noted that for $c = 1$ (i.e. when the affinity of both states towards the ligand is the same) and also when $L$ is negligibly small, the $\bar{Y}_F$ function simplifies to:

$$\bar{Y}_F = \frac{\alpha}{1+\alpha} = \frac{F}{K_R + F},$$

that is, to the Michaelis–Henri equation. The model therefore accounts for the homotropic co-operative effects which, as we pointed out, are almost invariably found with allosteric proteins. Let us now analyse the properties of the model

---

*) When $c$ is very small, eq. (2) simplifies to:

$$\bar{Y}_F = \frac{a(1+a)^{n-1}}{L + (1+a)^n}.$$

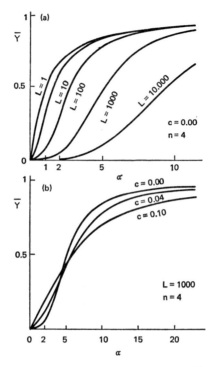

Fig. 1(a) and (b). Theoretical curves of the saturation function $\bar{Y}$ (eq. (2)) drawn to various values of the constants $L$ and $c$, with $n = 4$ (i.e. for a tetramer).

with respect to heterotropic interactions between different allosteric ligands. For this purpose, consider a system involving three stereospecific ligands, each binding at a different site. Assume that one of these ligands is the substrate (S) and, for simplicity, that it has significant affinity only for the sites in one of the two states (for example R). Assume similarly that, of the two other ligands, one (the inhibitor I) has affinity exclusively for the T state, and the other (the activator A) for the R state. Let $\bar{Y}_S$ be the fractional saturation of the enzyme with S.

According to the model, heterotropic effects would be due exclusively to displacements of the spontaneous equilibrium between the R and T states of the protein. The saturation function for substrate in the presence of activator and inhibitor may then be written as:

$$\bar{Y}_S = \frac{\alpha(1+\alpha)^{n-1}}{L'+(1+\alpha)^n},\tag{3}$$

where $\alpha$ is defined as above and $L'$ is an "apparent allosteric constant", defined as:

$$L' = \frac{\sum_0^n T_I}{\sum_0^n R_A},$$

where $\sum_0^n T_I$ and $\sum_0^n R_A$ stand respectively for the sum of the different complexes of the T state with I and of the R state with A. Following the same derivation as above, it will be seen that:

$$L' = L\frac{(1+\beta)^n}{(1+\gamma)^n},$$

with $\beta = \dfrac{I}{K_I}$ and $\gamma = \dfrac{A}{K_A}$, where $K_I$ and $K_A$ stand for the microscopic dissociation constants of activator and inhibitor with the R and T states respectively. Substituting this value of $L'$ in eq. (3) we have:

$$\bar{Y}_S = \frac{\alpha(1+\alpha)^{n-1}}{L\dfrac{(1+\beta)^n}{(1+\gamma)^n} + (1+\alpha)^n}. \tag{4}$$

This equation *) expresses the second fundamental property of the model, namely, that the (heterotropic) effect of an allosteric ligand upon the saturation function for another allosteric ligand should be to modify the homotropic interactions of the latter. When the substrate itself is an allosteric ligand (as assumed in the derivation of eq. (4)), the presence of the effectors should therefore result in a change of the *shape* of the substrate saturation curve. As is illustrated in fig. 2, the inhibitor increases the co-operativity of the substrate saturation curve (and also, of course, displaces the half-saturation point), while the activator tends to abolish the co-operativity of substrate (also displacing the half-saturation point). Both the activator and the inhibitor, as well as the substrate, exhibit co-operative homotropic effects.

The model therefore accounts for both homotropic and heterotropic inter-

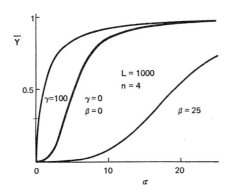

Fig. 2. Theoretical curves showing the heterotropic effects of an allosteric activator ($\gamma$) or inhibitor ($\beta$) upon the shape of the saturation function for substrate ($\alpha$) according to eq. (3).

---

*) A much more complicated, albeit more realistic equation would apply if the ligands were assumed to have significant affinity for *both* of the two states.

actions and for their interdependence. Its main interest is to predict these interactions solely on the basis of symmetry considerations. No particular assumption
has been, or need be, made about the structure of the specific sites or about
the structure of the protein, except that it is a symmetrically bonded oligomer,
the symmetry of which is *conserved* when it undergoes a transition from one to
another state. It is therefore a fairly stringent, even if abstract model, since co-
operative interactions are not only allowed but even required for any ligand
endowed with differential affinity towards the two states of the protein, and
heterotropic interactions are predicted to occur between any ligands showing
homotropic interactions.

## 3.   Application to the Description of Real Systems

### (a) *The kinetics of allosteric systems*

In fig. 3, results for the fractional saturation of haemoglobin by oxygen at
different partial pressures (Lyster, unpublished work) have been fitted to eq. (2).
While the fit is satisfactory, we feel that strict quantitative agreement is neither
sufficient nor necessary as a test of the basic assumptions of the model. It must
be borne in mind that in almost all enzyme systems, the saturation functions
with respect to substrate or effectors cannot be determined directly, and are only
inferred from kinetic measurements. (This of course does not apply to the case
of haemoglobin just cited.) Very often it is difficult to judge to what extent the
inference is correct, and the interpretation of kinetic results in terms of saturation
functions sometimes depends upon assumptions about the mechanism of the
reaction itself. It is to be expected, then, that most real systems will exhibit appreciable deviations from the theoretical functions, as indeed is very often the case

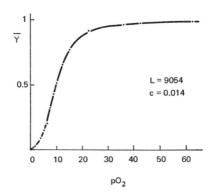

Fig. 3.   Saturation of haemoglobin with oxygen. Results (points) obtained by R. W. J. Lyster
(unpublished work) with horse haemoglobin (4.6%) in 0.6 M-phosphate buffer (pH 7) at 19 °C.
Solid line drawn to eq. (2) using the values of the constants $L$ and $c$ given on the graph.

for the much simpler Michaelis–Henri saturation law. We shall therefore discuss only the most characteristic qualitative predictions of the model in its application to real systems.

In any enzyme system, activating or inhibitory effects are measured in terms of variations of the two classical kinetic constants ($K_M$ and $V_M$), as a function of the concentrations of substrate (S) and effector(s) (F). Two classes of effects may then be expected in allosteric systems.

(a) "K *systems.*" Both F and S have differential affinities towards the T and R states (i.e. both F and S are *allosteric* ligands). Then evidently the presence of F will modify the apparent affinity of the protein for S, and conversely.

(b) "V *systems.*" S has the same affinity for the two states. Then there is no effect of F on the binding of S, nor of S on the binding of F. F can exert an effect on the reaction only if the two states of the protein differ in their *catalytic* activity. Depending on whether F has maximum affinity for the active or for the inactive state, it will behave as an activator (positive $V$ system) or as an inhibitor (negative $V$ system). It should be noted that this classification of allosteric systems is compatible with other mechanisms and does not depend upon the specific properties of the model. The following predictions, however, are based on the distinctive properties of the model.

(a) In an allosteric enzyme system, an *allosteric* effector (i.e. a specific ligand endowed with different affinities towards the two states) should exhibit co-operative homotropic interactions.

(b) In those systems in which an allosteric effector modifies the apparent affinity of the substrate, the substrate also should exhibit co-operative homotropic interactions.

(c) In those systems in which the effector does not modify the affinity of the substrate, the latter should not exhibit homotropic co-operative interactions.

As may be seen from inspection of table 1, where the properties of a number of systems have been summarized, all four classes of effects (positive and negative $K$ and $V$ systems) have been found among the dozen or so allosteric enzymes adequately studied. In inspecting table 1, it should be borne in mind that the published data concerning allosteric enzymes are very heterogeneous and often do not provide the kind of information which we are now seeking. Reasonably adequate kinetic data are available, however, for the systems numbered 1 to 8, 10, 13, 14, 16 ,18, 19 and 20. In all but two of these 15 systems, homotropic co-operative interactions of at least one of the ligands have been observed. Three of these systems (18, 19 and 20) show no $K$ effect of the inhibitor and no co-operative interactions of substrate, while the $K$ for systems 2 to 8 and 16 show

TABLE 1

Summary of properties of various allosteric systems *)

| Enzyme | Substrate | Inhibitor | Activator | V System | K System | Subunits | References |
|---|---|---|---|---|---|---|---|
| 1. Haemoglobin (vertebrates) (invertebrates) | Oxygen + | | | | + | + | Bohr, 1903; Wyman, 1963; Manwell, 1964 |
| 2. Biosynthetic L-threonine deaminase (E. coli K12) and (yeast) | L-Threonine + | L-Isoleucine + | L-Valine + | | + | (+) | Umbarger & Brown, 1958a; Changeux, 1961, 1962, 1963, 1964a,b; Freundlich & Umbarger, 1963; Cennamo et al., 1964 |
| 3. Aspartate transcarbamylase (E. coli) | Aspartate + Carbamyl phosphate | CTP + | ATP | | + | + | Gerhart & Pardee, 1962, 1963, 1964 |
| 4. Deoxycytidylate aminohydrolase (ass spleen) | dCMP + | dTTP + | dCTP + | | + | | Scarano et al., 1963, 1964; Scarano et al., 1967; Maley & Maley, 1963, 1964 |
| 5. Phosphofructokinase (guinea pig heart) | Fructose-6-phosphate ATP | ATP (+) | 3'-5' AMP | | + | | Passoneau & Lowry, 1962; Mansour, 1963, Vinuela et al., 1963 |
| 6. Deoxythymidine kinase (E. coli) | Deoxythymidine ATP + or GTP − | (dTTP) | dCDP | (+) | + | | Okazaki & Kornberg, 1964 |
| 7. DPN-isocitric dehydrogenase (N. crassa) | D-Isocitrate + DPN | (α-Ketoglutarate) | Citrate | | + | | Sanwal et al., 1963, 1964 |
| DPN-isocitric dehydrogenase (yeast) | D-Isocitrate + DPN | | 5' AMP | | + | | Hataway & Atkinson, 1963 |

| No. | Enzyme (source) | | | | | | | Reference |
|---|---|---|---|---|---|---|---|---|
| 9. | Homoserine dehydrogenase (*R. rubrum*) | Homoserine (−) Aspartate semialdehyde TPN–TPNH | L-Threonine | L-Isoleucine (+) L-Methionine | + | | + | Suram *et al.*, 1965, Datta *et al.*, 1964 |
| 10. | L-Threonine deaminase (*C. tetanomorphum*) | L-Threonine | ADP + | | | | + | Hayaishi *et al.*, 1963 |
| 11. | Acetolactate synthetase (*E. coli*) | Pyruvate (−) | L-Valine | | | | + | Umbarger & Brown, 1958b |
| 12. | "Threonine" aspartokinase (*E. coli*) | Aspartate (−) | L-Threonine | | | | + | Stadtman *et al.*, 1961 |
| 13. | L-Glutamine-D-fructose-6-P transaminase (rat liver) | L-Glutamine — D-Fructose-6-P | UDP-N acetyl-glucosamine − | | | | + | Kornfeld *et al.*, 1964 |
| 14. | Glycogen synthetase (yeast) (lamb muscle) | UDP-glucose — | Glucose-6-P — | | | | + | Algranati & Cabib, 1962; Traut & Lipmann, 1963 |
| 15. | Glutamate dehydrogenase (beef liver) | Glutamate | ATP GTP DPNH Oestrogens + Thyroxine | ADP Leucine + Methionine | + | (+) | + | (Ref. in Tomkins *et al.*, 1963) |
| 16. | Phosphorylase *b* (rabbit muscle) | Glucose-1-P + Glycogen P1 (+) | ATP | 5′ AMP + | + | | + | Helmreich & Cori, 1964; Madsen, 1964; Schwartz (personal communication); Ullmann *et al.*, 1964 |
| 17. | UDP-N acetyl-glucosamine-2-epimerase (rat liver) | UDP-N acetyl-glucosamine | CMP-N acetyl-neuraminic acid + | | | | + | Kornfeld *et al.*, 1964 |
| 18. | Homoserine dehydrogenase (*E. coli*) | Homoserine — Aspartate semialdehyde TPN-TPNH | L-Threonine + | | | | + | Patte *et al.*, 1963; Cohen *et al.*, 1963; Patte & Cohen, 1964 |
| 19. | "Lysine"-aspartokinase (*E. coli*) | Aspartate — ATP | L-Lysine + | | | | + | Stadtman *et al.*, 1961; Patte & Cohen, 1964 |

TABLE 1 (*continued*)

| Enzyme | Substrate | Inhibitor | Activator | V System | K System | Subunits | References |
|---|---|---|---|---|---|---|---|
| 20. Fructose-1-6-diphosphatase (frog muscle) (rat liver) | Fructose-1-6-diphosphate (−) | 5' AMP + | | + | | | Krebs, 1964; Salas et al., 1964; Taketa & Pogell, 1965 |
| 21. ATP-PRPP-pyro-phosphorylase (S. typhimurium) | ATP—PRPP | Histidine | | + | | + | Martin, 1962 |
| 22. "Tyrosine" 3-deoxy-D-arabinoheptulosonic-acid-7-phosphate synthetase (E. coli) | Phosphoenol-pyruvate—D-Erythrose-4-P— | L-Tyrosine | | + | | | Smith et al, 1962 |
| 23. "Phenylalanine" 3-deoxy-D-arabino-heptulosonic-acid-7-phosphate synthetase (E. coli) | Phosphoenol-pyruvate—D-Erythrose-4-P— | L-Phenylalanine | | + | | | Smith et al, 1962 |
| 24. Acetyl-CoA carboxylase (rat adipose tissue) | Acetyl CoA ATP, $CO_2$ | | Citrate + | | | + | Martin & Vagelos, 1962 |

*) The + and − signs against the name of the substrate(s) and effector(s) of each system indicate whether or not co-operative homotropic effects occur with the corresponding compound. A blank implies no relevant data, while (+) or (−) implies uncertainty. The + signs in the "$K$" and "$V$" columns indicate whether $K$ or $V$ effects have been observed. In the "subunit" column we have noted with a + those systems for which some evidence (direct or indirect) of the existence of subunits (not necessarily proved to be identical) has been obtained.

Note that (a) this summary is not claimed to be complete; (b) many of the systems listed have been described only recently and as yet incompletely; (c) the properties assigned to many systems represent our (rather than the original authors') interpretation of the data. We therefore assume responsibility for interpretative mistakes.

evidence of homotropic interactions for *both* substrate and effector(s), as predicted by the model. *)

It is somewhat difficult to judge whether systems 13 and 14 represent true exceptions or not. One of these (glycogen synthetase, no. 14) is a "positive $K$ system", where the occurrence of homotropic interactions might easily be missed. The other (glutamine—F6P transaminase, no. 13) is a "negative $K$ system" which has not yet been studied extensively. The possible significance of these exceptions will be considered in the general discussion.

Let us now examine some of the more specific predictions of the model. According to the theory developed above, the $V$ systems are described by the "function of state" ($\bar{R}$ or $1-\bar{R}$), assuming that the two states differ in their catalytic activity towards the substrate. We shall mostly discuss the properties of the $K$ systems, of which there are more examples and for which the predictions of the model are particularly interesting and characteristic.

According to the model, the complex kinetics of such systems simply result from displacements of the R $\rightleftharpoons$ T equilibrium, and their properties are described by eq. (3). We shall examine only a few typical experimental situations and compare them with predictions based on the model.

Consider first a $K$ system involving a substrate and an allosteric inhibitor. Assume that the R state binds the substrate, and the T state binds the inhibitor. We may expect that in any such system, the allosteric constant will be very different from 1. In other words, one of the two states (R or T) will be greatly favoured. Threonine deaminase of *E. coli* is a $K$ system, threonine being the substrate, and isoleucine the inhibitor (Umbarger, 1956). In the presence of inhibitor and substrate, the rate-concentration curve for *both* is S-shaped. In the *absence* of inhibitor, the substrate saturation curve is *still* S-shaped. According to the model, this indicates that the favoured state (in the absence of both ligands) is the one that has minimum affinity for threonine and maximum affinity for isoleucine. It is therefore expected that the saturation curve for inhibitor *in the absence of substrate* should be Michaelian, exhibiting no co-operative effect. This prediction has been verified experimentally (Changeux, 1946a). Moreover, as shown in fig. 4, the co-operativity of the inhibitor increases with the concentration of substrate.

More generally, in any $K$ system, we expect heterotropic effects to be expressed essentially as alterations of the homotropic "co-operativity" of any one allosteric ligand when in the presence of another. As a measure of homotropic effects, it is convenient to use the Hill approximation:

$$\bar{Y}=\frac{\alpha^{\underline{n}}}{Q+\alpha^{\underline{n}}}$$

* Attention must be directed to the fact that the homotropic effect of a ligand may not be expressed in the absence of an antagonistic ligand. For example, the co-operative interactions of G–1–P, in the case of phosphorylase *b*, are visible only in the presence of ATP (Madsen, 1964).

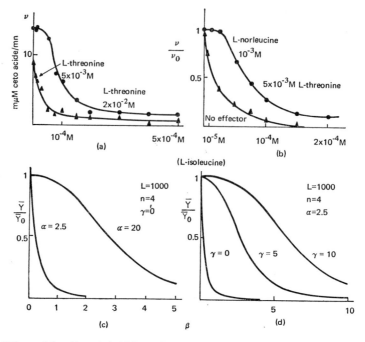

Fig. 4. Effects of the allosteric inhibitor L-isoleucine upon the activity of L-threonine deaminase. (a): In the presence of two different concentrations of the substrate (L-threonine). (b): At low concentration of substrate in the presence or absence of the allosteric activator L-norleucine. Compare with theoretical curves (c and d) describing similar situations according to eq. (3). Note that at low concentrations of substrate the co-operative effect of the inhibitor is scarcely detectable either in the theoretical or in the experimental curves. An increase of the concentration of substrate, or the addition of an activator, both reveal the co-operative effects of the inhibitor.

TABLE 2

Hill coefficients of homotropic interactions with respect to substrate ($n$), inhibitor ($n'$) and activator ($n''$) observed with dCMP deaminase

(From Scarano *et al.*, 1963; Scarano *et al.*, 1967)

|  |  | $n$ |
|---|---|---|
| Substrate (dCMP) | No effector | 2.0 |
|  | + dTTP    1.25 $\mu$M | 3.0 |
|  | „       2.25 $\mu$M | 4.1 |
|  | „      10.00 $\mu$M | 3.9 |
|  | + dCTP 100.00 $\mu$M | 1.0 |
|  |  | $n'$ |
| Inhibitor (dTTP) | Substrate concentration 4 mM | 3.4 |
|  |  | $n''$ |
| Activator (dCTP) | Substrate concentration 67 $\mu$M | 2.0 |

where $Q$ is a constant and $\underset{\sim}{n}$ (the Hill coefficient) is *not* the number of interacting sites (which we write $n$), but an interaction coefficient. It has been shown by one of us (Wyman, 1963) that under certain conditions the Hill coefficient can be interpreted as measuring the free energy of interaction between sites. As it may be seen from table 2, the Hill coefficients for the substrate of the allosteric system deoxycytidine deaminase are modified, in the expected direction, when the concentration of the other ligands (activator or inhibitor) varies. Another specific prediction of the model has been verified in the case of threonine deaminase, namely, the fact that a *true* competitive inhibitor (allothreonine), i.e. a substrate analogue (able to inhibit the enzyme by binding at the same site as the substrate), should exert the same effect as the substrate itself (L-threonine) as an antagonist of the allosteric inhibitor (isoleucine) (Changeux, 1964*a*). Another prediction, concerning the effect of analogues, is that at very low concentrations of substrate low concentrations of analogue should activate, rather than inhibit, the enzyme. This is observed with aspartic transcarbamylase (Gerhart & Pardee, 1963) (fig. 5) and also with threonine deaminase (Changeux, 1964*a*).

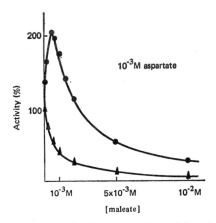

Fig. 5.    Effect of a substrate analogue (maleate) upon the activity of aspartic transcarbamylase at relatively low concentration of substrate (aspartate). Upper curve: native enzyme. Lower curve: desensitized enzyme. Note the large increase of activity at low maleate concentration which occurs with the native enzyme, but not with the desensitized enzyme (data from Gerhart & Pardee, 1963).

The effect of an allosteric *activator* in a $K$ system should be, according to the model, to decrease or abolish the substrate–substrate interactions. This has been observed in several different systems. As illustrated in fig. 6, the effect is particularly striking because, as expected, at saturating concentration of activator it results in converting the S-shaped rate–concentration curve for substrate into a Michaelian hyperbola. Moreover, of course, the presence of an activator should increase the co-operativity of an inhibitor, and conversely. Both effects are observed (see figs. 4 and 7).

It is clear from the model and the equations that the homotropic interactions of an allosteric ligand are independent of the absolute values of the microscopic dissociation constants. One may therefore expect that two sterically closely analogous ligands could bind to the same sites with the same interaction coefficient,

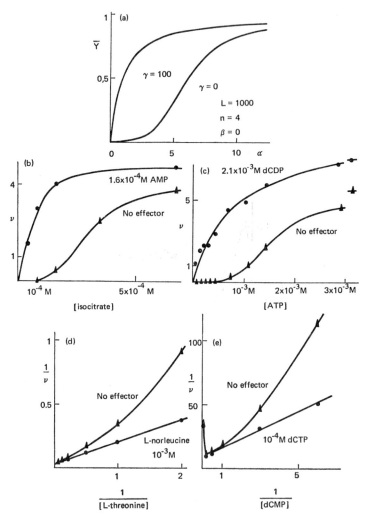

Fig. 6. Activity of various allosteric enzymes as a function of substrate concentration in the presence or absence of their respective activators. (a): Theoretical curve according to eq. (3). (b): DPN-isocitrate dehydrogenase from *Neurospora crassa* (results from Hataway & Atkinson, 1963). (c): Deoxythymidine kinase from *Escherichia coli* (results from Okazaki & Kornberg, 1964). (d): Biosynthetic L-threonine deaminase from *E. coli* (Lineweaver–Burk plot) (results from Changeux, 1962, 1963). (e): dCMP deaminase from ass spleen (Lineweaver–Burk plot) (results from Scarano *et al.*, 1963). Note that in all these instances, the presence of the allosteric activator abolishes the co-operative interactions of the substrate.

even though their affinities might be widely different. For example, with haemo-globin, the functionally significant steric features of the prosthetic groups must be virtually the same, whether the haems are bound to oxygen or to carbon monoxide. Therefore, although the affinity of carbon monoxide for the haem is known to be nearly 250 times that of oxygen, we should expect the interaction coefficients to be the same for both, as indeed they are (Wyman, 1948). When, however, the binding of two analogous ligands depends very much on steric factors, it may be expected that the *ratios* of the affinities of each ligand towards the two states of the protein (i.e. the constant *c* in eq. (1)) will be different. If so, the two ligands might bind to the same sites with widely different interaction coefficients. This appears to be the case, according to the observations of Okazaki & Kornberg (1964) for various triphosphonucleosides which act as phosphoryl donors in the deoxythymidine–kinase reaction. With ATP, for example, the rate–concentration curve is strongly co-operative, whereas with dATP the curves exhibit scarcely any evidence of homotropic effects. Furthermore, this enzyme, as

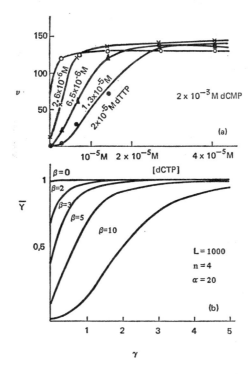

Fig. 7.  (a): Activity of dCMP deaminase as a function of the concentration of its allosteric-activator dCTP in the presence of substrate (dCMP) at near saturating concentration and at various concentrations of the allosteric inhibitor dTTP (results from Scarano, personal com-munication). (b): Theoretical curves of eq. (3) corresponding to a similar situation. Note that the co-operative effects of the activator are revealed only at relatively high concentration of the inhibitor.

shown by the same authors, is allosterically activated by CDP (fig. 6(c)). It is
easily seen that if this effect conforms to the model, activation should be observed
only with those substrates that show evidence of homotropic effects (ATP), and
not with those that do not (dATP). This, actually, is the observed result.

Since, again, the homotropic interactions are independent of absolute affinities,
certain conditions or agents may modify the affinity of an allosteric ligand without
altering its interaction coefficient. This is apparently the case for the Bohr effect
shown by haemoglobin: as is well known, the oxygen saturation curves obtained
at different values of pH can all be superimposed by a simple, adequately chosen,
change of the abscissa scale. In terms of the model, this would mean that the
binding of the "Bohr protons" does not alter the equilibrium between the two
hypothetical states of the protein. Hence also the Bohr protons themselves would
*not* be allosteric ligands, and their own binding is not expected to be co-operative.
This, again, appears to be the case, at least for human and horse haemoglobin
(Wyman, *loc. cit.*).

In the preceding paragraph, we have discussed only the more straightforward
predictions of the model. It should be pointed out that the model could also
account for more complicated situations, and for certain effects which were not
considered here. For example, it seems possible that, in some instances, the
phenomenon of inhibition by excess substrate might be due to an allosteric
mechanism (rather than to the classically invoked direct interaction between
two substrate molecules at the active site). This effect could be described on the
basis of our model by assuming two states with different affinities for the substrate,
the one with higher affinity being catalytically inactive. The equation for such
a situation would be of the form:

$$\frac{V}{V_m} = \frac{LS/K_a(1+S/K_a)^{n-1}}{L(1+S/K_a)^n+(1+S/K_I)^n}$$

with $K_I$ (dissociation constant of S with the inactive state) smaller than $K_a$ (dissociation constant with the active state).

(b) *Desensitization and dissociation*

One of the most striking facts about allosteric enzymes is that their regulatory
properties may be lost as a result of various treatments, without loss (indeed often
with increase) of activity (Changeux, 1961; Gerhart & Pardee, 1961). That it
should be so is understandable on the basis of the model, since conservation of the
interactions should depend upon the integrity of the whole native structure,
including in particular the inter-protomer binding, whereas conservation of
activity should depend only on the integrity of the active site. Also, according
to the model, the homotropic and heterotropic interactions should in general be
simultaneously affected, if at all, by alterations of protein structure. This was first

observed with threonine deaminase (Changeux, 1961) and ATCase (Gerhart & Pardee, 1962), and similar observations have since then been made with several other systems. These observations constituted the main initial basis for the assumption that regulatory interactions in general may be indirect (Changeux, 1961; Monod & Jacob, 1961; Monod *et al.*, 1963).

According to the model, loss of the interactions would follow from any structural alteration that would make one of the two states (R or T) virtually inaccessible. Now, one of the events most likely to result from various treatments of the protein is that quaternary (inter-protomer) bonds may be broken, completely or partially. One may therefore expect that:

(a)  Under any condition, or following any alteration, such that the protein is (and remains) dissociated, *both* types of interactions should disappear.

(b)  Conversely treatments, or mutations, which abolish the interactions should frequently be found to result in stabilization of a monomeric state.

These expectations are verified by observations made with at least two different systems (Gerhart & Pardee, 1963, 1964; Patte, Le Bras, Loviny & Cohen, 1963; Cohen & Patte, 1963).

Furthermore, since it is assumed from the model that in one of the two alternative states (R) the protomers are less constrained and therefore closer to the conformation of the monomer than in the other state (T), we expect that, under conditions where the protein is monomeric, it may exhibit high affinity for the ligand which stabilizes the R state, and little or no affinity for the ligand which stabilizes the T state. Hence, if the experiment can be performed, one may deduce *which* of the two states (R or T) is stabilized by a given ligand.

If conditions can be set up such that reversible dissociation of the protein actually occurs, one may expect that an allosteric ligand (i.e. any ligand exhibiting homotropic interactions) should now prove to act as a specific associative or dissociative agent. Actually, there is now clear evidence that under conditions where human haemoglobin shows a detectable amount of dissociation (low pH, high ionic strength), dissociation is favoured by oxygenation (Antonini, Wyman, Belleli & Caputo, unpublished experiments, 1961; Benesch, Benesch & Williamson, 1962; Gilbert & Chionione, recent unpublished experiments). Lamprey haemoglobin, in the oxygenated form, exists primarily as a monomer under all conditions, but when deoxygenated shows a strong tendency to polymerize (see table 3) (Briehl, 1963; Rumen, 1963). Myoglobin, which may be thought of as an isolated (and therefore relaxed) protomer of haemoglobin, has a much higher oxygen affinity, as would be expected on the basis of these two facts regarding human and lamprey haemoglobin.

Similarly, Changeux (1963) has found that in the presence of urea (1.5 M) threonine deaminase is reversibly dissociable. As expected, under these conditions, all three types of allosteric ligands active in the system, namely, the substrate

TABLE 3

Sedimentation coefficients of oxygenated and reduced lamprey haemoglobin (from Briehl, 1963)

| $t°$ C | pH | Haemoglobin concentration (E. 275) | Sedimentation coefficient ($S_{20,w}$) | |
|---|---|---|---|---|
| | | | Oxygenated | Reduced |
| 5.5 | 6.8 | 15.7 | 2.02 | 3.68 |
| 5.0 | 7.3 | 21.0 | 1.90 | 2.98 |

(threonine or analogue of threonine), the activator (valine) and the inhibitor (isoleucine) powerfully affect the dissociation, the inhibitor favouring the associated state, whereas both the substrate and the activator appear to stabilize the dissociated state. Hence, under normal conditions, the substrate and the activator presumably stabilize an R state, while the inhibitor favours a T state.

The observations of Datta, Gest & Segal (1964) on homoserine dehydrogenase from *Rhodospirillum rubrum* provide a further striking example of the effects of allosteric ligands upon dissociation of the protein. This enzyme is activated by both methionine and isoleucine, and inhibited by threonine. Both activators, as well as the substrate, promote dissociation of the protein, whereas the inhibitor favours an aggregated state.

We may conclude from the preceding discussions that the characteristic, unusual, apparently complex functional properties of allosteric systems can be adequately systematized and predicted on the basis of simple assumptions regarding the molecular symmetry of oligomeric proteins. In the next section, we shall examine the structural implications and the plausibility of these assumptions from a more general point of view.

## 4. Quaternary Structure and Molecular Symmetry of Oligomeric Proteins

### (a) Geometry of inter-protomer bonding

The first major assumption of the model is that the association between protomers in an oligomer may be such as to confer an element of symmetry on the molecule. The plausibility of this assumption has already been pointed out by Caspar (1963) and by Crick & Orgel (1964, and unpublished manuscript). We will analyse the implications of this assumption in terms of the possible or probable modes of bonding between protomers. Although next to nothing is known, from direct evidence, regarding this problem, the following statements would seem to be generally valid.

  (a)  A large number, probably a majority, of enzyme proteins are oligomers involving several *identical* subunits, i.e. protomers (see: Schachman, 1963;

Reithel, 1963; Brookhaven Symposium, 1964, *Subunit Structure of Proteins*, no. 17).

(b) In most cases the association between protomers in such proteins does not appear to involve covalent bonds.

(c) Yet most oligomeric proteins are stable as such (i.e. do not dissociate into true monomers, or associate into superaggregates), over a wide range of concentrations and conditions.

(d) The specificity of association is extreme: monomers of a normally oligomeric protein will recognize their identical partners and re-associate, even at high dilution, in the presence of other proteins (e.g. in crude cell extracts).

These properties indicate that within oligomeric proteins the protomers are in general linked by a *multiplicity* of non-covalent bonds, conferring both specificity and stability on the association. Clearly also the steric features of the bonded areas must play a major part.

Let us now distinguish between two *a priori* possible modes of association between two protomers. For this purpose we define as a "binding set" the spatially organized collection of all the groups or residues of *one* protomer which are involved in its binding to *one* other protomer. Considered together, the two linked binding sets through which two protomers are associated will be called the *domain of bonding* of the pair.

The two modes of association which we wish to distinguish may then be defined as follows.

(a) *Heterologous associations:* the domain of bonding is made up of two *different* binding sets.

(b) *Isologous associations:* the domain of bonding involves two *identical* binding sets.

These definitions imply the following consequences. *)

(1) In an isologous association (figs. 8 and 9), the domain of bonding has a two-fold axis of rotational symmetry. Along this axis, homologous residues (i.e. identical residues occupying the same position in the primary structure) face each other (and may form unpaired "axial" bonds). Anywhere else, within the domain of bonding, any bonded group-pair is represented *twice*, and the two pairs are symmetrical with respect to each other. Put more generally: in an isologous association, any group which contributes to the binding in one protomer furnishes precisely the same contribution in the other protomer. Isologous associations will therefore tend to give rise to "closed" i.e. *finite* polymers since, for example,

---

*) The validity of the statements that follow can be visualized and demonstrated best with the use of models. The interested reader may find it helpful to use a set of dice for this purpose.

an isologous dimer can further polymerize only by using "new" binding sets (i.e. areas and groups not already satisfied in the dimer). Note that this mode of association can give rise only to *even numbered* oligomers.

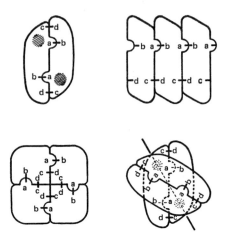

Fig. 8. Isologous and heterologous associations between protomers. Upper left: an isologous dimer. The axis of symmetry is perpendicular to the plane of the figure. Upper right: "infinite" heterologous association. Lower left: "finite" heterologous association, leading to a tetramer with an axis of symmetry perpendicular to the plane of the figure. Lower right: a tetramer constructed by using isologous associations only. Note that two different domains of bonding are involved.

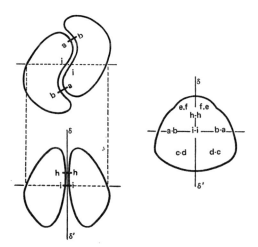

Fig. 9. Topography of the domain of bonding in an isologous association. Upper left: re-presented in a plane perpendicular to the axis of symmetry. Lower left: the same viewed in a plane of the axis of symmetry. Right: projection of the domain of bonding in a plane of the axis of symmetry. hh and ii, axial bonds; ab, ba, cd, dc, ef, fe, antiparallel bonds. It should be understood that in this figure the bonding residues a, b, c, etc. are supposed to project from under and from above the plane of the figure.

(2) In a heterologous association (fig. 8), the domain of bonding has no element of symmetry; each bonded group-pair is unique. Heterologous associations would, in general, be expected to give rise to polydisperse, eventually large, helical polymers except, however, in two cases.

    (a)  If polymerization is stopped at some point by steric hindrance, giving rise to a "hinged helix". Such aesthetically unpleasant structures should have less stability than "closed" structures.

    (b)  If a "closed" structure can be achieved such that any binding set which is used by one protomer is also satisfied in all the others. This is impossible of course in a dimer, but it is possible for trimers, tetramers, pentamers, etc., provided that the angles defined by the domains of bonding are right or nearly so. Such an oligomer would necessarily possess an axis of symmetry.

On the basis of these considerations, it is reasonable to assume that, if an oligomeric protein possesses a wide range of stability, it consists of a closed structure where all the protomers use the same binding sets; which implies, as we have just seen, that the molecule should possess at least one axis of symmetry.

Direct experimental evidence on this important problem is available for haemoglobin. As is well known, although made up of four subunits, haemoglobin is not, strictly speaking, a tetramer, since the $\alpha$ and $\beta$ chains are not identical. For our present purposes, however, we may consider the four subunits as equivalent protomers. The work of Perutz et al. (1960) has shown that these are associated into a pseudotetrahedral structure which possesses a twofold axis of symmetry.

Three further examples of oligomers possessing an element of molecular symmetry have recently been provided. Green & Aschaffenburg (1959) find that $\beta$-lactoglobulin (a dimer) has a dyad axis. Lacticodehydrogenase M4 (Pickles, Jeffery & Rossmann, 1964) and glyceraldehyde-phosphate-dehydrogenase (Watson & Banaszak, 1964), both tetrameric, appear to possess one (at least) axis of symmetry.

From the preceding discussion, and on the strength of these examples, it appears that oligomeric proteins are not only capable of assuming molecular symmetry, but also that this may be a fairly general rule.

Assuming this conclusion to be correct, it is of interest to enquire which mode of association (isologous or heterologous) may be most frequently used in Nature. For the reasons pointed out above, stable dimers, of which many examples are known, must represent isologous associations. Moreover, it may be pointed out that a symmetrical (isologous) dimer can further polymerize into a *closed* structure in two ways only.

(a) By again using isologous associations, thereby forming an isologous tetramer. Isologous polymerization, however, must stop at this point, since no further closed structure could be built by polymerization of such a tetramer.

(b) By using heterologous associations when the next closed structure would necessarily consist of three isologous dimers, and hence be an hexamer.

It follows from these remarks that (1) the exclusive use of isologous associations can lead to dimers and tetramers only; (2) the use of *both* isologous and heterologous domains of bonding should lead to even-numbered oligomers containing a *minimum* of six protomers; (3) the exclusive use of heterologous domains of bonding could lead to oligomers containing any number of protomers (except two). On this basis, the apparently rather wide prevalence of dimers and tetramers among oligomeric enzymes suggests rather strongly that the quaternary structures of these proteins are mostly built up by isologous polymerization.

### (b) *Protomer conformation: "quaternary constraints"*

The formation of stable, specific associations involving multiple bonds and strict complementarity between protein protomers is likely to imply in most cases a certain amount of re-arrangement of the tertiary structures of the monomers. Certain observations seem to confirm this assumption.

(1) The artificially prepared monomers of enzymes that are normally oligomeric generally exhibit functional alterations, suggesting that the structure of the active site in each protomer depends upon a conformation which exists only in the native oligomeric associated state (see Brookhaven Symposium, 1964, *Subunit Structure of Proteins*, no. 17).

(2) The rate of reactivation of oligomeric enzymes inactivated by dissociation into monomers is markedly dependent on temperature (alkaline phosphatase, Levinthal, Signer & Fetherolf, 1962; $\beta$-galactosidase, Perrin, manuscript in preparation; phosphorylase b, Ullmann, unpublished work). Since the association reaction does not involve the formation of covalent bonds, the temperature dependence of the rate is to be attributed, presumably, to a "conformational" transition state.

(3) The phenomenon of intra-cistronic complementation between different mutants of the same protein appears, as pointed out by Crick & Orgel (1964), to be due to a repair of altered structures which results from association between differently altered monomers of a (normally oligomeric) protein. Note that this interpretation necessarily implies, as pointed out by the authors, that the domain of bonding has an axis of symmetry.

It is reasonable therefore to consider that the conformation of each protomer in an oligomer is somewhat "constrained" by, and dependent upon, its association with other protomers. (An excellent discussion of this concept, as applied to haem proteins, is given by Lumry (1965).) In a symmetrical oligomer, all the protomers are engaged by the same binding sets and submitted to the same "quaternary constraints"; they should therefore adopt the same conformations. By contrast, in any non-symmetrical association, identical monomers would, as protomers, assume somewhat *different* conformations and cease to be truly equivalent. Thus, symmetry of bonding is to be regarded as a condition, as well as a result, of the structural equivalence of subunits in an oligomer.

These remarks justify the assumption that the specific biological properties of an oligomer depend in part upon its quaternary structure, and that the protomers will be functionally as well as structurally equivalent if, and only if, they are symmetrically associated within the molecule.

The last assumption of the model, namely, that in an allosteric transition the symmetry of quaternary bonding, and therefore the equivalence of the protomers, should tend to be conserved, may now be considered. Let us analyse the meaning and evaluate the possible range of validity of this postulate.

Consider a symmetrical oligomer (for simplicity, a dimer) wherein the conformation of each protomer is constrained and stabilized by the quaternary bonds (T). If these constraints were relaxed (i.e. the bonds broken) each protomer would tend towards an alternative conformation (R), involving certain tertiary bonds which were absent in the other configuration. The transition may be written

$$TT \xleftarrow{\quad \Delta F_1 \quad} X \xleftarrow{\quad \Delta F_2 \quad} RR,$$

where TT and RR stand for two symmetrical configurations and X for one (or several) non-symmetrical intermediate states. To say that symmetry should "tend to be conserved" is to imply that the occurrence of the R $\rightharpoonup$ T transition in one of the protomers should facilitate the occurrence of the same transition in the other. This would be the case of course if the intermediate state(s) X were less stable than either one of the symmetrical states; but it would also be the case, even if the X state were more stable than one of the symmetrical states, provided only that the $\Delta F$ of the first transition (from one of the symmetrical states to the intermediate state) were more positive than the second.

It is easy to see that the dissociation of a symmetrical oligomer should in general satisfy this condition. This may conveniently be symbolized as in fig. 10, where each subunit is represented as an arrow and only a minimum number of bonds is shown—actually two symmetrical (antiparallel) inter-protomer bonds (ab and ba) and one intra-chain bond (cb) the presence or absence of which is taken to characterize two distinct conformations (R and T) available to each subunit.

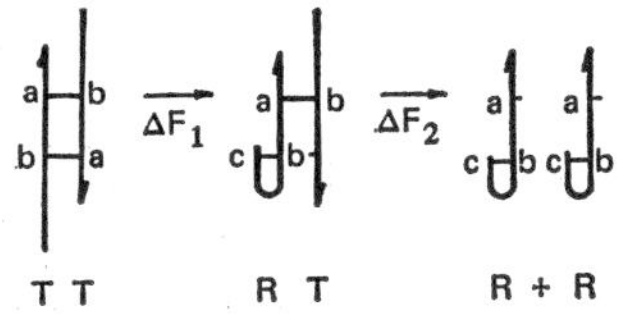

Fig. 10.

Although the symmetry of the protomers would not be conserved after dissociation into monomers, their equivalence would be, and the transition itself is symmetrical since it involves the breaking (or formation) of symmetrical bonds and symmetrical suppression (or creation) of identical quaternary constraints. The free energy of each of the two transitions may then be considered to involve two contributions: one ($\Delta F_b$), assignable to the breaking and formation of individual bonds, the other ($\Delta F_x$) associated with the freedom gained or lost by the protomers in respect to one another. By reason of symmetry, $\Delta F_b$ would be the same for both transitions, while $\Delta F_x$ would not. Since, in the example chosen, the second transition involves dissociation, the entropy gained in this step would be larger than in the first, and the sum of the two contributions would give $\Delta F_1 > \Delta F_2$, satisfying the condition of co-operativity. A ligand able to stabilize either the R or the T state would in turn exert homotropic co-operative effects upon the equilibrium.

There are examples in the literature of co-operative effects of this kind. The best illustration may be the muscle phosphorylase conversions which involve, as is well known, the formation of a tetrameric molecule (phosphorylase *a*) from the dimeric phosphorylase *b*. The conversion occurs when phosphorylase *b* is phosphorylated (in the presence of ATP and phosphorylase kinase). As expected since it is a tetramer, phosphorylase *a* contains four phosphoryl groups. Krebs & Fischer (personal communication) have observed that, when the amount of ATP used in the reaction is sufficient to phosphorylate only a fraction of the (serine) acceptor residues, a *stoichiometric amount* of fully phosphorylated tetrameric phosphorylase *a* is formed, while the excess protein remains dimeric and unphosphorylated. Another striking illustration of co-operative effects upon dissociation is provided by the work of Madsen & Cori (1956), who observed that phosphorylase *a* would dissociate into monomers in the presence of parachloromercuribenzoate, and showed that when the amounts of mercurial used were insufficient to dissociate all the protein, the remaining non-dissociated fraction did not contain any mercuribenzoate.

Reversible allosteric transitions however do not, in the majority of known cases, involve actual dissociation of the protomers. A transition between two undissociated symmetrical states of an oligomer would, nevertheless, be co-operative if it were adequately symbolizable, for example as in fig. 11, which expresses the assumption that one of the alternative conformations (T in this case) is stable

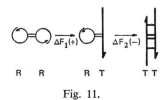

Fig. 11.

only when held by quaternary bonds which could be formed only at the price of breaking *symmetrically* certain tertiary bonds present in the other configuration (R).

In such a system, the $\Delta F$ of the first transition would be positive, the second negative, and the intermediate state (RT) therefore less stable than either one of the symmetrical states. Such a system could be very highly co-operative, and the strong homotropic interactions observed with many real systems *) suggests that they may conform to such a pattern.

However, fig. 12 symbolizes a much more general pattern of symmetrical transitions which is interesting to consider.

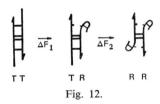

T T                    T R                R R

Fig. 12.

Here again the free energy assignable to the formation and breaking of individual bonds is the same in both transitions. Whether the RR ⇌ TT transition will be co-operative, non-co-operative, or anti-co-operative, should then depend entirely on the entropy term associated with the degrees of mutual freedom gained or lost by the protomers in each transition. If these entropy terms were equal in absolute value and of the same sign for both transitions (or if they were negligible) the system would be non-co-operative. In general, however, one would expect these two terms to be unequal and of significant magnitude in at least one of the transitions. The system would then be co-operative whenever the second entropy term was more positive than the first, and anti-co-operative in the reverse case.

The first possibility appears more likely on general grounds, since it seems reasonable to believe that certain degrees of mutual freedom, in a symmetrical dimer, may be held by either one of two (or more) symmetrical quaternary bonds, and liberated only when *both* are broken. Such a system would be closely comparable to a dissociating system, and it is interesting to note in this respect that in certain allosteric systems actual dissociation is observed under certain "extreme" conditions, whereas it is not seen under more normal conditions.

The possibility should also be considered that certain allosteric transitions might not involve a non-symmetrical intermediate. Such transitions would have to involve the initial breaking of axial bonds, eventually perhaps leading to, or allowing, the symmetrical breaking of symmetrical bonds as pictured in fig. 13. Such a mechanism would necessarily be co-operative.

It is impossible to say, at the present time, whether the co-operative homotropic

---

*) That is, when the Hill coefficient ($n$) approaches the value corresponding to the actual number of protomers.

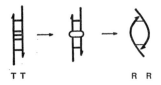

T T            R R

Fig. 13.

effects observed with real systems are better described by one or other pattern of symmetrical transitions. One might hope, however, to identify or eliminate some of these mechanisms by adequate thermodynamic and kinetic studies (using fast-mixing techniques) of the transition itself. It is clear in any event that none of these descriptions could apply to a non-symmetrically bonded oligomer, the protomers of which would have to assume different conformations and could not therefore undergo co-operatively the same transitions. On this basis, the fact that allosteric ligands appear invariably to exert co-operative homotropic effects may be taken as experimental evidence that the transitions which they stabilize occur in a symmetrical structure; indeed it was pointed out several years ago by one of us (Allen, Guthe & Wyman, 1950) that the symmetry properties of the oxygen saturation function for haemoglobin appeared to reveal the existence of elements of structural symmetry in the molecule itself. This inference was proved correct when the structure of haemoglobin was elucidated. Moreover, the recent work of Muirhead & Perutz (1963) and Perutz, Bolton, Diamond, Muirhead & Watson (1964) has shown that while the *quaternary* structure of haemoglobin is very significantly different in the oxygenated *versus* the reduced state, the molecular symmetry of the tetramer is conserved in both states. These observations would give a virtually complete illustration of the model if the X-ray pictures also showed some evidence of concomitant alterations of the tertiary structure of the protomers. This has not been observed; but it is reasonable to assume that a functionally significant allosteric transition need not involve more than a very small structural alteration of the protomers. In other words, given the very close and numerous intra-chain interactions, it would not be surprising that the quaternary constraints, even if strong, should not be expressed at the present level of resolution (5.5 Å) of the X-ray pictures. It is also possible that the quaternary constraints might not force any significant *sensu stricto* "conformational" alter-ation of the protomers, but only, for example, a (symmetrical) redistribution of charge within the molecule. We wish to point out that the assumptions of the model would remain valid also in such a case, and that the adjective "confor-mational" which we have used extensively (for lack of a better one) to qualify allosteric transitions, should be understood in its widest connotation.

## 5. General Discussion

In the preceding discussion we have tried to show, first that the functional properties of regulatory enzymes could be accounted for on the assumption that the quaternary structures of oligomeric proteins involve an element of symmetry in many, if not most, proteins made up of identical subunits (that is, presumably, in the majority of enzymes). We may now consider the problem in reverse and ask why molecular evolution should have so frequently favoured the appearance and maintenance of oligomeric globular proteins.

That it should be so must mean that there are functional advantages of some kind, inherent in the oligomeric state, and absent or difficult to achieve in the monomeric state. If most or all oligomeric proteins were endowed with the property of mediating allosteric interactions, especially homotropic interactions, we might believe that we had an answer to the question. Actually most of the enzymes known to be oligomeric are not, or at least are not known to be, allosteric. One should note, however, that the capacity to mediate physiologically significant interactions might be more frequent and widespread among proteins than has been realized so far. As we have seen, these properties are frequently very labile and may easily be lost during extraction and purification of an enzyme. Furthermore, it is conceivable that the effector for certain proteins may be an unknown or simply an improbable metabolite, if not, in some cases, another cellular protein (cf. Lehninger, 1964).

It probably remains true, however, that most oligomeric proteins are not endowed with specific regulatory functions. One must therefore presume that there are some other, more general, advantageous properties associated with the oligomeric state.

This problem may be related to the even more general question: Why should enzymes be so large, as compared with the size of their stereospecific sites? It seems reasonable to believe that two factors in particular contribute to determining a minimum size for enzymes. One is the requirement of fixing a very precise position in space for the several residues which together constitute the stereospecific site. Not only does this involve the necessity of a peptide chain with enough degrees of freedom (i.e. long enough) to allow the precise relative arrangement of these residues, but also the use of a further length of peptide to freeze these degrees of freedom, thereby conferring enough rigidity (i.e. specificity) upon the site. Another factor probably is the requirement that a given protein should *not* tend to associate more or less indiscriminately with other cellular proteins. As Pauling has pointed out, proteins are inherently "sticky", and the structure of enzymes must have been selected against the tendency to form random aggregates. Such a "purpose" may be, in part, fulfilled by decreasing the surface–volume ratio, and also by putting the polar groups on the surface, thereby increasing the solubility.

Now, association between monomers may evidently also contribute both to the fixation of an adequate structure and to a decrease in the surface–volume ratio, as well as to the covering-up of the hydrophobic areas of the monomers. Moreover, it is evidently more economical to achieve such results; whenever possible, by associating monomers rather than by increasing the unit molecular weight (i.e. the molecular weight per active centre).

These selective factors should therefore have favoured in general the appearance of closed (i.e. symmetrical) oligomers, since "open" structures (potentially infinite and polydisperse) would be disadvantageous in the case of most enzymes. Isologous (rather than heterologous) polymerization may have been frequently preferred for the same reasons, since this type of association leads to closed structures exclusively and, in the process of evolution from a monomeric to a polymeric state, it is evidently easier to start at the dimer stage (at which a heterologous association is still necessarily open), rather than right away at a higher stage.

However, the most decisive factor in the emergence and selective maintenance of symmetrical oligomeric proteins may have been the inherent co-operativity of their structure. To illustrate this point, consider schematically the events which may lead to the formation of a primitive dimer from a monomer.

On the surface of a protein monomer, any particular area contains a variety of randomly distributed groups, many of which may possess inherent chemical affinity for another one in the area. Since the *distance* between any two such groups is necessarily the same in two individual monomers, antiparallel association of the two pairs whenever possible would satisfy simultaneously two such valencies, creating a dimer involving two bonds and possessing a dyad axis. Furthermore, since this applies to *any* pair of groups capable of forming a bond, the monomers have a choice of *any one* of the mutually attractive pairs to achieve such a structure. Even so, the primitive dimer may not be formed, or might remain very unstable, because of the presence, within the area of contact of the protomers, of mutually repulsive groups. These pairs of groups would be distributed symmetrically about the dyad axis defined by the first two, mutually attractive, pairs. Therefore any mutation of *one* residue, conferring upon it the capacity to form a bond with its partner, would result in *two* new bonds being achieved in the dimer. Because of the interactions through "quaternary constraints" between the conformation of each protomer and the structure of their common domain of bonding, any such mutational event would affect symmetrically and co-operatively the functional properties of each of the two protomers. It is clear that, because of these reciprocal interactions, the same general reasoning applies to any mutation which might, even very discretely, affect the conformation of the protomers, including in particular the steric features of the domain of bonding which must of course play an important part in the stability of the association. Thus the structural and functional effects of single mutations occurring in a symmetrical oligomer, or allowing its formation, should be greatly amplified as compared with the effects of similar mutations in a monomer or in a non-symmetrical oligomer. In other words, because of the inherent co-operativity of their structure, symmetrical oligomers should constitute particularly sensitive targets for molecular evolution, allowing much stronger selective pressures to operate in the random pursuit of functionally adequate structures.

We feel that these considerations may account, in part at least, for the fact that most enzyme proteins actually are oligomeric; and if this conclusion is correct, the homotropic co-operative effects which seem at first to "characterize" allosteric systems should perhaps be considered only as one particular expression of the advantageous amplifying properties associated with molecular symmetry.

The same general argument may account for the fact that (apart from one or two possible exceptions) allosteric proteins have invariably been found to mediate *both* heterotropic and homotropic interactions, which implies of course that they are oligomeric. It should be clear from the discussion of the model that heterotropic interactions could *a priori* be mediated by a monomeric protein possessing two (necessarily different) binding sites, associated with two different "tautomeric" states of the molecule. If, for example, one of the states were stabilized by the substrate and the other by some other specific ligand, the latter would act as a competitive inhibitor. The saturation function $(\overline{Y}_s)$ would then simplify into:

$$\overline{Y}_s = \frac{\alpha}{L(1+\beta)+1+\alpha}$$

which we write only to indicate that, for $n=1$ (i.e. for a monomer) the model *formally* allows heterotropic effects to occur, but not of course, homotropic effects.

Just as the effect of a single amino-acid substitution will be greater in a symmetrical oligomer than in a monomer, the stabilization by a specific ligand of an alternative conformation, implying a significant increase of potential, may be possible in an oligomer when it would not be, for lack of co-operativity, in a monomeric protein. The fact that *both* heterotropic and homotropic interactions disappear when an allosteric protein is "desensitized" as a result of various treatments may be considered to illustrate this point, and actually constitutes one of the main experimental justifications of the model. It might be said in other words, that the molecular symmetry of allosteric proteins is used to amplify and effectively translate a very low-energy signal. *)

In addition, it is clear that the sigmoidal shape of the saturation curve characteristic of homotropic interactions may in itself offer a significant physiological advantage, since it provides the possibility of threshold effects in regulation.

---

*) Consider for example an allosteric system with an intrinsic equilibrium constant ($L = T_0/R_0$) of 1000. Assume, that the R state has affinity $1/K_R$ for a ligand F, and set $F/K_R = a$. In the presence of the ligand, the ratio of the two states will be: $\sum T/\sum R = 1000/(1 + a)^n$. Taking $a = 9$, for example, we would have, for a tetramer, $\sum T/\sum R = 0.1$. In order to reach the same value for the $T/R$ ratio with a monomeric system, the concentration of F would have to be more than one thousand times larger.

This property is of course essential in the case of haemoglobin, and it seems very likely that it has an important role in most, if not all, regulatory enzymes. Selection, in fact, must have operated on these molecules, not only to favour the structures which allow homotropic interactions, but actually to determine very precisely the energy of these interactions according to metabolic requirements.

The selective "choice" of oligomers as mediators of chemical signals therefore seems to be justified (*a posteriori*) by the fact that certain desirable physical and physiological properties are associated with symmetry, and therefore inaccessible to a monomeric protein.

We should perhaps point out here again that in the present discussion, as in the model, we accept the postulate that a monomeric protein or a protomer does not possess more than *one stereospecific site* able to bind a given ligand. That this postulate does apply to stereospecific sites is amply documented (cf. Schachman, 1963) and need not be discussed at length here. It is obvious of course that, lacking symmetry, a monomer or an individual protomer cannot present two or more *identical* elements of tertiary structure of any kind.

The postulate, however, does not apply to *group-specific*, as apposed to *stereo-specific*, ligands. Homotropic interactions of various kinds (not necessarily co-operative) may therefore occur in the binding of group-specific ligands (such as SH reagents, detergents, ions, etc.) whether the protein is monomeric or not. As is well known, the vast literature on the denaturation of proteins is replete with descriptions of multimolecular effects exerted by various group-specific reagents. It may be worth noting in this respect that in the last analysis, the co-operative effects of such reagents are accounted for by the simultaneous attack of numerous bonds occupying functionally similar (although not geometrically symmetrical) positions in the molecule.

The significance of this generalization may be made clear by considering the melting of double-stranded DNA. This is a typically co-operative phenomenon the co-operativity of which is evidently dependent upon and expresses the (helical) symmetry of the "domain of bonding" between the two strands in the Watson–Crick model. In the last analysis therefore, the axial symmetry requirement for homotropic co-operative effects to occur with a globular protein, when *stereo-specific* ligands are concerned, reflects essentially the fact that, in general, only one stereospecific site able to bind such a ligand exists on a protein monomer or protomer.

Gerhart (1964) and Schachman (1964) have recently reported the successful separation, from crystalline aspartic transcarbamylase, of two different subunits, one of which bears the specific receptor for aspartate, and the other the receptor for CTP. It is very tempting to speculate on the possibility that this remarkable and so far unique observation may in fact correspond to a general rule, namely, that a protein should contain as many different subunits (peptide chains) as it bears stereospecifically different receptor sites. The emergence and evolution of such structures, by association of primitively distinct entities, would be much easier to understand than the acquisition of a new stereospecific site by an already existing and functional enzyme made up of a single type of subunit.

We have so far not discussed one of the major assumptions of the model, namely, that allosteric effects are due to the displacement of an equilibrium between discrete states assumed to exist, at least potentially, apart from the binding of a ligand. The main value of this treatment is to allow one to define, in terms of the allosteric constant, the contribution of the protein itself to the interaction, as distinct from the dissociation constants of the ligands. This distinction is a useful and meaningful one, as we have seen, and its validity is directly justified by the fact that the affinity of a ligand may vary widely without any alteration of its homotropic interaction coefficient. But it should be understood that the "state" of the protein may not in fact be exactly the same whether it is actually bound, or unbound, to the ligand which stabilizes it. In this sense particularly, the model offers only an over-simplified first approximation of real systems, and it may prove possible in some cases to introduce corrections and refinements by taking into consideration more than two accessible states.

We feel, however, that the main interest of the model which we have discussed here does not reside so much in the possibility of describing quantitatively and in detail the complex kinetics of allosteric systems. It rests rather on the concept, which we have tried to develop and justify, that a general and initially simple relationship between symmetry and function may explain the emergence, evolution and properties of oligomeric proteins as "molecular amplifiers", of both random structural accidents and of highly specific, organized, metabolic interactions.

This work has benefited greatly from helpful discussions and suggestions made by our friends and colleagues Drs R. Baldwin, S. Brenner, F. H. C. Crick, F. Jacob, M. Kamen, J. C. Kendrew, A. Kepes, L. Orgel, M. F. Perutz, A. Ullmann. We wish to thank Mr. F. Bernède for his kindness in performing many calculations with the computer.

The work was supported by grants from the National Institutes of Health, National Science Foundation, Jane Coffin Childs Memorial Fund, Délégation Générale à la Recherche Scientifique et Technique and Commissariat à l'Energie Atomique.

## References

D. W. Allen, K. F. Guthe & J. Wyman, *J. Biol. Chem.* **187** (1950) 393.

I. Algranati & E. Cabib, *J. Biol. Chem.* **237** (1962) 1007.

R. E. Benesch, R. Benesch & M. E. Williamson, *Proc. Nat. Acad. Sci., Wash.* **48** (1962) 2071.

C. Bohr, *Zentr. Physiol.* **17** (1903) 682.

R. W. Briehl, *J. Biol. Chem.* **238** (1963) 2361.

D. L. D. Caspar, *Advanc. Protein Chem.* **18** (1963) 37.

C. Cennamo, M. Boll & H. Holzer, *Biochem. Z.* **340** (1964) 125.

J. P. Changeux, *Cold Spr. Harb. Symp. Quant. Biol.* **26** (1961) 313.

J. P. Changeux, *J. Mol. Biol.* **4** (1962) 220.

J. P. Changeux, *Cold Spr. Harb. Symp. Quant. Biol.* **28** (1963) 497.

J. P. Changeux, Thèse Doctorat ès Sciences, Paris. *Bull. Soc. Chim. Biol.* **46** (1964*a*) 927, 947, 1151; **47** (1965) 115, 267, 281.

J. P. Changeux, *Brookhaven Symp. Biol.* **17** (1964*b*) 232.

G. N. Cohen & J. C. Patte, *Cold Spr. Harb. Symp. Quant. Biol.* **28** (1963) 513.

G. N. Cohen, J. C. Patte, P. Truffa-Bachi, C. Sawas & M. Doudoroff, Colloque International C.N.R.S. *Mécanismes de régulation des activités cellulaires chez les microorganismes.* Marseille: (1963) 243.

F. Crick & L. L. Orgel, *J. Mol. Biol.* **8** (1964) 161.

P. Datta, H. Gest & H. Segal, *Proc. Nat. Acad. Sci., Wash.* **51** (1964) 125.

R. M. Ferry & A. A. Green, *J. Biol. Chem.* **81** (1929) 175.

M. Freundlich & H. E. Umbarger, *Cold Spr. Harb. Symp. Quant. Biol.* **28** (1963) 505.

J. C. Gerhart, *Brookhaven Symp. Biol.* **17** (1964) 232.

J. C. Gerhart & A. B. Pardee, *Fed. Proc.* **20** (1961) 224.

J. C. Gerhart & A. B. Pardee, *J. Biol. Chem.* **237** (1962) 891.

J. C. Gerhart & A. B. Pardee, *Cold Spr. Harb. Symp. Quant. Biol.* **28** (1963) 491.

J. C. Gerhart & A. B. Pardee, *Fed. Proc.* **23** (1964) 727.

D. W. Green & R. Aschaffenburg, *J. Mol. Biol.* **1** (1959) 54.

J. A. Hataway & D. E. Atkinson, *J. Biol. Chem.* **238** (1963) 2875.

O. Hayaishi, M. Gefter & H. Weissbach, *J. Biol. Chem.* **236** (1963) 2040.

E. Helmreich & C. F. Cori, *Proc. Nat. Acad. Sci., Wash.* **51** (1964) 131.

S. Kornfeld, R. Kornfeld, E. F. Neufeld & P. J. O'Brien, *Proc. Nat. Acad. Sci., Wash.* **52** (1964) 371.

H. Krebs, *Proc. Roy. Soc.* B, **159** (1964) 545.

A. Lehninger, Centenaire Société de Chimie Biologique, *Bull. Soc. Chim. Biol.* **46** (1964) 1555.

C. Levinthal, E. R. Signer & K. Fetherolf, *Proc. Nat. Acad. Sci., Wash.* **48** (1962) 1230.

R. Lumry, unpublished work.

N. B. Madsen, *Biochem. Biophys. Res. Comm.* **15** (1964) 390.

N. B. Madsen & C. F. Cori, *J. Biol. Chem.* **223** (1956) 1055.

F. Maley & G. F. Maley, *Science,* **141** (1963) 1278.

G. F. Maley & F. Maley, *J. Biol. Chem.* **239** (1964) 1168.

T. E. Mansour, *J. Biol. Chem.* **238** (1963) 2285.

C. Manwell, *Oxygen in the Animal Organism* London: Pergamon Press (1964).

D. B. Martin & P. R. Vagelos, *J. Biol. Chem.* **237** (1962) 1787.

R. G. Martin, *J. Biol. Chem.* **237** (1962) 257.

J. Monod, J. P. Changeux & F. Jacob, *J. Mol. Biol.* **6** (1963) 306.

J. Monod & F. Jacob, *Cold Spr. Harb. Symp. Quant. Biol.* **26** (1961) 389.

H. Muirhead & M. F. Perutz, *Nature* **199** (1963) 633.

R. Okazaki & A. Kornberg, *J. Biol. Chem.* **239** (1964) 269.

J. Passoneau & O. Lowry, *Biochem. Biophys. Res. Comm.* **7** (1962) 10.

J. C. Patte & G. N. Cohen, *C. R. Acad. Sci. Paris,* **259** (1964) 1255.

J. C. Patte, G. Le Bras, T. Loviny & G. N. Cohen, *Biochim. biophys. Acta* **67** (1963) 16.

M. F. Perutz, W. Bolton, R. Diamond, H. Muirhead & H. C. Watson, *Nature* **203** (1964) 687.

M. F. Perutz, M. G. Rossmann, A. F. Cullis, H. Muirhead, G. Will & A. C. T. North, *Nature* **185** (1960) 416.

B. Pickles, B. A. Jeffery & M. G. Rossmann, *J. Mol. Biol.* **9** (1964) 598.

F. J. Reithel, *Advanc. Protein Chem.* **18** (1963) 124.

N. M. Rumen, *Fed. Proc.* **22** (1963) 681.

M. Salas, E. Vinuela, J. Salas & A. Sols, *Biochem. Biophys. Res. Comm.* **17** (1964) 150.

B. Sanwal, M. Zink & C. Stachow, *Biochem. Biophys. Res. Comm.* **12** (1963) 510.

B. Sanwal, M. Zink & C. Stachow, *J. Biol. Chem.* **239** (1964) 1597.

E. Scarano, G. Geraci, A. Polzella & E. Campanile, *J. Biol. Chem.* **238** (1963) 1556.

E. Scarano, G. Geraci & M. Rossi, *Biochem. Biophys. Res. Comm.* **16** (1964) 239.

E. Scarano, G. Geraci & M. Rossi, *Biochem.* **6** (1967) 192.

H. Schachman, *Cold Spr. Harb. Symp. Quant. Biol.* **28** (1963) 409.

H. Schachman, Sixth Intern. Congress Biochem. (1964) New York: Pergamon Press.

L. C. Smith, J. M. Ravel, S. Lax & W. Shive, *J. Biol. Chem.* **237** (1962) 3566.

E. R. Stadtman, G. N. Cohen, G. Le Bras & H. de Robichon-Szulmajster, *J. Biol. Chem.* **236** (1961) 2033.

E. Sturani, F. Datta, M. Hugues & H. Gest, *Science* **141** (1963) 1053.

K. Taketa & B. M. Pogell, *J. Biol. Chem.* **240** (1965) 651.

G. M. Tomkins, K. L. Yielding, N. Talal & J. F. Curran, *Cold Spr. Harb. Symp. Quant. Biol.* **28** (1963) 461.

R. Traut & F. Lipmann, *J. Biol. Chem.* **238** (1963) 1213.

A. Ullmann, P. R. Vagelos & J. Monod, *Biochem. Biophys. Res. Comm.* **17** (1964) 86.

H. E. Umbarger, *Science* **123** (1956) 848.

H. E. Umbarger & B. Brown, *J. Biol. Chem.* **233** (1958*a*) 415.

H. E. Umbarger & B. Brown, *J. Biol. Chem.* **233** (1958*b*) 1156.

E. Vinuela, M. L. Salas & A. Sols, *Biochem. Biophys. Res. Comm.* **12** (1963) 140.

H. C. Watson & L. J. Banaszak, *Nature* **204** (1964) 918.

J. Wyman, *Advanc. Protein Chem.* **4** (1948) 407.

J. Wyman, *Cold Spr. Harb. Symp. Quant. Biol.* **28** (1963) 483.

*Theoretical Physics and Biology*, © 1969, *North-Holland Publ. Co., Amsterdam*

# L'ORIGINE DE LA SYMETRIE BILATERALE CHEZ L'EMBRYON

ETIENNE WOLFF

*Laboratoire d'Embryologie expérimentale, Collège de France, Nogent-sur-Marne, France*

Les organisateurs de ce colloque m'ont demandé de parler de questions de symétrie. J'ai choisi comme sujet: l'origine de la symétrie bilatérale chez certains embryons. Je ne sais pas bien ce qui est à l'origine de cette symétrie, mais je sais qu'elle existe et qu'on la voit apparaître à un stade défini et précoce. Après Jacques Monod qui a parlé de la symétrie à l'échelle moléculaire, je voudrais parler de la symétrie à l'échelle *molaire*, c'est-à-dire à l'échelle des grands ensembles.

Il est très remarquable de constater que, chez la plupart des espèces animales, les premières déterminations ne sont pas des déterminations matérielles, mais des déterminations d'axes, de polarité, de plans de symétrie. La nature, si je puis m'exprimer ainsi, procède comme le peintre ou le dessinateur (pas le surréaliste ou le non figuratif) qui met en place les grandes directions de son plan, avant de couvrir sa toile de couleurs. On dirait que des déterminations abstraites précèdent les déterminations concrètes. Ce n'est bien entendu qu'une interprétation anthropomorphique. Mais il est cependant très frappant de constater qu'au stade de l'oeuf non incubé des oiseaux ou à celui de l'oeuf fécondé des amphibiens, on pourrait représenter l'individu virtuel par un cercle ou une sphère homogène, traversée par un vecteur orienté qui représente à la fois l'axe céphalo-caudal et le plan de symétrie bilatérale (fig. 1). Rien autre n'est déterminé. Il faut toutefois ajouter à ce schéma vide la dénomination poulet, ou grenouille, ou oursin, ou ascidie, car ce fantôme irréel contient en lui des possibilités d'avenir concrètes et bien définies. Mais je veux seulement rappeler qu'à un stade précoce, rien n'est encore déterminé et qu'un territoire quelconque est capable de multiples prestations, sinon de toutes. Ainsi le blastoderme d'oiseau, ou de mammifère, peut être fragmenté en autant de parties qu'on le veut: chacune peut donner

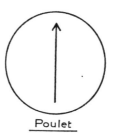

Poulet

Fig. 1.  Représentation schématique d'un embryon d'une espèce quelconque de Vertébré.

naissance à un embryon complet. Cela veut dire aussi qu'un territoire qui, dans le développement normal, donnerait, par exemple, la région du tronc peut aussi bien donner la tête que la queue, si on morcelle l'oeuf ou le blastoderme.

## 1. L'oeuf d'amphibien

Considérons l'oeuf d'amphibien avant la fécondation. C'est un système mon-axone, à symétrie radiaire. Il est homogène suivant tous ses méridiens. Il est hétérogène en latitude, par suite d'une répartition inégale du vitellus et du pigment superficiel (fig. 2A). Cette hétérogénéité latitudinale définit un axe déjà orienté, qui correspond approximativement au futur axe céphalo-caudal, la tête se trouvant du côté le moins chargé en vitellus et le plus riche en pigment. Cette détermination n'est du reste pas irréversible, elle peut être modifiée par différentes interventions expérimentales. Mais à ce stade il n'y a encore aucun plan de symétrie. Au contraire, dès que l'oeuf a été inséminé, (dès que le spermatozoïde a touché la surface de l'ovule), le plan de symétrie bilatérale apparaît, il se manifeste par des phénomènes morphologiques (croissant clair (fig. 2B)) et dynamiques (rotation

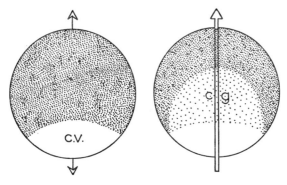

Fig. 2. Passage de la symétrie axiale à la symétrie bilatérale dans l'oeuf d'amphibien. (A): Oeuf non fécondé. Les deux flèches symbolisent l'axe de symétrie de l'oeuf. C.V., champ vitellin. (B): Oeuf fécondé. Formation du croissant gris C.G., du futur côté dorsal. La flèche symbolise le plan de symétrie bilatérale.

de la pellicule ovulaire superficielle). C'est le point de pénétration du spermatozoïde qui définit le plan de symétrie bilatérale et, en même temps la polarité dorso-ventrale; le côté ventral se trouve du côté du point d'entrée du spermatozoïde.

Le point de pénétration du spermatozoïde, donc le plan de symétrie, n'est pas préétabli. On peut le déterminer à volonté, comme le montrent les expériences de fécondation dirigée, suivant la méthode de Roux et d'Ancel et Vintemberger, qui consiste à faire pénétrer le spermatozoïde en un point quelconque, en guidant le sperme vers ce point (fig. 3). On retrouve toujours la traînée de pigment qu'en-traîne la tête du spermatozoïde dans le plan de symétrie bilatérale.

Mais il y a d'autres moyens d'agir de l'extérieur sur la position du plan de symétrie et même de contrarier l'action du spermatozoïde. On peut faire agir la pesanteur, comme l'ont montré Ancel et Vintemberger. L'oeuf de grenouille possède un gradient de densité, correspondant à sa charge en vitellus. Le pôle vitellin placé en haut retombe toujours vers le bas comme un ballon lesté. Il décrit dans sa chute un arc correspondant à un certain méridien (fig. 4). Si l'angle correspondant à cet arc est assez grand, c'est précisément ce méridien qui définira le plan de symétrie; cette rotation, si elle est suffisante, peut annuler l'action du spermatozoïde. En pratique, la rotation efficace doit être de 360°, ce qui veut dire qu'il faut faire effectuer au pôle vitellin deux chutes successives suivant le même plan: les rotations successives cumulent leurs effets. D'autres facteurs, tels l'activation par des agents expérimentaux, la compression, peuvent avoir les mêmes effets.

Il est donc possible de faire passer le plan de symétrie bilatérale par un point quelconque de l'oeuf. Il s'ensuit que tout le plan d'organisation de l'individu futur est sous la dépendance d'une intervention extérieure à l'oeuf, que ce soit le spermatozoïde ou la rotation d'orientation. C'est dans un plan et du côté choisis à l'avance par l'expérimentateur que se situera par exemple le tube nerveux

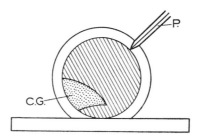

Fig. 3.   Schéma d'une insémination dirigée. P, pincette amenant les spermatozoïdes au contact de l'oeuf. Le croissant gris C.G., se forme du côté opposé au point d'entrée du spermatozoïde. (d'après Ancel et Vintemberger).

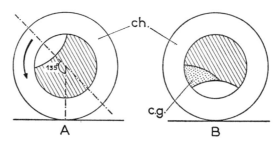

Fig. 4.   Bascule du pôle végétatif sous l'action de la pesanteur. (A): Le pôle végétatif est placé sur un axe faisant avec l'axe vertical un angle de 135°. En raison de sa densité, il tombe vers le bas, décrivant un arc de 135°. (B): Cette manoeuvre ayant été répétée 2 ou 3 fois, le croissant gris se forme du côté où le pôle végétatif a basculé. (d'après Ancel et Vintemberger).

et la chorde dorsale. La situation future de tous les organes dépend de cette première détermination.

Si l'on regarde les choses de plus près, on s'aperçoit que les deux facteurs de symétrisation ont pour effet de déclencher une réaction de membrane, c'est-à-dire une contraction et une rotation de la fine pellicule cytoplasmique qui entoure l'ovule. Il semble que, chez l'oeuf d'amphibien, la couche superficielle de l'ovule ne soit pas absolument solidaire des couches plus profondes, et que, sous l'influence de différents facteurs, elle ait tendance à tourner par rapport au substratum. Le croissant gris permet même de mesurer l'ampleur de cette rotation: elle est d'environ 30°. C'est ce déséquilibre entre la périphérie et l'intérieur qui crée une hétérogénéité, d'où résultent la polarité dorso-ventrale et la symétrie bilatérale.

## 2. L'oeuf d'oiseau

Si l'on considère l'oeuf de poule, c'est aussi un mécanisme extérieur qui impose à l'embryon son axe céphalo-caudal et son plan de symétrie bilatérale. On sait que la position de l'axe céphalo-caudal est définie chez les embryons d'Oiseaux par la règle bien connue de Von Baer. Si un observateur place l'oeuf devant lui de manière que le petit bout de la coquille soit tourné du côté droit, l'axe de l'embryon est perpendiculaire à l'axe de rotation de la coquille, et l'extrémité céphalique est située du côté opposé à l'observateur. Cette règle est approximative. Elle se vérifie dans 70 % des cas environ chez l'oeuf de poule et de cane, la proportion est beaucoup plus importante et voisine de 100 chez le pigeon. Une telle relation entre l'orientation de l'embryon et les enveloppes inertes de l'oeuf permet de soupçonner que des facteurs externes doivent exercer leur action sur l'orientation de l'embryon. Effectivement Vintemberger et Clavert (1953–1960) ont montré que celle-ci est en rapport avec l'enroulement des chalazes, c'est-à-dire avec le sens de rotation de l'oeuf dans l'oviducte (fig. 5). A un enroulement

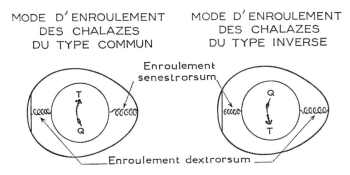

Fig. 5.   Représentation schématique des deux modes principaux d'enroulement des chalazes. TQ, direction céphalo-caudale. (T, tête; Q, queue). (d'après Vintemberger et Clavert).

dextrorsum de la chalaze gauche, correspond une position typique de l'axe de l'embryon; à un enroulement senestrorsum correspond une orientation inversée de 180°.

D'autre part, ces auteurs ont montré que l'orientation définitive de l'embryon est acquise dans l'utérus. Il existe un stade critique au-delà duquel aucune intervention ne change plus la détermination de l'axe céphalo-caudal. Mais on peut la modifier en changeant la position de l'oeuf entier avant le stade critique, par une opération simple, faite sous le contrôle des rayons X.

Si le grand axe de la coquille coïncide avec l'axe de l'utérus, le petit bout de l'oeuf étant tourné vers le cloaque, l'embryon suit la règle de von Baer: son orientation est conforme. Si au contraire, le gros bout de l'oeuf est tourné vers le cloaque, on obtient une orientation inverse de l'embryon. Si l'oeuf est placé de telle manière que son grand axe soit oblique par rapport à l'axe de l'utérus, l'orientation de l'embryon est intermédiaire entre l'orientation normale et l'orientation inverse (fig. 6).

Ceci prouve que la détermination de l'axe céphalo-caudal et du plan de symétrie

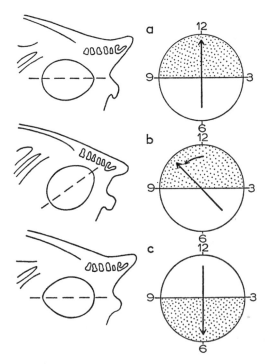

Fig. 6.   Relation entre l'orientation de l'axe de l'oeuf utérin et l'orientation du plan de symétrie bilatérale. a: axe horizontal, petit bout dirigé vers le cloaque (présentation normale); le plan de symétrie bilatérale est conforme à la règle de Von Baer. b: grand axe de l'oeuf oblique par rapport à l'axe de l'utérus, orientation intermédiaire de l'axe de l'embryon. c: axe horizontal, gros bout dirigé vers le cloaque; orientation inversé de l'embryon.

bilatérale dépend encore de facteurs externes, on peut même dire de facteurs mécaniques, tels que le mouvement de rotation et le frottement de l'embryon contre ses enveloppes. Vintemberger et Clavert ont montré que l'on peut aussi bien faire tourner le chorion autour de l'oeuf de grenouille que faire tourner celui-ci à l'intérieur de son enveloppe. Le résultat est le même, ce qui met en cause, non l'action de la pesanteur, mais le frottement de l'oeuf contre son enveloppe ou réciproquement. Mais, ce qui est plus complexe que dans le cas des amphibiens, c'est que ces facteurs externes agissent non sur une cellule, mais sur un blastoderme comprenant déjà un grand nombre de cellules. Cela laisse supposer que des transformations simultanées se produisent dans toutes les cellules du blastoderme, qui acquièrent simultanément les mêmes propriétés.

On peut donc déterminer expérimentalement et à volonté l'emplacement du plan de symétrie bilatérale. Comment expliquer ces résultats? Nous n'aurons pas la naïveté de croire qu'une intervention aussi simple qu'un contact, une rotation, un déplacement, joue un rôle essentiel dans un processus aussi complexe que la mise en place de toute l'organisation embryonnaire: car c'est bien de cela qu'il s'agit. L'emplacement de toutes les parties de l'organisme dépend de ces manoeuvres très simples. On ne peut s'empêcher de comparer leur effet à celui d'un déclic, d'une minuterie. La symétrie bilatérale existe à l'état potentiel, une action mécanique la fait passer à l'état réel. Cela ne veut pas dire grand'chose. Mais comment ne pas imaginer que l'orientation d'un ensemble est faite de l'orientation des parties, du groupement de nombreuses molécules qui jusqu'alors étaient réparties sans ordre. Sous l'influence d'un facteur extérieur, les molécules s'orientent parallèlement les unes aux autres, comme de la limaille de fer sous l'influence d'un aimant. Il doit en être ainsi dans la couche superficielle de l'ovule de grenouille, dans toutes les cellules du blastoderme de poulet. Il s'agit vraisemblablement de molécules protéiques. Mais il reste à démontrer cette hypothèse — à préciser quelles sont les molécules qui s'orientent, comment et pourquoi elles s'orientent. C'est vers la physique théorique et expérimentale que nous nous tournons pour expliquer ces phénomènes. La symétrie à l'échelle molaire est justiciable d'une explication à l'échelle moléculaire.

*Theoretical Physics and Biology,* © 1969, *North-Holland Publ. Co., Amsterdam*

## DISCUSSIONS

S. E. BRESLER: I would like to make a short comment on Dr. Monod's lecture. As he showed the allosteric enzymes have a cooperative non linear kinetics which is revealed by a sigmoid curve. This can be explained not only by the interaction of subunits. It was shown that if you have two kinds of ligands, two kinds of substrates which are used by the enzyme, they may interact directly or indirectly through the tertiary structure of the enzyme. Then the binding constant of one of the substrates will depend on the concentration of the second substrate. This seems to be a rather widespread situation. For instance we have studied an enzyme called phosphorylase B where you have this case because the binding of one of the substrates say phosphate, depends very strongly on the presence of the second substrate—glycogen. The change is five times if you go from zero concentration to saturation. So we thought we had here an anti-Monod case, a case where the sigmoid character has another origin than the interaction of subunits. It came out that this idea was wrong. We were able to make straightforward spectrophotometric measurements of the concentration of the complex of the enzyme with the first substrate in the absence of the second substrate. When we have done these measurements in the presence of the allosteric inhibitor, we obtained nice sigmoid curves, cooperative binding curves which showed that in this case of course, you have to do with the interaction of subunits. Because you don't have the second substrate at all.

M. KOTANI: I should like to ask Dr. Monod a question, which may be a minor point in your lecture. Dr. Monod seems to assume two discrete conformations for an allosteric protein molecule. It seems to me, however, more natural to assume continuous sequence of possible confirmations. To describe this situation, I use a continuous parameter $S$, which defines conformation, keeping the symmetry. In my mind the essence of Dr. Monod's theory is described in the following way. For simplicity I assume only one kind of ligand.

This protein molecule is assumed to consist of $n$ protomers, in your language, so that it can bind up to $n$ ligands. When $m$ ligands are attached to this protein molecule, where $0 \leqslant m \leqslant n$, and the conformation of the molecule corresponds to a given value of $S$, then the molecule has a certain free energy $f(m, S)$. Molecules without ligands have a distribution in conformation in thermal equilibrium as given by

$$\left[ \exp - \frac{f(0, S)}{kT} \right],$$

which corresponds to curve $m = 0$ in the figure. Now, I believe that the essence of Dr. Monod's theory is represented in the following form of $f(m, S)$ for general $m$:

$$f(m, S) = m\phi(S) + f(0, S).$$

Since $f(m, S)$ is linear in $m$, no direct interaction between binding sites is assumed. The distributions in conformation of these molecules are essentially given by

$$x^m \left[ \exp - \frac{f(m, S)}{kT} \right],$$

where $x$ denotes the activity of the ligand. Perhaps $\phi(S)$ and $f(0, S)$ will have minima at different $S$, and the distribution will shift with the increase of $m$, as shown qualitatively in the figure. This corresponds to what Dr. Monod has shown in his beautiful theory.

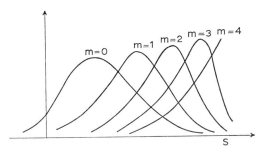

Fig. 1.

However in Dr. Monod's formulation, as far as I understand, the function $f(m, S)$ is assumed to have minima of delta function type at two discrete points. My question is whether you have a special argument for having only two discrete conformations for these protein molecules.

J. MONOD: If I understand well, it seems to me that your way of putting the interaction would lead to cinetics of the Adair type. I.e., it would lead to predicting that there would be a gradual change in the affinities for the substrate for the $n$ binding sites, i.e., when you have one substrate molecule on the enzyme, then the other sites would have higher affinity, when you have two, the affinity would still be higher, and so on. So that it would lead to an Adair model.

There are a number of actual observations which are not in accord with the Adair model. One is the type of analysis which I showed at the end, which indicates that there is no increase in the binding constant. The binding constants are the same throughout. But I think the best answer is given by Dr. Eigen and Kirschner's experiments. And perhaps Dr. Eigen himself might comment on that.

M. EIGEN: I should perhaps say a few words about the two possible models because one may see how one can differentiate between these models. A particularly simple example would be an enzyme consisting of several (e.g., four) identical subunits. Binding of a substrate usually would be governed by the simple law of mass action and characterized by a hyperbolic binding curve. Similarly, steady-state kinetics of catalytic turnover according to Michaelis would lead to a correspondingly shaped rate curve.

On the other hand, binding of substrate by cooperative interaction of the subunits would change the character of these curves. As is known from binding of oxygen to hemoglobin, those binding curves can become sigmoidal, indicating "cooperation" in binding. This means low affinity at low oxygen concentration and high affinity at concentrations at which hemoglobin becomes partly saturated.

Several theories have been proposed to describe this behaviour. Adair assumed a set of four different and successively increasing binding constants. This certainly can describe empirically any sigmoidal curve. However, it might be possible that the number of intrinsic parameters can be reduced, similar to the treatment of helix coil transformations where the $n$-subunit system ($n$ being a large number) could be represented by only two or three intrinsic parameters.

Such a model was proposed by Monod, Wyman, and Changeux, and is represented in a generalized form in the figure. This model requires only three parameters. R and T denote two isomeric states of the enzyme that can bind the substrate. The affinity for the substrate (D) in the R-state is high, whereas in the T-state it is low (or even negligible), but T is the preferred state in the absence of binding. The subunits in either the R- or the T-state are "degenerate", so that only two intrinsic binding constants, one for any subunit in the R-state and one for any subunit in the T-state, must be assumed. In addition to these constants, we need one parameter to describe the transformation from R- to T-state—a conformation change—which is assumed to be extremely cooperative, i.e., all-or-none (all subunits are either in the R or the T conformation). The two simplifying assumptions—that is, the complete degeneracy of R-states or T-states for binding and all-or-none transformation from T to T derived by Monod from symmetry

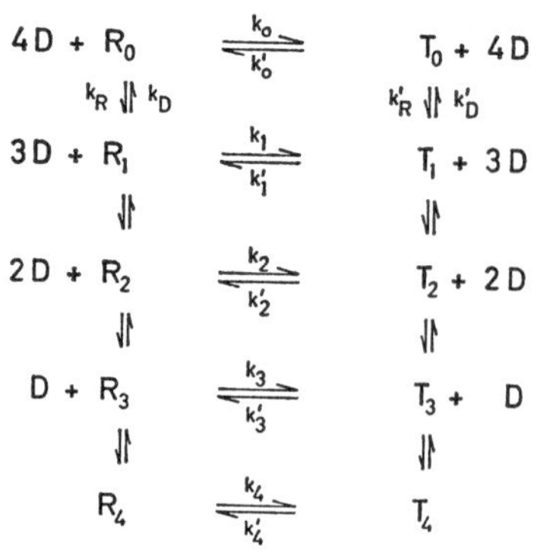

$$
\begin{array}{ccc}
4D + R_0 & \underset{k'_0}{\overset{k_0}{\rightleftharpoons}} & T_0 + 4D \\
{}_{k_R}\Updownarrow{}^{k_D} & & {}_{k'_R}\Updownarrow{}^{k'_D} \\
3D + R_1 & \underset{k'_1}{\overset{k_1}{\rightleftharpoons}} & T_1 + 3D \\
\Updownarrow & & \Updownarrow \\
2D + R_2 & \underset{k'_2}{\overset{k_2}{\rightleftharpoons}} & T_2 + 2D \\
\Updownarrow & & \Updownarrow \\
D + R_3 & \underset{k'_3}{\overset{k_3}{\rightleftharpoons}} & T_3 + D \\
\Updownarrow & & \Updownarrow \\
R_4 & \underset{k'_4}{\overset{k_4}{\rightleftharpoons}} & T_4
\end{array}
$$

Fig. 2.

considerations—have been subject to criticism. It is obvious that a decision about the mechanism cannot be derived from equilibrium studies. However, relaxation spectrometry can provide the tool for elucidation of such complex reaction mechanisms. Instead of looking at the system after all steps have equilibrated and characterizing it by one integrant number, we expand the rapidly perturbed equilibrium function on the time axis, and thus we may see each step equilibrating with its own characteristic rate.

What do we expect for a system such as that depicated in the figure? If the system is characterized by three intrinsic parameters it should yield three characteristic time constants. They would have a particularly simple form, if the all-or-none conformation change ($R \leftrightarrow T$) is the slowest step. Let us assume this condition and see how the system would "relax" after perturbation. At the first instant, the two conformations would be frozen in; each would start to bind substrate with a characteristic rate. Usually one of them—often the more affine form—will show a shorter relaxation time. There will be only one detectable time constant for the binding at all R-subunits. This can be represented by a second-order reaction between "free" substrate molecules (D) and unoccupied R-binding sites ($4R_0 + 3R_1 + 2R_2 + R_3$). The reverse process is a first-order dissociation from all occupied R-sites ($R_1 + 2R_2 + 3R_3 + 4R_4$). According to considerations in the preceding chapters, the inverse relaxation time ($1/\tau_I$) should be a linear function of the sum of the concentrations of free substrate and unoccupied binding sites. The slope will yield the "intrinsic" rate constant of recombination to R-states, and the intercept will yield the intrinsic rate constant of dissociation from R-states.

Similarly, the binding by the less affine form (T) should yield an analogous dependence for the inverse relaxation time ($1/\tau_{II}$), i.e., an increase with the sum of free substrate plus unoccupied T-binding site concentrations. The latter term, however, must be corrected by a factor that results from coupling to the more rapidly established R-state equilibrium. Both equilibria are coupled via the substrate concentration; in other words, while T-states react with the substrate, the change of substrate concentration modifies the more rapidly equilibrating R-state binding. These factors (which would also modify $1/\tau_I$ if both relaxation times were of the same order of magnitude) can easily be calculated. Again, for the T-state binding there is only one detectable relaxation time for all four subunits, and it yields the intrinsic rate constants of recombination and dissociation (the ratio of which also yields the intrinsic binding constant).

The third relaxation time now must be characteristic of the first-order conformation change. Nevertheless, it may be substrate-concentration dependent, as the substrate binding equilibria will rapidly rearrange during the transformation: $R \leftrightarrow T$. At very low concentrations of substrate we expect no binding, so the system will be in the $T_0$- and $R_0$-state only. Similarly, at very high concentrations of D, where R and T are saturated, the system will be in the $R_4$- and $T_4$-state only.

In both cases we expect first-order transformations and therefore concentration-independent $1/\tau_{III}$ values. In general, however, these are different for both cases. Sigmoidal characteristics will result only if (at low concentration) the $T_0$-state is more stable than the $R_0$-state, whereas at high concentrations of D the system favors the $R_4$-state more than the $T_4$-state. This is equivalent to the fact that the rate of $R_0 \rightarrow T_0$ is larger than that of $T_0 \rightarrow R_0$, but the rate of $R_4 \rightarrow T_4$ is smaller than that of $T_4 \rightarrow R_4$. If we raise the concentration of D to a level at which R starts to saturate, $1/\tau_{III}$ will decrease from the constant value (given by $k_{RT_0} + k_{TR_0}$) to another constant value $k_{TR_0} + k_{RT_4}$ (but usually $k_{RT_4} \ll k_{TR_0}$). If we then reach the level at which T saturates, $1/\tau_{III}$ will again change to $k_{TR_4} + k_{RT_4} \approx k_{TR_4}$. It might well be that saturation of the less affine form does not change the rate constants $k_{TR}$ detectably. If so, this latter step in $1/\tau_{III}$ should not be detectable. In any case, $1/\tau_{III}$ should remain constant at higher concentrations. Furthermore, $1/\tau_{III}$ should be independent of the enzyme concentration throughout the whole concentration range (at constant "free"-substrate concentration).

With such a relaxation spectrum we should have an excellent tool to test the nature of this simple type of allosteric control mechanism. The existence of the first two relaxation effects, for which $1/\tau$ increases linearly with concentration, should give detailed information on binding in both states. From a comparison at low and high concentrations (low and high saturation), we should learn whether the simplifying assumptions of degeneracy, or intrinsic binding constants, are fulfilled. The third relaxation time should inform us about the existence of a conformation change. If $1/\tau$ decreases from one constant level to another, we should expect two conformations with first-order transformations. The sharpness of the decrease of $1/\tau$ decides about the cooperativity. An all-or-none transformation of a four-subunit system requires fourthpower terms in substrate concentration. On the other hand, any model that assumes only one binding form would never yield this type of $1/\tau$ relation. For such a model, all $1/\tau$ values should more or less increase with the "free" substrate concentration. Moreover, the non-existence of any of the simplifying assumptions in the RT-model (degeneracy, all-or-none transformation) would result in a higher complexity of the relaxation spectrum, e.g., more than three relaxation times, or curvature changes, in $1/\tau$ relations. If, on the other hand, a relaxation spectrum like the one described above could be detected, one could well conclude that Monod's model is the simplest to account for the experimental facts and that—even if a more complex mechanism is involved—it is still a good approximation.

Obviously, quite some luck is required for one to encounter a system agreeable enough to show three relaxation times with concentration dependencies that can be analyzed explicitly in such a simple manner. Dr. Kirschner in our group had such luck with the enzyme glyceraldehyde-phosphate-dehydrogenase, a key enzyme in the glycolysis pathway. He also was fortunate enough to isolate this enzyme from yeast instead of from muscle—fortunate because only the former allows the complete analysis in a concentration range that is easily accessible to relaxation spectrometry. Details about these investigations are published elsewhere. The behaviour described in this chapter is exactly reflected by the enzyme.

From these studies Monod's assumptions about symmetry seem to be met completely, although according to Koshland, Némethy, and Filmer, nonsymmetric models in which the simplifying assumptions of Monod are not fulfilled should not be excluded a priori. It may well be that both models are possible in principle, but that symmetric models had a better chance to survive during evolution. Multisubunit enzymes seem to be products of "adaptation" or "learning". The binding and cooperation of subunits is a product of successive mutuation and selections. Any symmetric model would have advantages, for improvement of subunit interaction becomes effective at more than one side. Although it would be premature to draw general conclusions from only one example, two conclusions are possible: (1) Monod's model represents a useful working hypothesis; (2) techniques now available (such as relaxation spectometry) allow a complete analysis of such molecular models of control function.

H. C. LONGUET-HIGGINS: Two points. About the nerve membrane, the Hodgkin-Huxley theory gives a complete account of the conduction phenomenon assuming that the permeability of the membrane to potassium (and to sodium) is a function of the membrane potential. And this is enough, in fact, to give the phenomenon without assuming a cooperative activity inside the membrane.

Can I make the other point, which was about the tendency for like to associate with like, which is at the basis of Dr. Monod's idea for interpreting the allosteric phenomenon in terms of symmetry. If one disregards the somewhat embarrassing exception of the alpha-beta interaction in the haemoglobin molecule, then it is suggested that one thinks of this tendency to preserve symmetry as in some way related to the phenomenon of the crystallisation of pure crystals from mixtures.

It really is very remarkable that one can, from a complicated mixture of non-polar or slightly polar molecules obtain in almost all cases pure crystals (rather than mixed crystals, mixed crystals being the exception. This observation is of course in no sense a hypothesis as to *why* like should pair with like, although one can hazard the guess that it must have something to do with the necessity of matching the optical spectra of the members of the pair because an exact match between their optical spectra produces the maximum Van der Waals between two molecules, other things being equal. The problem of the purity of crystals seems to be at least related to Dr. Monod's problem about the connection between symmetry and allostery.

A. KATCHALSKY: The point raised by Dr. Longuet-Higgins deserves discussion in this group. It is such an intriguing and basic physiological fact that in the excitation of the nerve membrane the permeability increases very drastically—at least by two orders of magnitude. Rojas from South America has proved that not only electrolytes but also non-electrolytic substances which have nothing to do with potassium-sodium exchange become very permeable through the nerve membrane upon stimulus.

Hodgkin and Huxley have taken notice of this fact and have incorporated it into the phenomenological theory by assuming that the permeability coefficient changes with potential—which is a correct statement but does not explain why permeability should change with potential.

Indeed the permeability does not change continuously but increases very rapidly in a kind of exponential manner presumably with an exponent of 4. It is very interesting that there is an exponential increase. For a physico-chemist or a biophysicist could expect a cooperative phenomenon or that the very change in potential release a change in structure. Something structurally changes which opens larger holes in the membrane system. What is the simplest assumption. There is a group of neuro-physiologists who are now tending to believe that what happens is a kind of phase transition of the type which was presented by Monod. I want to make it absolutely clear that this does not in anyway contradict the finding of Hodgkin and Huxley but tries to provide it with a physicochemical basis.

F. LYNEN: Y a-t-il des questions au sujet de l'exposé fait par le M. Wolff?

H. C. LONGUET-HIGGINS: You will forgive, perhaps, a very naive question asked in ignorance. It is of course very interesting that one obtains a development of symmetry in a large organism. But perhaps what is even more interesting is that on the whole one obtains organisms, or populations of organisms, which are *not* symmetrical; most of us are right-handed and we all have our hearts on the left. There must be some point at which the small-scale "handedness" of our constituent molecules develops into a large-scale handedness. I would like to know the usual view about this problem.

E. WOLFF: Il est déjà très difficile de comprendre la question de la symétrie. En ce qui concerne la question de l'asymétrie, il n'est pas plus facile d'avoir une idée des facteurs qui la déterminent. Aussi ne suis-je pas en mesure de répondre à votre question.

M. MAGAT: Il a été parlé de l'apparition d'une symétrie, au moment de la perturbation. En réalité, il y a une diminution de symétrie, parce qu'il y a d'abord une symétrie sphérique; ensuite, on obtient un plan de symétrie. Donc, le degré de symétrie est beaucoup plus bas. Et c'est quelque chose de beaucoup plus facile à comprendre, du point de vue physicochimique, que l'apparition d'une symétrie.

E. WOLFF: Le point de vue du physicien est certainement très judicieux. Pour le biologiste, il n'est pas très réaliste. L'apparition simultanée d'un plan de symétrie bilatérale, d'une polarité dorso-ventrale, d'une orientation céphalo-caudale est l'étape capitale de la transformation d'un œuf indifférencié en un système différencié, c'est-à-dire en un *embryon*. Je vais poser un problème, à mon tour : il est relatif aux Echinodermes. L'Echinoderme commence, comme à peu près tous les autres organismes, par avoir une symétrie axiale. Et puis, au cours du développement embryonnaire, il acquiert une symétrie bilatérale, qui se marque en particulier par les deux sacs coelomiques. Un peu plus tard, au moment de la métamorphose de la larve de l'Echinoderme, il se reconstitue une symétrie radiaire, une symétrie pentaradiaire, assez rare chez les organismes animaux. Il y aurait donc une augmentation de la symétrie, au sens des Physiciens. Ce n'est donc pas toujours dans le sens d'une diminution de la symétrie que se fait le développement. Dans certains cas, il peut y avoir une acquisition, une augmentation de la symétrie.

M. MAGAT : Je m'excuse... Je n'ai peut-être pas bien compris la situation... Vous avez d'abord une symétrie cylindrique. Cela signifie que vous avez un nombre infini de plans, qui sont tous des plans de symétrie.

Ensuite : l'apparition d'un plan dissymétrique. Donc, votre symétrie baisse. Puis, quand vous introduisez encore cinq autres directions, votre symétrie diminue encore. Ainsi, vous avez, je crois, là aussi, une diminution de symétrie ! Il faudrait voir la chose avec des mathématiciens, qui connaissent mieux la question.

E. WOLFF : Il me semblait que dans la symétrie bilatérale, il n'y avait qu'un plan de symétrie, tandis que dans la symétrie pentaradiale, il y en a cinq.

S. L. SOBOLEV : Peut-être ai-je mal compris... La symétrie de rotation et la symétrie de réflexion sont deux opérations qui ne se confondent pas. Vous pouvez avoir une symétrie de rotation sans avoir une symétrie de réflexion. De laquelle s'agit-il ? (Dans ce que M. Wolff a dit, c'est-à-dire une apparition de nouvelles quantités). On voit toujours, avec une symétrie de rotation, qu'il s'agit de molécules à symétrie hélicoïde, pour laquelle la réflexion est impossible. Alors, vous voyez que ce n'est pas tout à fait simple !

Je peux, maintenant, soutenir un peu ce que M. Wolff disait, car l'apparition du plan avec une symétrie de réflexion est une chose nouvelle. N'est-ce pas ?

E. WOLFF : Au moment où s'établit le plan de symétrie bilatérale, il y a des changements macroscopiques et microscopiques dans la pellicule qui entoure l'ovule. C'est une pellicule très mince, dans laquelle se produisent des transformations continuelles, considérables, telles que des mouvements de particules. Ces mouvements, qui ont été bien étudiés, sont d'abord des mouvements fluctuants. Ensuite, toute la pellicule est entraînée, mais pas d'une manière égale. Elle effectue une rotation de 30° environ, en se contractant vers la future face dorsale. Il s'agit certainement de jeux de molécules extrêmement complexes.

L. Rosenfeld: Mais qui se stabilisent ensuite.

E. Wolff: Oui. Mais tout est en mouvement, à un certain moment.

H. C. Longuet-Higgins: May I put my previous point again? It seems to me to be a point of great interest. There are three things that might happen in the evolution of an organism from a chromosome in a suitable environment. The first is that it might develop mirror symmetry on a large scale, and this we have heard about. Another thing that it might do is what *we* do, and that is to grow up all right-handed, all with the same "chirality" on a large scale ("chirality" means "handedness"—the word is due to Eddington). The third thing that might happen is that you might get some individuals who were left-handed and others right-handed, at random.

It is much more easy to understand the first and the third of these possibilities, when they arise, than the second. Because the second possibility—everybody's heart being on the left—implies that the "chiral" control must be exerted right through the morphogenetic process, right from the molecular level up to the level of the organism as a whole. And if one looks at the growing organism at the level of cellular structure one does *not* notice any very conspicuous chirality at that level. Therefore it seems to me very odd that we should not be a more or less equal mixture of people with their hearts on the right hand side and others with their hearts on the left.

H. S. Bennett: If one wishes to invoke d-1 molecular asymmetries to account for the asymmetric distribution of certain organs and the asymmetric development of certain portions of the nervous system with respect to the saggital plane at the center of the body, one has to explain the occasional appearance of what we call "situs inversus". That is a condition characterized by the heart and other organs on the side opposite to their usual location. Situs inversus occurs in people whose molecular components gave rise to the same sense rotation as do those of normal people. This suggests that there is not a simple or strict relationship between the molecular sense of rotation of molecules in our bodies and the asymmetries of our organs.

E. Wolff: Il est certain que l'asymétrie, droite ou gauche, est quelque chose de remarquable. On l'a étudiée, du point de vue génétique, dans certaines espèces animales, chez les escargots et chez d'autres espèces de gastéropodes. Il y a des races dextres et des races sénestres. De sorte que l'asymétrie a des bases génétiques.

A. Tavares de Sousa: M. Wolff vient de montrer sur l'œuf de Batracien et sur l'œuf d'Oiseau que la symétrie bilatérale de l'embryon est sous la dépendance de conditions d'ordre mécanique (point de pénétration du spermatozoïde, rotation de l'œuf).

Ma remarque c'est que, dans le cas de l'œuf de Grenouille les remaniements du matériel formatif se font à l'intérieur d'une seule cellule, alors que dans le cas de l'œuf de Poule il s'agit d'un blastoderme dont les cellules sont limitées par des membranes. Je demande à M. Wolff si l'on peut concevoir une interprétation qui soit valable, au fond, pour les deux cas et quelles modifications faudrait-il lui faire subir pour l'adapter à chacun des cas.

E. WOLFF: C'est une intervention extérieure qui provoque la symétrisation. Et ce sont tout de même des interventions analogues dans les deux groupes. Faisons abstraction de l'action du spermatozoïde. J'ai parlé du cas des Amphibiens où la rotation de l'œuf—ne disons pas la pesanteur,—peut provoquer la symétrisation. Or, qu'est-ce que la rotation de l'œuf? La rotation de l'œuf, c'est la descente de l'ovule, avec frottement contre ses membranes. Il y a d'abord une forte adhérence, qui cesse après l'insémination, au moment de la rotation de l'œuf dans son chorion, mais il y a cependant encore un certain frottement entre la paroi de l'œuf et le chorion.

Clavert et Wintenberger ont fait une expérience qui, au lieu de faire tourner l'œuf dans son enveloppe, consiste à faire tourner celle-ci autour de l'œuf. Ce qui aboutit exactement au même résultat, c'est-à-dire à la symétrisation. Dans l'œuf de l'Oiseau, nous avons le même processus: une rotation de l'œuf entier à l'intérieur de ses enveloppes. La bascule de l'œuf dans l'utérus entraîne un frottement de l'ovule, c'est à dire, du jaune par rapport à son enveloppe inerte l'albumine. Voilà ce qui est commun entre les deux espèces.

Ce qui est différent, c'est que, comme vous l'avez fait remarquer dans l'œuf de batracien, il s'agit d'une cellule unique, une cellule dans laquelle toutes les molécules peuvent entrer en relation les unes avec les autres. Il y a des passages possibles entre les différents points de la surface et de l'intérieur de l'œuf—tandis que dans l'œuf de Poule, il s'agit d'un blastoderme comportant déjà un grand nombre de cellules séparées par des cloisons. Alors, dans chaque cellule, doit se réaliser le même processus dans la cellule unique de l'œuf de Batracien. C'est quelque chose de difficile, mais non impossible, à concevoir. Le résultat est la détermination du plan de symétrie par un nombre élevé de cellules. Les mêmes transformations qui se produisent dans la cellule unique de l'œuf des Amphibiens, doivent s'effectuer dans chacune des cellules d'un même blastoderme.

---

# MECANISMES PHYSIOLOGIQUES

*1ère séance*

PRÉSIDENT P. LÖWDIN

---

B. B. LLOYD

The Concept of Regulation in Physiology

M. LINDAUER AND H. MARTIN

Special Sensory Performances in the Orientation of the Honey Bee

Discussions

*Theoretical Physics and Biology,* © 1969, *North-Holland Publ. Co., Amsterdam*

# THE CONCEPT OF REGULATION IN PHYSIOLOGY

B. B. LLOYD

*Magdalen College and University Laboratory of Physiology, Oxford, England*

The rapporteurs of this conference have been asked to set a broad framework for subsequent discussion and not to cover topics in detail and entirety. After a brief discussion of homeostasis I shall therefore concentrate on some quantitative and speculative aspects of respiratory regulation, describing controller and material equations, and briefly discuss frequency of breathing and the effect of high altitude. The problem of physiological set point is then related to the respiratory regulation of pH. Regulation in exercise and voluntary interference with ventilation are then briefly discussed.

Energy, material and information pass through the individual animal at various and varying rates during its existence. Body temperature and posture are typical resultants of a balance between energy input and output, and body weight a typical resultant of a material balance, and are examples of what Cannon [1] called homeostasis, which may be defined as the maintenance of a variable that is subject to throughput within a tolerable and usually fairly narrow range of an average value. Regulation may be defined as the sum total of devices and processes by which homeostasis is achieved.

The information throughput is different from those of energy and matter. Wiener [2, p. 121] has likened the activity and development of a human brain during its life to a single run on a computer: memory and learning, whether neural or immunological, are scarcely examples of homeostasis. And the transmission and multiplication of genetic material by reproduction is largely restrained by the environment, and not subject, except perhaps in civilized man, to homeostatic regulation [see 3 for a different view], which is based on negative feedback, whereas much sexual behaviour, such as mating and parturition, provides examples of positive feedback, which is otherwise rather rare in normal biology, though perhaps not in pathology.

The variables which show homeostasis are legion and the ones most easily and frequently studied are chemical concentrations in the blood. These may be classified into respiratory, ionic, acid-base, nutritional, excretory and hormonal, each of these classes showing more or less overlap with the others; and the list of hundreds of normal values in any reference book of medical chemistry [4] is a list of homeostases, which are also seen in growth and organ size, in the hydrostatic problems of the circulation of the blood, and in the regulation of the frequency, amplitude and wave form of cyclic processes, such as sleeping and waking and the sexual rhythms, which often profoundly influence supposedly steady levels.

Claude Bernard "Le plus illustre physiologiste de notre époque" (the most illustrious physiologist of his epoch) according to Vulpian [5, 1, vii], once wrote [5, 1, 340] that "Tout acte d'un organisme vivant a sa fin dans l'enceinte de cet organisme" (every act of a living organism has its aim within the ramparts of that organism), which may have prompted J. S. Haldane [cited in 1 & 3] to state that "No more pregnant sentence was ever framed by a physiologist" about Bernard's [5, 1, p. 113] proposition that "La fixité du milieu intérieur est la condition de la vie libre, indépendante". This most famous of physiological generalizations comprises three notions of great importance, the free life, the internal milieu, and the latter's constancy. Physiologists have spent little time debating whether Bernard's 'La vie libre' referred to the mobility of a mammal or to the political freedom of man, but in the last century there has been immense and intense physiological investigation of the identity, composition and regulation of the internal environment, of which Bernard [5, 2, p. 5] stated 'C'est le sang; non pas à la vérité le sang tout entier, mais la partie fluide du sang, le plasma sanguin, ensemble de tous les liquides interstitiels, source et confluant de tous les échanges élémentaires.' (It is the blood; not indeed the whole blood, but the fluid part of the blood, the plasma, including all the interstitial liquids, the source and confluence of the elementary exchanges.)

This concept of the internal environment is quite complex. Our skin exists in atmospheric air, but the alveolar air in chemical contact with blood in the lungs contains much less oxygen (14 as opposed to 21 %) and far more $CO_2$ (6 as opposed to 0.03 %) than atmospheric air. We can thus be said to live in an atmosphere containing a partial pressure of oxygen, $Po_2$, of 100 torr, and a $Pco_2$ of 40 torr, which at once differentiates us from the creatures in the sea, where both these pressures are usually lower.

The arterial blood exchanges gases with the alveolar air, so that for most purposes in the normal man we can accept the equality of the alveolar and arterial partial pressures ($P_A$ and $P_a$) of oxygen, nitrogen, $CO_2$ and water, but, just as we really live in alveolar, not atmospheric, air, our tissues live in venous rather than arterial blood, with a $Po_2$ less than 50 and a $Pco_2$ above 40 torr. The immediate environment of cells is the tissue fluid, which tends to be slightly more venous than venous blood, with lower $Po_2$ and higher $Pco_2$, and the intracellular fluids, in which the cellular components such as mitochondria function, may have $Po_2$'s of only a few torr, and at the actual points in the cell where oxygen is converted to water its concentration tends to zero. The chemical reduction of oxygen supplies energy to the cell, and the continuous but fluctuating energy demands of the cell depend on an appropriate transport of $O_2$ from air to the mitochondrion, and of $CO_2$ from the enzymes of the Krebs cycle to the air. The maintenance of appropriate flows of $CO_2$ and $O_2$ is the primary task of ventilation and circulation, and in discussing the concept of regulation in physiology I shall speak mainly about this task. This needs no apology, for not only did Bernard [5, **2**, p. 141] state that 'La respiration est le phénomène le plus caracteristique de la vitalité, c'est-à-dire de l'être en activité vitale', (Respiration is the most characteristic phenomenon of vitality, that is to say of the being in vital activity) but the physiology students being examined in Oxford a fortnight ago were asked to discuss the proposition that 'Respiratory physiologists are the most quantitative neurophysiologists'. Respiration has indeed been much investigated, as its inputs and outputs are relatively easy to measure; it operates over a 30-fold range of activity, and it provides the most characteristically quantitative examples of chemoneural interactions.

We are still in doubt, owing to inadequate experimentation, as to what chemical variables directly affect the ventilation, but it is generally agreed that the $H^+$ concentration of the blood is effective at the chemoreceptors in the aorta and the fork of the carotid artery, and that there is an effective $H^+$ concentration inside the cranium. It is also accepted that the partial pressure of oxygen in the arterial blood, $P_{a}o_2$, has a potent effect on the arterial chemoreceptors. The experimentally observed relations between chemical factors and ventilation $\dot{V}$ have been summarized in various equations, of which the following is a recent, though still controversial, example [6].

$$\dot{V} = h\{\psi(\lambda + \log H_a^+/H_{a\theta}^+)/(P_a o_2 - \gamma) + \mu + \log H_c^+/H_{c\theta}^+\}. \tag{1}$$

$\dot{V}$ represents the total air breathed out by a subject in unit time (l./min.) and is easily measured by means of a gas meter, while $h$ is partly a size parameter (an elephant will breathe more than a mouse having the same blood chemistry), and partly a measure of overall sensitivity.

The first term inside the curly bracket is meant to represent the activity of the chemoreceptors. These small bodies, situated on the aorta and the carotid artery, fire more nerve impulses to the brain when the pH or $P_{o_2}$ falls in the arterial blood reaching them. The term shows a linear function of the logarithm of blood hydrogen ion, $(\lambda + \log H_a^+/H_{a\theta}^+)$, multiplying a reciprocal function of oxygen $\psi/(P_a o_2 - \gamma)$, and represents the notion, which is still being investigated, that the product of these terms generates the impulse traffic in the chemoreceptor nerves. This traffic adds to two central terms, $\mu$ representing a residual nervous effect independent of chemical stimuli, and a final term containing $H_c^+$, a central hydrogen ion concentration, perhaps in cerebrospinal fluid or brain interstitial fluid. When $H^+$ falls below the threshold value $H_\theta^+$ in either log term the effect is taken to be zero rather than negative.

We now turn from these informational relationships to the metabolic or material relationships that normally exist between ventilation and the respiratorily important variables. A simple example is the equation relating $P_{co_2}$ with ventilation:

$$\dot{V}(P_A\,co_2 - P_I\,co_2) \simeq j\dot{V}co_2. \tag{2}$$

This states that the product of ventilation and the difference between alveolar and inspired $P_{CO_2}$ is approximately proportional to the metabolic production of $CO_2$, so that if $\dot{V}$ is doubled at constant $\dot{V}_{CO_2}$, $P_{ACO_2} - P_{ICO_2}$ is halved, and so on, a relationship we term the metabolic hyperbola. This means that if $\dot{V}$ changes for any reason, such as a change in one of the terms in the right-hand side of equation (1) there will usually be a change in $P_{ACO_2}$, which is itself a main determinant of $H_a^+$ and $H_c^+$ and hence of $\dot{V}$. The form of the equations shows that when $\dot{V}$ rises, $P_{ACO_2}$ approaches $P_{ICO_2}$ and hence $H_a^+$ and $H_c^+$ tend to fall, causing a depression of ventilation. This negative feedback is an example of the myriad of negative feedback arrangements in physiology. A similar negative feedback applies to oxygen, for which we may write

$$\dot{V}(P_{I\,O_2} - P_{A\,O_2}) \simeq j\dot{V}_{O_2}. \tag{3}$$

When $\dot{V}$ rises, $P_{IO_2} - P_{AO_2}$ falls, so that, at constant $P_{IO_2}$, $P_{AO_2}$ rises. This reduces $1/(P_{aO_2} - \gamma)$ in equation (1), so that the stimulus is reduced and $\dot{V}$ tends to fall. In short, knowing $P_{IO_2}$, $P_{ICO_2}$ and $\dot{V}_{O_2}$ ($\dot{V}_{CO_2}$ is usually close to it in value) and the basal characteristics of the blood and c.s.f. of the subject, we should be able from equations (1), (2) and (3) to predict the steady-state values of $\dot{V}$, $H_a^+$, $H_c^+$ and $P_{aO_2}$. The non-steady state is more complicated, and a masterly treatment of the relationships corresponding with equations (2) and (3) in the non-steady state for lungs, blood, tissues, brain and c.s.f. is to be found in the paper of Grodins, Buell and Bart [7].

Their paper, like our discussion so far, treats ventilation as a steady process, represented by a steady current of air passing over the lung surface, but breathing, as Yamamoto and Raub [8] amusingly point out in an intriguing theoretical paper, is of course rhythmic, involving $V_T$, the tidal volume or amplitude, and a frequency $f$, such that

$$\dot{V} = fV_T. \tag{4}$$

Hey et al. [9] have shown that over a wide range of ventilations with a variety of respiratory stimuli (increased body temperature is an exception, causing $m$ to increase) there is a linear relation between $\dot{V}$ and $V_T$ of the form

$$\dot{V} = m(V_T - k). \tag{5}$$

The parameter $m$ has the dimensions of a frequency, min$^{-1}$, and it is highly correlated with the sensitivity-size parameter $h$ of equation (1), which has the dimensions l.min$^{-1}$. This high correlation implies some important overlap between chemical and neural phenomena, and raises the question of the origin of the rhythm of ventilation and of its optimal frequency. The latter is related to work of breathing, to the volume and resistance of the respiratory tubes that do not exchange gases with blood, and with the notion of length-tension appropriateness [cf. 10, 11]. Mathematical ideas of optimality are of growing importance in biology [12].

Although equations such as (1), (2) and (3) provide a basis for respiratory regulation, they are far from adequate as explanations of every important respiratory phenomenon. Altitude and exercise provide the most striking examples of physiological respiratory alteration, and both provide interesting examples of regulatory peculiarities.

At altitude $P_{IO_2}$ and hence $P_{AO_2}$ are reduced, and equations (1), (2) and (3) permit a prediction of $\dot{V}$ in the steady-state which is reasonably borne out by experience. The increase in $\dot{V}$ without an increase in $P_{ICO_2}$ will by equation (2) lower $P_{ACO_2}$, and this will cause $H_a^+$ and $H_c^+$ to fall below their threshold values $H_{a\theta}^+$ and $H_{c\theta}^+$, so that the log terms are effectively zero. Whereas an abnormal alkalinity of the blood is only slowly rectified by renal excretion of $HCO_3^-$, an abnormally alkaline c.s.f. is rapidly rectified by a secretory lowering of $HCO_3^-$ so that the pH of c.s.f. is soon restored to its usual figure of about 7.32, at which $H_c^+$ is sufficiently high to promote ventilation and hence a raised $P_{AO_2}$. This extremely interesting phenomenon, by which c.s.f. pH is maintained and $P_{AO_2}$ considerably augmented at altitude, was hinted at by Kellogg [13] and confirmed by Mitchell, Severinghaus and their collaborators [14], when they heroically analysed their own c.s.f.'s at an altitude of 3800 m.

A well-known effect of altitude is the increase in red cells per unit volume of blood. This is yet another example of negative feedback, the hypoxia calling up a response in the blood-forming system which tends to negative the effect of hypoxia. There are, however, two disadvantages of polycythaemia, the obvious one being the rise in blood viscosity, which can overload the heart and circulation. Secondly, and this is less obvious, because haemoglobin is the main blood buffer polycythaemia depresses the slope of the blood log $H^+$, log $P_{CO_2}$ line and thus tends to cause a reduction in ventilation which could lead to further hypoxia and hence more haemoglobin, a positive feedback to which there is no defence except blood-letting or a return to lower altitudes. This is a speculative contribution to the aetiology of the undesirably high polycythaemia of chronic mountain sickness, but brings out the point that where an element in a normally effective homeostatic mechanism develops positive feedback with an open-loop gain greater than 1, disease and perhaps death ensue. This question is discussed in Milhorn's [15, pp. 102, 358] comprehensive text, in which control theory is systematically expounded in a physiological context, at much greater length than in the useful shorter book by Bayliss [16].

The phenomenon by which c.s.f. pH tends to be jealously regulated near 7.32 without a steady-state error, by a combination of ventilation, secretion, brain bloodflow and metabolism, raises the whole question of set-points in physiological regulation. Why is the blood pH 7.4, the c.s.f. pH 7.318? On the former, Priban and Fincham [17] have made the ingenious and stimulating suggestion that the normal pH of blood is set at 7.4 because this is the pH at which haemoglobin is most effective as a physiological acid-base regulator. Haemoglobin is a buffer system which accepts protons and lessens pH change as $CO_2$ diffuses into blood from the tissues, but this simple physicochemical buffering is further supplemented by the Bohr-Haldane effect, by which oxyhaemoglobin is a stronger acid than reduced haemoglobin, and oxygen is driven off haemoglobin by an increase of $CO_2$ or $H^+$ concentration. Priban regards this effect as being most marked at a pH of about 7.4, and suggests that this property of haemoglobin acts on the ventilation to bring blood pH back to the figure at which haemoglobin is most effective, using the device known to control engineers as hill-climbing to provide a mechanism in detail.

This suggestion is open to serious criticism as a full explanation of respiratory regulation, but it contains the important idea that the set-point to which a regulatory system returns may be embodied in the properties, physicochemical or otherwise, of a large molecule evolved in more primitive evolutionary conditions. Once evolved the large molecules can tend to improve their biological effectiveness by promoting the selection of an internal environment which enhances the biological value of their physicochemical peculiarities. Priban's idea is also close to the notion that an enzyme could during the course of evolution bring its intra- or extracellular environmental pH to the value at which it is most effective as a catalyst.

A further suggestion as to the respiratory regulation of blood pH comes from the work of Rahn [18], who has shown in cold- and warmblooded animals that over a range of body temperatures the pH to which the blood is normally brought by regulation varies with temperature in parallel with the pH of neutral water, so that the pH of blood is about 0.6 above that of physicochemical neutrality. Albery [19] has argued that this may be related to the change with temperature of the pK of some biologically important buffer system.

Yet another aspect of blood pH regulation is the possibility that it is set to give a minimum effective total of $[H^+] + [OH^-]$, both $H^+$ and $OH^-$ being powerful hydrolytic catalysts for biologically important polymers such as proteins, esters and polysaccharides. If we can argue that because $H^+$ is about twice as mobile as $OH^-$, and that the ideal compromise pH is reached when $[H^+]$ is $0.5 [OH^-]$, we should expect at 37° a biological pH of 6.95, 6.8 being the pH of neutral water. This seems rather far from that of blood, but is probably close to intracellular pH's [20].

The point emerging at this stage of our discussion is that the pH's of blood, interstitial fluid and cells undoubtedly show homeostasis and are subject to regulation, but that the set point to which they regulate is not some standard pH tucked away in a little compartment to which reference is continually made, or a pointer reading as in a chemical engineering plant, but may be embodied in the physicochemical or biochemical properties of key macromolecules. The links between these properties and the regulatory devices have yet to be worked out, and until this is done in detail this notion remains speculative.

In discussing the pH regulation of the blood it may be convenient to think of factors tending to disturb it and of those tending to correct it. A low-protein vegetarian diet tends to cause alkalinity, and a high-protein diet acidity by the oxidative production of sulphuric acid: these are relatively slight effects dealt with in the long run by the kidney. The ingestion of any sort of food leads to the secretion of acid in the stomach, and Dodds [21] showed that the resulting alkalinity of the blood is dealt with by a depression of ventilation, a consequential retention of metabolic $CO_2$, and a partial restoration of pH to the norm. Haldane [22] showed that 10 or 15 g of ammonium chloride could be ingested daily without apparent ill effect and that it gave an effect very similar to that of the administration of hydrochloric acid. This causes acidity of the blood, an increase in ventilation, a consequential lowering of the $CO_2$ concentration of the blood and hence a partial restoration of the pH to normal.

As long as the daily ingestion of 10 g of ammonium chloride continues, the blood pH remains at about 7.3, this giving the steady-state error of 0.1 pH needed to drive the kidney to dispose of the 10 g of ammonium chloride into an acid urine: similarly the oral ingestion of sodium bicarbonate can lead to a steady-state blood pH of 7.45, this being the pH at which the kidney will dispose of the bicarbonate into an alkaline urine. This dietarily imposed range of 0.15 pH is quite large, and implies that the kidney operates as a proportional controller with a steady-state error, that is as if the rate of excretion of acid is a direct function of the acidity of the blood, and there is no evidence, from the persistence of abnormal pH's when the diet is abnormal, that integral control operates to bring the final steady-state error to zero. The kidney seems also to lack derivative control responding to the rate of change of blood [H+], but this is not a defect, for the most precipitate change in blood pH comes during and after the most violent exercise, such as the running of 400 m. in 45 seconds. Energy for this comes partly from oxidation and partly from the conversion of some 40 g of glycogen to lactic acid in the muscles. The concentration of lactic acid in the blood may rise from 10 to 150 mg per 100 ml, and, if there were no ventilatory adjustments, the pH would fall to 6 or less. As we all know from personal experience there is during or immediately after violent exercise a rise in ventilation (partly predictable from equation (1)) which drives off $CO_2$ and moderates the fall in pH so that the lowest value reached is about 7, and then during recovery there is a steady and rapid return to normal as lactic acid is removed from the blood, mainly by the liver. If the kidney responded immediately to the low pH of exercise, we should be in grave difficulties. Lactic acid, a valuable source of calories under aerobic conditions, would be lost in the urine, and the metabolic cost of violent exercise would rise ten-fold, a situation of no advantage to primitive man and of benefit only to the obese of the twentieth century (if they could be persuaded to make use of it).

This short-term and very effective ventilatory regulation of blood pH depends on the volatility of the main acid product of metabolism, $CO_2$. By hyperventilation we can lose roughly one molecule of $CO_2$ for each molecule of lactic acid added, though once again, as with the longer-term kidney, there will be a steady-state error, showing the the ventilation-blood-$H^+$ system seems to work as a proportional controller without integral control. The importance of the blood $H^+$-$CO_2$-ventilation system is its rapidity of action, which is second only to that of the first line of defence, the physicochemical buffers of the body fluids, of which haemoglobin and bicarbonate are the most important.

When $P_{A CO_2}$ is raised by inhalation of $CO_2$, the blood supply to the brain generally rises through the agency of a roughly proportional control mechanism by which the diameter of the brain arterioles is a direct function of $P_{A CO_2}$, so that $P_{c.s.f. CO_2}$ rises less than $P_{A CO_2}$. This means that the stimulus to raise c.s.f. $[HCO_3^-]$ by secretion is smaller than it would otherwise be. If this regulation of brain bloodflow did not exist, raising blood $CO_2$ would entail a secretory rise in $HCO_{c.s.f.}$, a change which could lead to serious underventilation if the extra $CO_2$ were suddenly removed. The raising of brain bloodflow in response to high $P_{A CO_2}$ is thus a valuable short-term regulation by a rapidly reversible process, similar in function to the respiratory as opposed to renal adjustment of blood pH during the lactacidaemia caused by violent exercise. It should be noted, however, that the final regulation of c.s.f. pH to 7.32 in a wide variety of conditions [14: but see 23] may imply integral control without a steady-state error.

Exercise is the most normal and least understood stimulus to ventilation. It can raise $\dot{V}_{CO_2}$ and $\dot{V}_{O_2}$ of equations (2) and (3) 25-fold above their resting values, and this must cause an increase in either or both of $\dot{V}$ and $(P_{A CO_2} - P_{I CO_2})$. It is obviously possible in principle to calculate $\dot{V}$ in exercise from equations (1), (2) and (3) and the appropriate subsidiary equations relating $H^+$ to $P_{O_2}$ and $P_{CO_2}$, but it is found in practice that $\dot{V}$ is nearly always greater than the prediction, and indeed greater than the value predicted when values actually measured in exercise are substituted in the controller equation. It is therefore suggested that exercise increases $\mu$, possibly by impulses coming from active limbs, and we [24] have found in man, following in the wake of M. Dejours [see e.g. 25] that the increase in $\dot{V}$ at the beginning of exercise is independent of the pre-existing $P_{O_2}$ and $P_{CO_2}$ in the chemical background, which argues strongly that this initial exercise effect is an increase in $\mu$. We [26] are currently investigating short-term and steady-state aspects of the hypoxia parameters ($\psi$ and $\gamma$) in exercise, which now appears to evoke a change in the form of equation (1). The hyperventilation seen in exercise over and above the prediction of equation (1) is obviously of great interest, and has been attributed to feed forward, showing up as an increase in $\mu$. The absence of changes (steady-state error) in $P_{a CO_2}$ and $P_{a O_2}$ in moderate exercise points also to integral control.

Professor Dejours has always grasped the most interesting nettles, seldom if ever being stung, of respiratory physiology, and I see from the programme of the June 1967 meeting of French-speaking physiologists that he is now working on respiratory aspects of speech. Speech is the most human of motor activities, using one of the most primitive of quasi-autonomic activities, respiration, for its purposes: and it has for some time been known [27] that during speech the respiratory muscles are used primarily not for maintaining the homeostasis of blood pH or brain $Po_2$, but for keeping a constant pressure of gas against the vocal cords. This capacity for over-ruling the usual chemoneural respiratory rhythms is of immense biological importance for drinking, swimming, talking, spitting, yawning, grimacing, laughing, sobbing, cleaning one's teeth, snorting, singing, diving, sneezing, sniffing, coughing, blowing, sucking, eating, drinking and for rejecting some hostile atmospheres and inhaling others, and it is interesting to speculate as to where the will to hold one's breath, presumably cortical in origin, conflicts with the chemical drive which ultimately forces one to breathe. Measurements during and after breath-holding provide, albeit at a low level, objective and reproducible data on the will and its pharmacology [28, 29], and it is of interest that Dr Bhattacharyya has found in our laboratory that at the breath-holding breaking point the movement of the chest is affected by the blood gases, but the effects, though fairly consistent for an individual, differ widely between subjects.

By definition, but by nothing else, the gaps in this presentation cannot be its most salient features. By largely confining it to respiratory regulation we have been able to go into some detail, though the treatment of the only controller equation we have discussed has been brief and dogmatic. The unsolved problems in regulatory physiology are the details, largely requiring physiological experimentation, of these control equations, the conceptual problems of the non-steady-state, in which Grodins has given an admirable lead, and the search for the physical counterpart of the set point to which many homeostatic systems return, with or without steady-state error.

The problem of ventilation in exercise remains partially unsolved, and the central nervous links between the chemical input and the rhythmic neuromuscular output remain largely unexplored. There is no doubt that the physiology of regulation has already benefited from the limited application of the techniques and ideas of mathematics, physics, engineering and physical chemistry that has so far been made. During the next twenty years a systematic extension of these applications to the scores of regulations which are now merely verbally described will transform this branch of physiology into a rigorous quantitative discipline. Regulation was made for man and his free life, not man for regulation. May this neo-Bernardism bring many more physically trained theorists into biology!

## References

[1]  W. B. Cannon, *The Wisdom of the Body*, (Norton, 1939).

[2]  N. Wiener, *Cybernetics, or Control and Communication in the Animal and the Machine*, 2nd ed., (M.I.T. Press, Wiley, 1961).

[3]  L. L. Langley, *Homeostasis*, (Chapman & Hall, 1966).

[4]  K. Diem, Ed. *Documenta Geigy Scientific Tables* 6th ed., (Geigy, 1962).

[5]  C. Bernard, *Leçons sur les phénomènes de la vie*, 2 vols, sec. ed., (Baillière, 1885).

[6]  B. B. Lloyd, *The interactions between hypoxia and other ventilatory stimuli*, pp. 146–165 in: *Proc. int. Symp. cardiovasc. respir. Effects Hypoxia, Kingston, Ont. 1965*. Eds. J. D. Hatcher and D. B. Jennings, (Karger, 1966).

[7]  F. S. Grodins, J. Buell and A. J. Bart, Mathematical analysis and digital simulation of the respiratory control system, *J. Appl. Physiol.* **22** (1967) 260–276.

[8]  W. S. Yamamoto and W. F. Raub, Models of the regulation of respiration in mammals, Problems and promises, *Computers Biomed. Research* **1** (1967) 65–104.

[9]  E. N. Hey, B. B. Lloyd, D. J. C. Cunningham, M. G. M. Jukes and D. P. G. Bolton, Effects of various respiratory stimuli on the depth and frequency of breathing in man, *Respiration Physiology* **1** (1966) 193–205.

[10]  J. G. Widdicombe, *The regulation of bronchial calibre*, pp. 48–82 in: *Advances in Respiratory Physiology*, Ed. C. G. Caro, (Arnold, 1966).

[11]  J. B. L. Howell and E. J. M. Campbell, Eds. *Breathlessness*, (Blackwell, 1966).

[12]  R. Rosen, *Optimality Principles in Biology*, (Butterworth, 1967).

[13]  R. H. Kellogg, *The role of $CO_2$ in altitude acclimatization*, pp. 379–395 in: *The Regulation of Human Respiration*, Eds. D. J. C. Cunningham and B. B. Lloyd, (Blackwell, 1963).

[14]  R. A. Mitchell, *Cerebrospinal fluid and the regulation of respiration*, pp. 1–47 in: *Advances in Respiratory Physiology*, Ed. C. G. Caro, (Arnold, 1966).

[15]  H. T. Milhorn, *The application of control theory to physiological systems*, (Saunders, 1966).

[16]  L. E. Bayliss, *Living control systems*, (English Un. Press, 1966).

[17]  I. P. Priban and W. F. Fincham, Self-adaptive control and the respiratory system, *Nature* **208** (1965) 339–343.

[18]  H. Rahn, *Gas transport from the external environment to the cell*, pp. 3–23 in *Development of the lung*, Eds. A. V. S. de Reuck and R. Porter, (Churchill, 1967).

[19]  J. Albery and B. B. Lloyd, *Variation of chemical potential with temperature*, pp. 30–33 in: *Development of the lung*, Eds. A. V. S. de Reuck and R. Porter, (Churchill, 1967).

[20]  C. McC. Brooks, F. F. Kao and B. B. Lloyd, Eds. *Cerebrospinal Fluid and the Regulation of Respiration*, (Blackwell, 1965).

[21]  E. C. Dodds, Variations in alveolar carbon dioxide pressure in relation to meals, *J. Physiol.* **54** (1921) 342–348.

[22]  J. B. S. Haldane, Experiments on the regulation of the blood's alkalinity, II. *J. Physiol.* **55** (1921) 265–275.

[23]  J. R. Pappenheimer, The ionic composition of the cerebral extracellular fluid and its relation to control of breathing, *Harvey Lectures* **61** (1967) 71–94.

[24]  D. J. C. Cunningham, B. B. Lloyd and D. Spurr, The relationship between the increase in breathing during the first respiratory cycle in exercise and the prevailing background of chemical stimulation, *J. Physiol.* **185** (1966) 73–75P.

[25]  R. Lefrançois and P. Dejours, Etude des relations entre stimulus ventilatoire gaz carbonique et stimulus ventilatoires neurogéniques de l'exercice musculaire chez l'homme, *Revue Etud. Clin. Biol.* **9** (1964) 498–505.

[26]  D. J. C. Cunningham, D. Spurr and B. B. Lloyd, *The drive to ventilation from arterial chemoreceptors in hypoxic exercise*, pp. 301–323 in: *Arterial Chemoreceptors*, Ed. R. W. Torrance, (Blackwell, 1968).

[27]   M. H. Draper, P. Ladefoged and D. Whitteridge, Expiratory pressures and air flow during
       speech, *Brit. Med. J.* **1** (1960) 1837–1843.
[28]   R. C. Stroud, Combined ventilatory and breath-holding evaluation of sensitivity to
       respiratory gases, *J. Appl. Physiol.* **14** (1963) 353–356.
[29]   B. B. Lloyd and D. J. C. Cunningham, *A quantitative approach to the regulation of human
       respiration*, pp. 331–349 in: *The Regulation of Human Respiration*, Eds. D. J. C. Cunningham
       and B. B. Lloyd, (Blackwell, 1963).

## Additional references

P. Dejours, S. Wagner, M. Dejager and M.-J. Vichon, *Ventilation et gaz alvéolaire pendant
le langage parlé*, Proceedings of the 35th Réunion of the Association des Physiologistes,
(1967). *J. Physiol.* **59** (1967) 386.

J. D. Horgan and R. L. Lange, Digital computer simulation of respiratory response to
cerebrospinal fluid $Pco_2$ in the cat, *Biophys. J.* **5** (1965) 935–945.

H. H. Loeschcke, Homoiostase des arteriellen $CO_2$-Drucks und Anpassung der Lungen-
ventilation und den Stoffwechel als Leistungen eines Regelsystems, *Klin. Wchschr.* **38**
(1960) 366–376.

*Theoretical Physics and Biology*, © 1969, *North-Holland Publ. Co., Amsterdam*

# SPECIAL SENSORY PERFORMANCES IN THE ORIENTATION OF THE HONEY BEE

M. LINDAUER and H. MARTIN

*Zoologisches Institut, University of Frankfurt, Germany*

Adaptation, regulation and homeostasis, these fundamental performances of the living organism, are only possible when the latter can gain information about conditions and changes in its environment. Every animal, including human beings, can however register only a diminutive fraction from the *objective* environment through its "interpreters"—the sense organs. In order to understand the behaviour of a living organism it is important to recognize the *subjective* part of its environment.

Three examples, drawn from the orientation behaviour of the bee, may us reveal a section of a world, inaccessible to our sensory system.

## 1. The suncompass-orientation

Bees use the sun as a compass. It is important to realize, that they use the sun not only for their private orientation but mainly as a point of reference in their mutual communication: When a forager bee in the darkness of the hive indicates the location of a food source by the tail-wagging dance [1] the angle between sun and goal is transposed into the field of gravity. Astonishingly enough—our bees communicate by dances even if the sun is hidden behind clouds and if they can see only a patch of blue sky. That is possible because the light that comes from the blue sky is polarized showing definite relations to the sun's position with both the direction of vibration and the intensity of polarization. V. Frisch discovered this twenty years ago and today we know that there exist true analyzers for polarized light in the bee's eye. The fine structure of the rhabdomeres gives the prerequisites for the dichroic absorption of polarized light [2, 3, 4, 5].

Another question is whether the bees are familiar with the whole pattern of polarization on the blue sky and its regular arrangement around the sun. If you keep young unexperienced bees in a closed room under artificial light and let them fly out for the first time after three weeks than they are unable to use the sun as a compass. It takes at least 5 days, or 500 foraging flights until they learn to use the moving sun and the pattern of polarization on the blue sky as a compass. Another experiment demonstrates that this suncompass is not innate but must be learnt by the bees: I brought a bee colony from Ceylon to Poona (India) and from Ceylon to Munich. The bees behaved in the northern hemisphere as if they would be still under the Ceylonese sky e.g. they calculated the sun movement anticlockwise instead of clockwise; it took 40 days until they got familiar with the orientational cues on the new sky [6, 7].

A prerequisite for suncompass-orientation is an exact time memory. The bees have to know the azimuth of the sun every minute of the day for calculating the changing angles between the moving sun and the fixed position of the food source. Beside the findings of Beling [8], Wahl [9] ,and Renner [10, 11] the following experiment may surprise every one how precisely the time sense works in sun-compass-orientation: In specific situations [6, 12] you can induce "marathon-dances"; in this case the bees continue dancing for many hours without leaving the hive. Therefore they cannot control in the meantime the changing angle between sun and goal; time memory and knowledge of the sun movement exclusively have to save this critical dilemma. Even at night with artificially induced dances the bees show where on the sky the sun could be found; and − the human observer hardly can believe it − they don't calculate the sun's movement at an average speed: they take into account the seasonal variations of the change of azimuth of the sun [13].

## 2.  Problems of orientation during building activity

A masterpiece of the regulated communal work in the beehive is the building of the honeycomb, a performance of orientation marvelled at since long time not only by biologists but mainly by mathematicians, and physicists, in front of all by REAUMUR. In this situation the m e c h a n i c a l  s e n s e s are alone responsible. The honeycomb must be built in the darkness of the hive; without the guidance of optical clues the builders have to measure the overall dimensions of the space within the cells and have to control the angles between the comb-cell walls as well as their inclination towards the foundation. Of the different control mecha-nisms in the building activity we know only two:

1.)  The bristle fields in the neck of the bees are used as gravity receptors to orient the cells in the gravity field. If you glue these sensory bristles with paraffin in all bees of a colony not a single comb cell can be built (even during 4 weeks); but if you remove the paraffin, the bees start with construction of combs immediately.

2.)  A group of highly specialized pressure receptors on the antennal tip controls the t h i c k n e s s  o f  t h e  c e l l  w a l l, which is strictly $72\,\mu$ in worker cells and $95\,\mu$ in drone cells (fig. 1a/b). These receptors work on the following principle: after the foundations of the wall are laid down the builders move along one side of the wall, continually making pressing movements with the rounded edges of their mandibles. The cell walls get indented about $5\,\mu$ deep under controlled pressure of about 4 dyn − just as when one draws two fingers over a tightly stretched curtain. The tips of the antennae − by bipolar scanning movements − register the change dynamically during the indenting and return movements. At the same time they perceive through the sensory hairs on their tips any counter

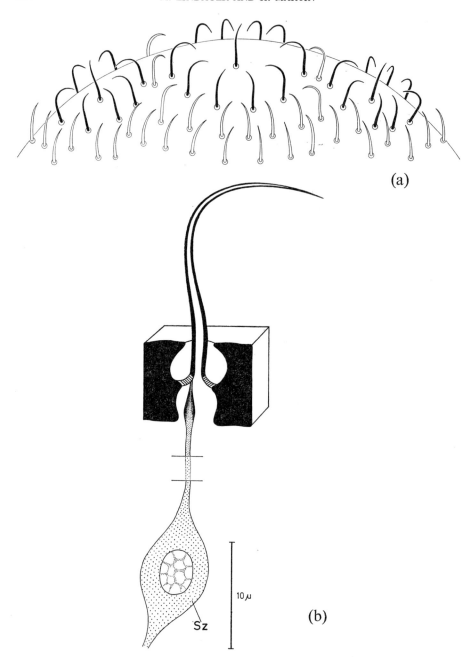

(a)

(b)

Fig. 1.   Three groups of specialized sensory hairs on the tip of a bees' antenna; in the center of each is a straight bristle surrounded by 7–10 hooked bristles in a circle which measure the counterpressure of the wax wall, when it is dented by the mandibles. Note in fig. 1b that the chitinous membrane on the basis of the bristle allows a downward movement of the hair for less than 5 $\mu$. This is exactly the range in which the builders indent the cell walls.

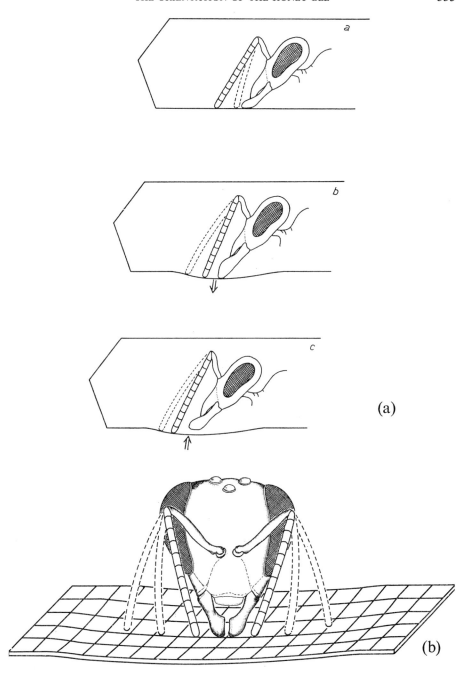

Fig. 2. (a): Pressing movements of the mandibles (a, b), the elastic aperiodic returning movement of the wall (c) and the simultaneous control of the counterpressure by the tips of the antennae
provide the information to measure the thickness of the wall.
(b): The tips of the antennae measure three parameters relevant to the thickness of the comb-
wall: counterpressure, change in shape and its speed.

pressure (fig. 2a/b). This control mechanism—in order to be effective—has two prerequisites:

1.) One has to use always the same raw material; in fact, this is the case: the building material is intrinsic wax (palmitic acid ester and myricil-alcohole) which is transformed into an amorphic stage by kneading them together with the secretion of the mandibular glands.

2.) The temperature must be kept exactly at 35° C because in this temperature range only the wax wall—after mandibular pressure—shows "aperiodic displace-

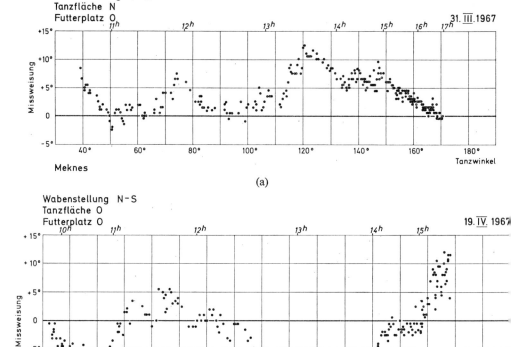

(a)

(b)

Fig. 3. Bees visiting a feeding table 400 m eastwards of the hive announce by their dances on the vertical comb the angle between sun and goal (transposed into the gravity field). The abscissa gives the changing angle between goal and the advancing sun throughout the day. All spots on the 0-line correspond to this angles performed by the dancing bees indicating the direction to the goal without mistake. All spots above or below the 0-line are deviations. The diurnal curve of these deviations depends on the position of the dance floor in the earth magnetic field. (In fig. 3a the dance floor faced northwards, in fig. 3b eastwards.)

ment". In fact since long time it is known that the temperature in the building cluster is kept day and night exactly at 35° C [14].

### 3. Orientation in the gravity field influenced by the earth's magnetic field

In the last ten years studying of orientation of dancing bees we encountered much trouble when the bees apparently made small genuine "mistakes". But there was a system in these errors: As I have mentioned already the optical angle between sun and goal is transposed into the gravity field; the wagging line of the dancer keeping the same angle with the perpendicular. When one measures, for a whole day long, this dance angle on the vertical comb, than—as the angle between the advancing sun and the goal changes the dance angle deviates in a characteristic course up to 25° from the expected normal (fig. 3a). The recruits in spite of this

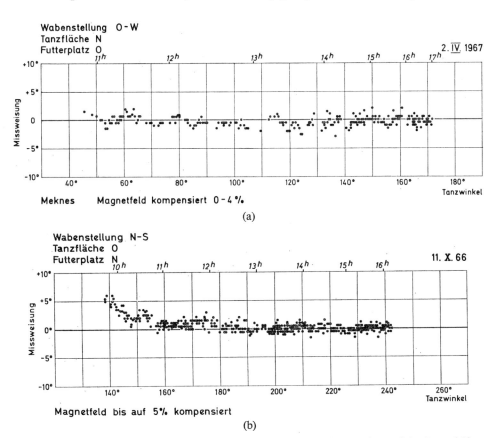

(a)

(b)

Fig. 4. The experimental situations in fig. 4a and 4b are identical to those of fig. 3a and 3b respectively. In both cases the dance floor was put in the center of a Helmholtz-drum and the earth's magnetic field was compensated up to 4 percent. The bees dance almost without deviation —they transpose the angle between goal and sun exactly into the gravity field throughout the day.

indicated deviation search for the right direction without misorientation. The communication system therefore in the "bee language" is correct, it demands however the right interpretation. To our astonishment we found the last year, that this diurnal „deviation" curve ran a quite different course when the vertical "dance floor" was put at different position in the earth magnetic field (fig. 3a/b). In the same way this deviation curve alters when the bees from Frankfurt are taken to Marocco where the earth's magnetic field has a different intensity and inclination. (Frankfurt: total intensity 0,43 Oersted; inclination 64°. Marocco: total intensity 0,41 Oersted; inclination 49°.) The deviations disappear almost completely when the hive, with its dancing bees, is put under a Helmholtz drum and the magnetic field is compensated (fig. 4a/b). The results will be discussed in detail elsewhere [15].

From this it seems that the bees are receptive to the fields of the earth magnetic force. We do not yet know the receptive mechanism for this nor its exact biological meaning. We hope the bees will give us some chance to enter in this unknown field step by step.

## References

[1]  K. v. Frisch, *Tanzsprache und Orientierung der Bienen*, (Springer, 1965).

[2]  K. v. Frisch, *Gelöste und ungelöste Rätsel der Bienensprache*, Naturwissenschaften **35** (1948) 12–23 u. 38–43; *Die Polarisation des Himmelslichtes als orientierender Faktor bei den Tänzen der Bienen*, Experientia **5** (1949) 142–148.

[3]  K. v. Frisch, *Die Sonne als Kompass im Leben der Bienen*, Experientia **6** (1950) 210–221.

[4]  T. H. Goldsmith, *Fine structure of the retinulae in the compound eye of the honeybee*, J. Cell. Biol. **14** (1962) 489–494.

[5]  T. H. Goldsmith, *The visual system of Insects*, in: *The Physiology of Insetcs*, Ed. M. Rockstein, 397–462, Vol. 1, New York-London (1964).

[6]  M. Lindauer, *Sonnenorientierung der Bienen unter der Aequatorsonne und zur Nachtzeit*, Naturwissenschaften **44** (1957) 1–6.

[7]  M. Lindauer, *Angeborene und erlernte Komponenten in der Sonnenorientierung der Bienen*, Z. Vergl. Physiol. **42** (1959) 43–62.

[8]  I. Beling, *Über das Zeitgedächtnis der Bienen*, Z. Vergl. Physiol. **9** (1929) 259–338.

[9]  O. Wahl, *Neue Untersuchungen über das Zeitgedächtnis der Bienen*, Z. Vergl. Physiol. **16** (1932) 529–589.

[10] M. Renner, *Neue Versuche über den Zeitsinn der Honigbiene*, Z. Vergl. Physiol. **40** (1957) 85–118.

[11] M. Renner, *Über ein weiteres Versetzungsexperiment zur Analyse des Zeitsinnes und der Sonnenorientierung der Honigbiene*, Z. Vergl. Physiol. **42** (1959) 449–483.

[12] M. Lindauer, *Time-compensated sun orientation in bees*, Cold Spring Harbor Symposia on quantitative Biology **25** (1960) 371–377.

[13] H. Beier, I. Medngorac and M. Lindauer, *Synchronisation et dissociation de l'horloge interne des abeilles par des facteurs externes*, Ann. Epiphyties **19** (I) (1968) 133–144.

[14] H. Martin and M. Lindauer, *Sinnesphysiologische Leistungen beim Wabenbau der Honigbiene*, Z. Vergl. Physiol. **53** (1966) 372–404.

[15] M. Lindauer end H. Martin, *Die Schwereorientierung der Bienen unter dem Einfluss des Erdmagnetfeldes*, Z. Vergl. Physiol. **60** (1968) 219–243.

*Theoretical Physics and Biology*, © 1969, *North-Holland Publ. Co., Amsterdam*

# DISCUSSIONS

H. C. LONGUET–HIGGINS: Perhaps one could be told how big the measurements of the vertical squares are. The deviations shown are deviations between the angle of the dance and the correct angle for locating the flowers?

M. LINDAUER: It is just at the 0 line.

H. C. LONGUET–HIGGINS: Your interpretation, then, is that the bees make a composite function of the earth's magnetic field and the polarization of the light from the sky?

Dr. MENDELSSOHN: Just one short question. It looks immediately of course, since the acting mechanism is a polarization of the light and is affected by magnetic field, that there should exist somewhere in the mechanism of the bee a rotation of polarization by a magnetic field. Has it been tried to see what happens if you do not just compensate the magnetic field but put a rather strong magnetic field in another direction, I mean, can you force the bee that way.

M. LINDAUER: It has been done. But I mentioned that when they go from Francfort to Meknes, the total intensity is different. Then the curve is different.

F. HALBERG: La figure 1 est intéressante d'un point de vue historique. Elle a été publiée à Leipzig, en 1840, par Gauss et Weber dans "Resultate aus den Beobachtungen des Magnetischen Vereins im Jahre 1839". Cette figure représente des courbes de déclinaison magnétique. Gauss et Weber ont utilisé des informations recueillies en différents points du globe, et, en particulier, à Milan, à Berlin et à Uppsala.

   Malgré le déplacement géographique Milan–Berlin–Uppsala, les courbes sont presque identiques entre elles. Vous voudrez bien remarquer que ces graphiques sont publiés sans analyses complémentaires, par exemple, sans faire appel à la méthode des moindres carrés. Ce qui est intéressant est que cette publication a été sanctionnée par Karl F. Gauss, le "roi des mathématiques" et père de la méthode des moindres carrés, bien que les résultats n'aient pas été soumis à une critique statistique. En effet, la figure ne comporte pas d'étude de la variation statistique des valeurs représentées. Une telle étude était réellement superflue même pour ne discuter que de "l'harmonie des courbes", c'est à dire, sans vouloir essayer de quantifier les composantes du phénomène.

   Les aspects statistiques du géomagnétisme ont été discutés d'une façon élégante

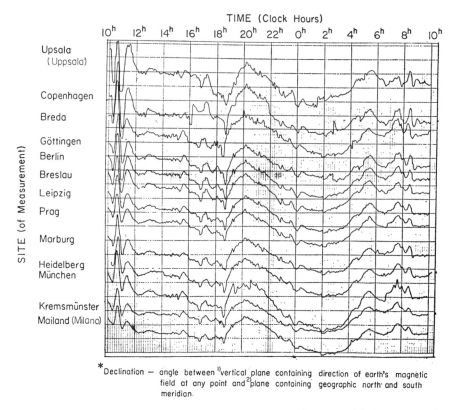

<sup></sup>*Declination — angle between [1]vertical plane containing direction of earth's magnetic field at any point and [2]plane containing geographic north and south meridian.

Fig. 1. The "harmony of curves" in itself can definitely be documented in instances such as these without any inferential statistics. [cf also Gauss, Carl Friedrich u. Wilhelm Weber. (Resultate aus den Beobachtungen des magnetischen Vereins im Jahre 1839. Herausgegeben von Carl Friedrich Gauss und Wilhelm Weber. Mit 4 Steindrucktafeln. Leipzig, im Verlage der Weidmannschen Buchhandlung, 1840)].

par MM. S. Chapman et J. Bartels dans "Geomagnetism, Volume II, Analysis of the Data and Physical Theories, Oxford Univ. Press, 1940".

M. Lindauer — comme MM. F. Brown et D. Beischer aux U.S.A. — a eu le grand mérite d'avoir pris en considération des équivalents biologiques des phénomènes représentées à la figure 1 et aussi les effets biologiques d'autres variations géomagnétiques. Ces effets biologiques ne sont peut être pas tout à fait en relation de causalité classique.

Dans ces conditions, je pose au M. Lindauer la question suivante: quelle est l'erreur qui peut dépendre de l'échantillonnage? Ce qui m'amène à poser deux questions complémentaires: 1) y a t-il une ressemblance entre les effets du géomagnétisme mesuré par vous en des emplacements différents et les phénomènes rapportés par Gauss et Weber? 2) Avez-vous une notion des récepteurs physiologiques sensibles aux phénomènes magnétiques dont vous faites état?

M. LINDAUER: Les différences quotidiennes sont peut-être les mêmes. Mais la déclinaison entre, par exemple, Francfort et un autre point, serait de 12 degrés. Il y a des différences de 60 degrés et d'autres de 49 degrés par rapport à Francfort.

F. HALBERG: Quel est l'ordre des déviations?

M. LINDAUER: Les déviations sont de l'ordre de 100 gammas; les différences de l'inclinaison sont supérieures à 1000 gammas.

W. ELSASSER: I would like to comment only on a quantitative relationship. I think there is a misunderstanding here. These daily variations of the magnetic field are very small. If that was not so no sailor could use a compass for navigation.

The deviations that you showed were of the order of 10–15°. So I don't think there is any kind of contradiction. Evidently, what the bees see is not the magnetic daily variation, it's not anything produced by the external atmospheric field which is so small that it can be measured only with very delicate instruments.

W. REICHARDT: In relation to Dr. Lindauer's paper I would like to add a remark concerning dichroitic absorption in the rhabdomeric structures of the insect compound eye. We have recently studied dichroitic absorption in rhabdomers of the housefly (Diptera) using polarized light. If polarized light is thrown onto the distal end of a single rhabdomer and the vector of polarization rotated, a degree of 50 % modulation is found in the light beam leaving the rhabdomeric structure. This finding is consistent with the hypothesis that the pigment molecules are located and randomly distributed in the membranes of the tubuli of which the rhabdomers are made up. Therefore dichroitic absorption in the insect eye seems to be entirely due to the geometry of tubuli oriented perpendicular with their long axis to the Poynting vector of the stimulating light.

E. D. BERGMANN: If Dr. Lindauer permits me, I would like to add to his beautiful statement some data which show that insects seem to be as well versed in chemistry as they are in physics. These data refer to the well-known phenomenon of sex attraction in insects.

It is known that male insects are attracted to the females over distances of many kilometers, a fact which has puzzled the scientists for a long time. One has recently isolated several such sex attractants and has determined their structure. I would like to take the data of Butenandt for the virgin silk moth. The quantity required to attract the males is only $10^{-10}$ micrograms. If one wants to translate this figure into something more meaningful, one can calculate from the vapor pressure of the compound that 1 cm$^3$ of air over a thin layer of this compound contains 192 individual molecules. The total quantity of sex attractant which is produced by one female silk moth is $10^{14}$ individual molecules. Thus one can see that, given a normal current of air, one can supply a considerable volume of gas with the concentration of the sex attractant which I mentioned.

This is, indeed, a very small quantity; let us compare it with other data which we possess, pertaining to the sensitivity to smell, the sensitivity of the olfactory receptors. There are two data which I remember. One is that Alsatian dogs, which are very sensitive to butyric acid, require 190 individual molecules/cm$^3$ of air in order to feel uncomfortable. The other figure is that 42 individual molecules only of secondary butyl mercaptan/cm$^3$ of air are sufficient to make it impossible for a human being to remain in this atmosphere.

So we are coming down now to figures which are in the neighbourhood of one individual molecule per chemoreceptor. This, I think, is a very remarkable feat of nature. It also shows that the chemoreceptor and the active chemical compound, e.g., the sex attractant, must be structurally very well fitted to each other in order to give this extraordinary sensitivity.

P. Dejours: Mon intervention a trait à la présentation du M. Lloyd. M. Fessard nous a dit: "Vous, les physiologistes de la respiration, vous avez bien de la chance, car, au moins, vous pouvez parler en termes quantitatifs". M. Lloyd, aujourd'hui, nous a donné un exemple de la façon dont les physiologistes respiratoires utilisent ces termes quantitatifs: vous avez vu un certain nombre d'équations. En réalité, nous rencontrons d'énormes difficultés pour nous exprimer quantitativement. Lorsque nous utilisons le symbole $\dot{V}$, on veut dire volume ventilé par unité de temps ou débit ventilatoire. Mais la ventilation est un phénomène rythmique, périodique, produit d'un volume et d'une fréquence. Il est bien évident que l'utilisation du seul terme $\dot{V}$ dans nos équations est tout à fait insuffisante, car ce terme n'est pas représentatif du phénomène ventilatoire et ne peut être considéré que comme un index imparfait. Il arrive souvent que dans deux circonstances on mesure un même débit ventilatoire, mais que dans un cas le volume est faible et la fréquence élevée, et que dans l'autre cas ce soit le contraire. Il est bien évident qu'il n'est pas permis de dire que les régimes ventilatoires sont identiques dans ces deux circonstances. Il est évidemment impossible d'utiliser une seule quantité pour rendre compte d'un phénomène périodique; mais tous les jours nous nous heurtons à ce problème car, en physiologie respiratoire, l'expression mathématique complète de la fonction ventilatoire serait trop complexe pour représenter une expression, facilement maniable, du phénomène. J'aimerais connaître votre opinion, M. Lloyd, sur ce point.

Vous avez beaucoup parlé des changements de respiration en différentes circonstances, par exemple au cours de la parole, du cri, du chant, etc. Cela est très important, car la plupart des physiologistes de la respiration, à la différence des neurophysiologistes, les physiologistes de la respiration ont la très mauvaise habitude de toujours rechercher un régime stationnaire qui leur parait plus facile à analyser. C'est une déplorable habitude. Car, dans la nature nous sommes soumis à des sollicitations multiples qui modifient notre ventilation. On peut dire, si l'on veut, que le système opérationnel qu'est notre organisme est soumis à différentes stimulations qui peuvent être des fonctions échelon, rectangulaires, sinusoïdales ou de n'importe quelle forme, et que ce système réagit. Sans cesse, de nouvelles sollicitations s'opposent à un régime stationnaire de la ventilation. A cet égard, le régime stationnaire de repos est artificiel: il n'existe pas dans la nature, on devrait parler de repos forcé ou imposé.

B. B. LLOYD: Dr. Dejours' two questions are of course connected. He asked why we don't pay more attention to the fact that the ventilation is a function in time of the amplitude. Luckily ventilation is almost exactly sinusoidal, so one can fit a simple sine curve, which represents ventilation really rather well:

$$V = a + \tfrac{1}{2} V_T \sin 2\pi ft.$$

There are three things you can alter in this pattern. First you can alter the setting of the point $a$, the average volume of the chest: this is an important notion, in terms of buoyancy or of oscillations in arterial $P_{CO_2}$. Secondly you can alter amplitude, or tidal volume $V_T$, and thirdly the frequency $f$. We know that these three things alter. The linear plot of ventilation $fV(=\dot{V})$ against $V_T$ shows that up to a certain limit if you know $\dot{V}$ or $V_T$, you know the other. You could rewrite all the equations that I gave you in terms of $V_T$ as a function of $H^+$ and $P_{O_2}$ and you could also write frequency as a similar function. So I am not absolutely convinced that when we are doing steady-state experiments we need to bother much about the frequency and the amplitude, at any rate in man. In some animals it's quite different because ventilation regulates temperature. As soon as you start transient analysis, you have to be careful about the phase of the ventilation when you impose the transient. This is a situation where we undoubtedly need the help of people who know how to analyse these rhythmic phenomena.

It is, however, interesting that the chemical variables and the amplitude-frequency parameters are related. Dr. Bhattacharyya working in our laboratory has been looking at the movement of the chest after breathholding. The subject is suddenly told to hold his breath at a given point and then after a certain time he has to breathe again. Obviously the direction in which he starts again will depend on what part of the cycle he started holding his breath in. If he is holding his breath on low oxygen he tends to breathe in even when his chest is on the full side. On the other hand, if you make him hold his breath on high $CO_2$ then in general there is a tendency for him to expire.

L. ROSENFELD: Simply as a matter of curiosity, I should like to point out that there is a case in which men did nearly as well as bees in the way of orientation. The Vikings are known as hardy navigators and they were as ignorant of physics as the bees. They were able to find their bearings in circumstances in which there was no sun to guide them. In a saga there is an allusion to a "sun stone" by which they could achieve this. Now, it is well known that in Iceland there are several strongly anisotropic minerals, in particular the famous Iceland-spar which led Huygens to the discovery of the polarization of light. So it is an easy induction that they used such anisotropic stones to analyse the polarization of the light from the sky, and that they had found empirically the correlation between the appearances they observed and the position of the sun.

B. B. LLOYD: We have mechanisms such as those in the foetus which enable us to grow up and become adults. Then if we go to altitude we may find that these mechanisms let us down. This I suppose is the origin of some disease which cannot be attributed to the invasion of an outside organism such as a virus or a bacterium. Eventually our homeostatic negative feedbacks start letting us down and in various regions of physiology go over into positive feedback and ultimately disaster. There are other parallels with this hypertension effect at high altitude; for example there is an increase of red cells in the blood at high altitude. This can be advantageous, but beyond a certain point it becomes disadvantageous and some subjects have either to be bled or come down to a lower altitude if they wish to survive. This takes us away from the physiology of regulation into its pathology, which is enough for at least three other conferences.

*Journée du 30 juin 1967*

# MECANISMES PHYSIOLOGIQUES

*2ème séance*

PRÉSIDENT G. CARERI

## F. HALBERG
Chronobiologie, Rythmes et Physiologie Statistique

Discussions

*Theoretical Physics and Biology*, © 1969, *North-Holland Publ. Co., Amsterdam*

# CHRONOBIOLOGIE;

## RYTHMES ET PHYSIOLOGIE STATISTIQUE *)

FRANZ HALBERG

*Chronobiology Laboratories, Department of Pathology,*
*University of Minnesota Medical School, Minneapolis, Minnesota, U.S.A.*

## 1. Introduction

Au cours de cette réunion, nous avons discuté surtout jusqu'à maintenant de la Biologie moléculaire que l'on essaie de transformer en Biologie *"sub-moléculaire"*. Nous avons examiné non seulement ses molécules, mais aussi l'organisme global, plus ou moins considéré comme une structure uniquement ou exclusivement spatiale.

Une autre dimension des organismes doit être étudiée: celle du temps, de manière à connaître leur structure temporelle. Certains aspects de haute fréquence, d'une structure temporelle ont été examinés par des chercheurs, tels que M. A. Fessard, il y a une vingtaine d'années [1, 2]. Fessard traitait, dans une monographie publiée en 1936, de l'activité rythmique spontanée du système nerveux et des systèmes excitables. Ce sera mon privilège de vous soumettre ici des documents sur le degré d'activité rythmique spontanée de processus biologiques dont les périodes ont une durée d'un jour, d'une semaine et même d'un an (tableau 1).

Les recherches relatives à ces périodes introduisent certaines "propriétés" du *temps* ou de la *durée* dans la représentation des phénomènes biologiques: (fig. 1). Celle-ci est habituellement donnée en terme de *structure spatiale* (siège anatomique, histologique et même moléculaire d'un processus) et en terme d'action et de réaction biophysique et (ou) biochimique. Ce faisant, on cherche à préciser la nature (le "quoi") d'un ensemble de processus et *où* il se situe dans l'espace de l'organisme. L'analyse des variations rythmées de cet ensemble de processus doit nous permettre de le situer dans le temps. On cherche alors *quand*, dans l'échelle du temps, peut être située telle ou telle variation intéressante du processus considéré. Il est bien évident que les réponses aux trois questions: *où?*, *quoi?* et *quand?* sont liées et que la réponse à la question *quand?* doit apporter aux deux autres un complément indispensable [2].

*) Les travaux ici discutés ont été subventionés par l'U.S. Public Health Service (5-K6-GM-13981) et la N.A.S.A. (NAS 2-2738 and NGR-24-005-006). C'est un plaisir de remercier Alain Reinberg, Maître de Recherches, C.N.R.S., pour une coopération inestimable dans plusieurs lignes de recherches presentés dans cet essai. De plus, lui et Hélène Astier, Docteur des Sciences à l'Université de Montpellier, ont eu l'amabilité d'examiner le style d'une partie de ce texte.

TABLEAU 1

Domaines du Spectre des Rythmes Biologiques

| DOMAINE* | Haute fréquence $\tau < 0.5h$ | Moyenne fréquence $0.5h \leqslant \tau \leqslant 6j$ | Basse fréquence $\tau > 6j$ |
|---|---|---|---|
| REGION | $\tau \sim 0.1$ s<br>$\tau \sim 1$ s<br>et cetera | ULTRADIENNE $(0.5 \leqslant \tau < 20)$<br>CIRCADIENNE $(20 \leqslant \tau \leqslant 28h)$<br>INFRADIENNE $(28 < \tau \leqslant 6j)$ | CIRCASEPTIDIENNE $(\tau \sim 7j)$<br>CIRCAVIGINTIDIENNE $(\tau \sim 20j)$<br>CIRCATRIGINTIDIENNE $(\tau \sim 30j)$<br>CIRCANNUELLE $(\tau \sim 1$ année$)$ |
| RYTHME | Electroencéphalographique<br>Cardiaque<br>Respiratoire | Sommeil rapide (REM)<br>Veille – sommeil<br>Repos-activité<br>Réaction aux médicaments<br>Constituants sanguins<br>Variables urinaires<br>Processus métaboliques en général | Menstruation<br><br>Excrétion des 17-céto-stéroïdes avec composantes spectrales dans toutes les régions montrées ci-dessus et dans d'autres domaines |

\* Domaines et régions [(nommées selon des critères de fréquence (f)] sont partagées selon la fréquence réciproque c'est à dire la période, $\tau$, de la fonction utilisée pour l'approximation d'un rythme. s = seconde, h = heure, j = journée.

Pour les variables examinées jusqu'à maintenant on trouve des composantes statistiquement significatives dans plusieurs domaines spectrales.

C'est à l'étude des réponses à la question "quand" que s'adresse la chronobiologie—une discipline *in statu nascendi* s'occupant *inter alia*, de l'activité rythmique.

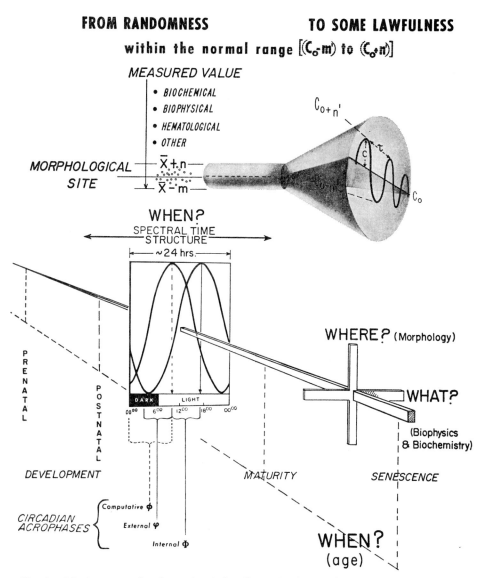

Fig. 1.   Rhythmometry for the study of circadian and other rhythms adds a new domain to organisms and yields new endpoints such as the circadian acrophases shown at the bottom of the figure.

FRANZ HALBERG

TABLEAU 2

La rythmométrie, complément de la biométrie classique

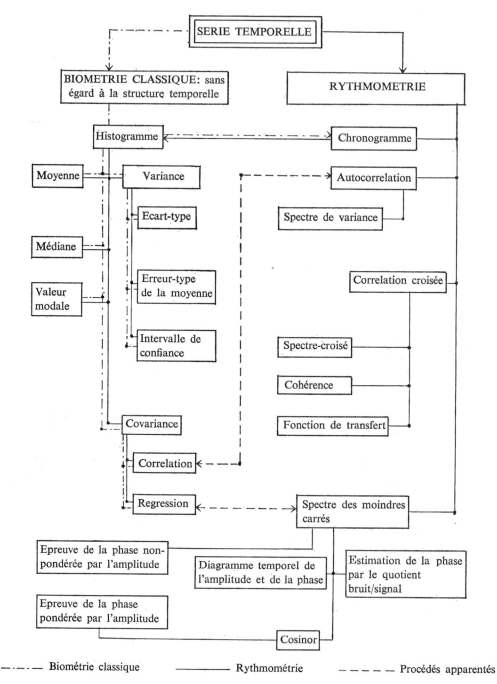

—·—·— Biométrie classique    ——— Rythmométrie    — — — — Procédés apparentés

## 2. La chronobiologie

La chronobiologie a pour but d'inventorier, de définir, de caractériser et de mesurer par des moyens appropriés (e.g., tableau 2 et fig. 2) les caractéristiques temporelles des phénomènes biologiques, tels que l'activité rythmique qui apparaît comme une des propriétés fondamentales de la matière vivante. La chronobiologie, illustrée par des données relatives aux rythmes circadiens [2–11] comprend déjà les branches ci-après:

a) *La chronophysiologie* étudie, chez le sujet sain, les caractéristiques temporelles de tout un ensemble de processus physiologiques—métaboliques, endocriniens, nerveux, etc.... Le chronophysiologiste étudie certains rythmes après une manipulation (définie le mieux possible) de l'organisme—par exemple après une ablation glandulaire (fig. 3a), et une substitution hormonale ou après des modifications de l'ambiance (fig. 4) suivant des horaires modifiés avec ou sans vols transmeridiens ou dans des situations expérimentales aussi différentes et particulières que le séjour dans une grotte (figs. 5 et 6) ou dans l'espace extra-terrestre.

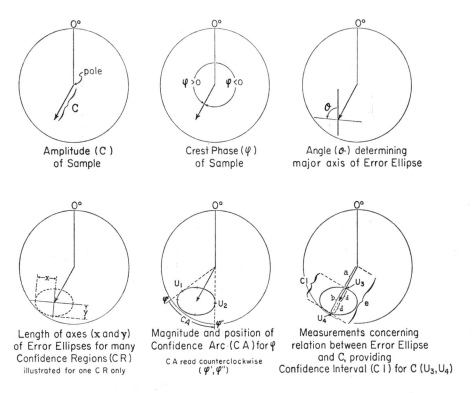

Fig. 2a. Numerical estimates provided by routines for the drawing of a cosinor.

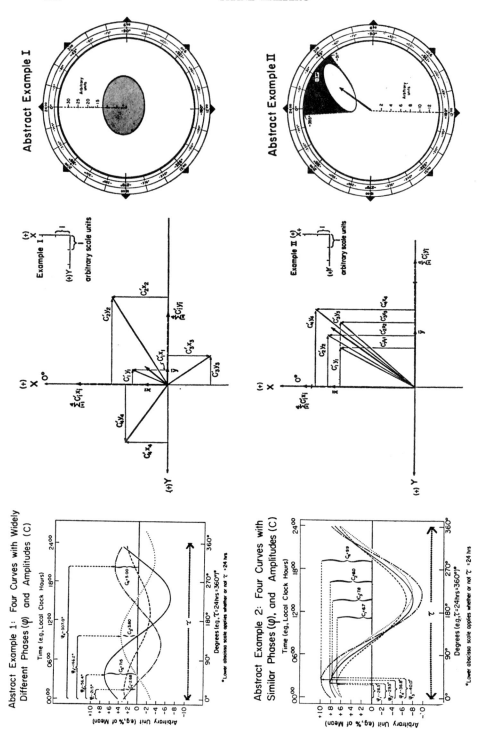

Fig. 2b.   Abstract illustrations of cosinor procedure. From the curves displayed in the time domain on the left, one arrives via a vectorial transformation in the middle to the cosinor plots shown on the right [cf. 5]. Display of cosinor results by electromechanical plotter [discussed in ref. 5].

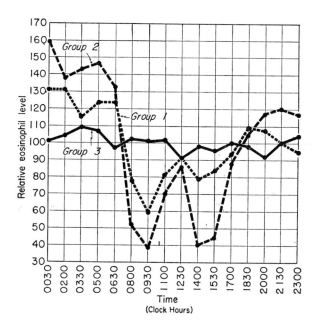

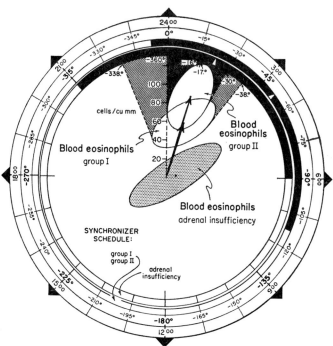

Fig. 3a.  Chronogram of 3 sets of transverse data, on the top, summarized by cosinor at the bottom. Note failure to detect a rhythm in Addison's disease (adrenocortical insufficiency).

Group   I — healthy, limited activity;
Group  II — healthy, unrestricted activity;
Group III — adrenal insufficiency, limited activity.

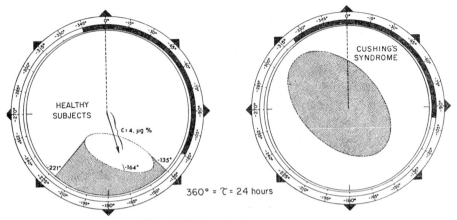

No pole overlap if confidence coefficient ≤ .998
.95 confidence interval of C: 2.7₂ to 7.0₈ µg %

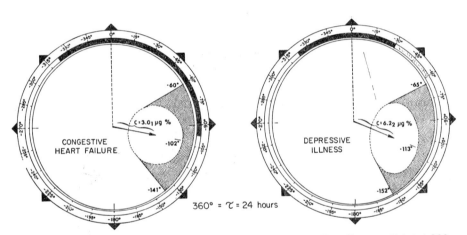

No pole overlap if confidence coefficient ≤ .995
.95 confidence interval of C: 1.1₁ to 4.9₁₁ µg %

No pole overlap if confidence coefficient ≤ .992
.95 confidence interval of C: 2.5₉ to 9.85 µg %

Fig. 3b. Cosinor evaluation of circadian rhythm of plasma 11–hydroxycorticosteroids in congestive heart failure and depressive illness—rhythm not detected in Cushing's syndrome. Data of M. S. Knapp, P. M. Keane and J. G. Wright (British Medical Journal **2**: 27–30, 1967). Failure to detect a rhythm in plasma 11-hydroxycorticosteroids in patients with Cushing's syndrome.

b) *La chronopathologie* étudie les altérations temporelles des biorythmes (fig. 3) de sujets malades: 1) altérations qui résultent de la maladie et, ce qui semble plus important encore, 2) altérations qui peuvent jouer un rôle déterminant dans l'état pathologique [2].

c) *La chronopharmacologie* étudie les effets de substances actives (médicamenteuses ou autres): 1) sur les caractéristiques temporelles biologiques (figs. 7–9);

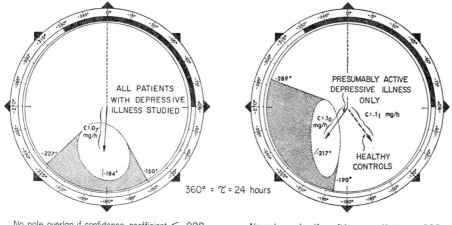

No pole overlap if confidence coefficient ≤ .998
.95 confidence interval of C: .0₃ to .l₂ mg/h

No pole overlap if confidence coefficient ≤ .998
.95 confidence interval of C: .0₄ to .l₆ mg/h

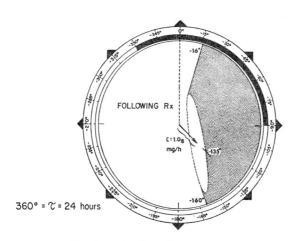

No pole overlap if confidence coefficient ≤ .996
.95 confidence interval of C: .0₄ to .l₂ mg/h

Fig. 3c. Possible circadian dyschronism of 17-ketosteroid excretion in untreated patients with depressive illness. Original data of M. Sakai (Yokohama Medical Bulletin II: 352–367, 1960). Synchronizer schedule extrapolated from timing of urine collection spans [cf. also 14 and 15].

et 2) en fonction de leur distribution dans le "temps biologique" déterminée par les mesures d'acrophase (sommet de la fonction périodique) (fig. 10). Dès lors, en augmentant le rapport: "activité thérapeutique/activité toxique" d'un médicament, la chronopharmacologie peut apporter des informations complémentaires dans le domaine de la pharmacologie classique (réduction des effets secondaires et meilleure tolérance, par exemple). En outre, dans la mesure ou une altération rythmique peut jouer un rôle déterminant dans une maladie (désyn-

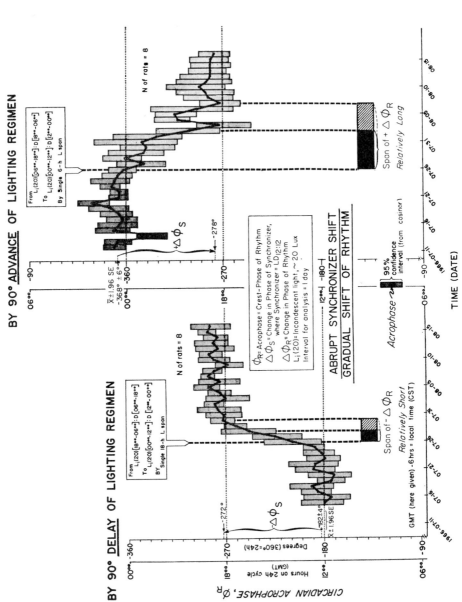

Fig. 4. "Polarity" in the circadian system of the rat—more rapidly delayed than advanced [cf 59–62]. Phase shifting of 24-hour-synchronized circadian rhythm in intraperitoneal temperature of female MSD rats [35, 36].

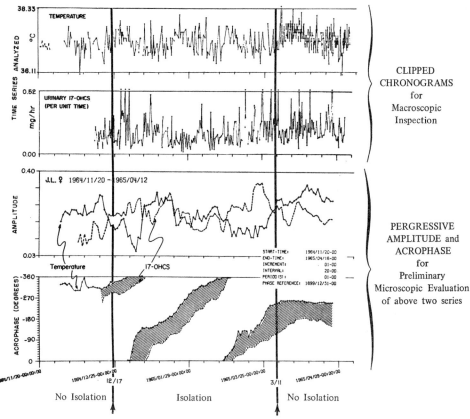

Fig. 5. Phase relations of circadian rhythms in the urinary excretion of 17-hydroxycortico-steroids and in rectal temperature of a woman during isolation in a cave and following re-synchronization with a 24-hour-cyclic societal routine. Clipped chronogram of the time series shown on top. (For clipping, the mean and standard deviation were repeatedly computed and all values above and below the mean $\pm$ 3 standard deviations were repeatedly equated to the nearer of these limits until the result of this iterative procedure was no longer associated with a change in mean and standard deviation to the nearest four decimal places. Extreme values thus "clipped" are indicated by a dot on top of the corresponding chronogram).

A macroscopic inspection of these time plots is barely contributory. Long-period changes of some regularity—corresponding presumably to the menstrual cycle [69]—are apparent for rectal temperature in particular; changes with shorter period also are suggested by the record, yet it seems unjustified on the basis of inspection alone even to attempt to ascribe a precise period to a circadian rhythm and certainly it would be cumbersome, to say the least, to discuss the phase relations to each other of circadian components in the two time series.

By contrast, the display of acrophase in the bottom row—part of the microscopic approach—indicates first that the rhythms of both functions changed their period during isolation—only to be resynchronized with a 24-hour-cyclic routine thereafter; second, that the rectal temperature acrophase lagged behind that for 17-hydroxycorticosteroid excretion during isolation, as well as following resynchronization; and third, that resynchronization of body temperature occurred considerably faster than that of 17-hydroxycorticosteroid excretion. The latter finding may be related at least in part to the circumstance that the $\phi$ of rectal temperature was nearer its usual temporal placement in relation to the synchronizer than the $\phi$ of 17-OHCS, on the day of emergence from the cave.

From the pergressive amplitude diagram—third row from top—it can be seen that the amplitude, notably that of the rhythm in 17-OHCS excretion during isolation, showed no indication of damping as a conditioned reflex phenomenon might be anticipated to do. If there was a difference between the amplitude at the end of isolation and that upon resynchronization, this measure of the extent of circadian periodic change in 17-OHCS excretion indicated a more marked rhythm at the end of isolation than following resynchronization.

chronisation circadienne, par exemple), de nouvelles applications thérapeutiques peuvent être mises en évidence pour des drogues déjà connues (par exemple l'ACTH pour une resynchronisation) ou de nouvelles drogues peuvent faire l'objet d'applications intéressantes et, nous l'espérons, importantes.

d) *La chronotoxicologie* étudie les effets non désirés et/ou dangereux d'agents chimiques, physiques ou autres, y compris ceux de poisons, de substances polluantes ou contaminantes, et l'éventualité de surdosages de médicaments utiles. Ces effets sont étudiés: 1) en relation avec les altérations caractéristiques temporelles (fig. 11) et, 2) en fonction de leur distribution dans le temps biologique (figs. 12 et 13) [26–33].

Toute recherche dans l'une de ces branches entre dans le domaine de la chronobiologie, puisque la validité du travail ainsi réalisé dépend étroitement du degré de quantification des rythmes et des changements de phase lorsqu'on étudie les interactions et les interrelations de ces rythmes entre eux.

### 3.  Analogie physique

Un spectre de rythmes physiologiques comportant des fréquences caractéristiques, différentes entre elles, peut être comparé au spectre des radiations électromagnétiques. Le dernier fait apparaître la distribution de l'intensité spectrale des oscillations composantes pour différentes fréquences, comprenant, entre autres domaines, celui de la lumière visible. Les spectres physiologiques révèlent dans quelle mesure les "intensités" ou les amplitudes d'oscillations biologiques, pour différentes fréquences, contribuent à la variabilité globale rencontrée dans les mesures biologiques, effectuées en séries temporelles. Comme les spectres électromagnétiques, les spectres physiologiques couvrent de larges domaines de fréquences.

On peut pousser plus loin cette analogie: de même que la lumière blanche peut être perçue directement à l'œil nu lorsque l'intensité des fréquences composantes se situe au-dessus du seuil de la vision, de même par le report des données expérimentales dans l'échelle des temps, certains des rythmes biologiques, tels que certains rythmes circadiens, peuvent également être clairement discernés sans l'emploi de procédés spectraux—si et seulement si, ces rythmes sont suffisamment accentués.

Certains autres rythmes biologiques peuvent ne présenter qu'une faible amplitude, et—ce qui est souvent le cas—la contribution d'un rythme aux changements observés sur un diagramme, où les résultats expérimentaux sont simplement représentés en fonction du temps, peut être estompée par des interférences appelées "bruit"; de tels rythmes sont alors brouillés, ou même situés au-dessous du "seuil de visibilité", dans ce type de diagrammes—le chronogramme. Leur détection relève alors des techniques spéciales. Entre autres procédés, les spectres de variance, les spectres des moindres carrés et la méthode dénommée "cosinor" nous permettent de discerner directement ces mêmes rythmes en dépit d'un rapport: "signal (rythme)/bruit" peu favorable, tout comme on peut discerner les diverses couleurs si l'on interpose un prisme dans un faisceau de lumière blanche. De telles comparaisons n'ont pas le dessein de signaler de quelconques interactions [65] entre les rythmes biologiques et les oscillations électromagnétiques, mais elles permettent de souligner la nécessité de méthodes spéciales si l'on désire identifier toutes ces oscillations [5, 10]; des objectifs chronobiologiques spécifiques [2] peuvent être atteints par ces méthodes.

## 4.   L'Analyse de données chronobiologiques

Pour commencer par un exemple relativement simple, un certain nombre d'étapes ont conduit au développement de ce que l'on nomme maintenant la méthode cosinor (figs. 2–4, 7–9, 13, 14, 20) [5]. Par ce procédé, on arrive éventuellement à réaliser l'étalement, sous forme de représentation en coordonnées polaires, de l'amplitude (C) d'un rythme, de sa phase (∅) et de l'ellipse d'erreur correspondante—puis à procéder ultérieurement au développement d'un "spectre cosinor". Du cosinor on peut tirer, par exemple, des renseignements relatifs à l'amplitude C et la phase ∅ de certains rythmes de sujets en bonne santé (fig. 14) [13]. Par la même méthode, on peut voir apparaître également des altérations de C et/ou de ∅ dépendant d'un état pathologique tel que par exemple un cancer [11] une psychose dépressive [14, 15], la maladie d'Addison (fig. 3a) et le syndrome de Cushing (fig. 3b) [16–18]. En effet, même un résultat négatif, par le cosinor, peut faire l'objet d'une constatation positive pour un rythme donné; autrement dit, il peut révéler un état pathologique (fig. 3). C'est ce fait qui a conduit au choix de l'analogie entre la "microscopie" de rythmes au moyen d'analyses faites par un ordinateur électronique et l'examen des tissus à l'aide d'un microscope qui peut révéler les altérations d'un tissu—modifications qui sont souvent plus fines et qui peuvent éventuellement se révéler plus importantes que ne l'est un effet terminal global, tel qu'un manque de cellules dans un tissu nécrotique.

En outre, de même que l'anatomopathologiste emploie plus d'une coloration et plus d'un outil (il peut même recourir, à l'occasion, à la microscopie électronique), le chronophysiologiste et chronopathologiste doivent employer plus d'une série de méthodes d'analyses par un ordinateur et plus d'un diagramme (tableau 2). A part la méthode du Cosinor, il doit appliquer des méthodes qui servent à détecter un changement de fréquence—tel que la désynchronization d'un rythme circadien provoqué par la cécité—lorsque la période de ce rythme devient égale p. ex. à 23.5 heures au lieu de 24 heures (fig. 15) [21]. Il doit pouvoir disposer des instruments nécessaires à mesurer une transposition de variance, telle que celle qui intervient dans le développement normal de l'enfant (fig. 16) ou dans une réponse caractéristique de sujets adultes à des drogues telles que la réserpine [22].

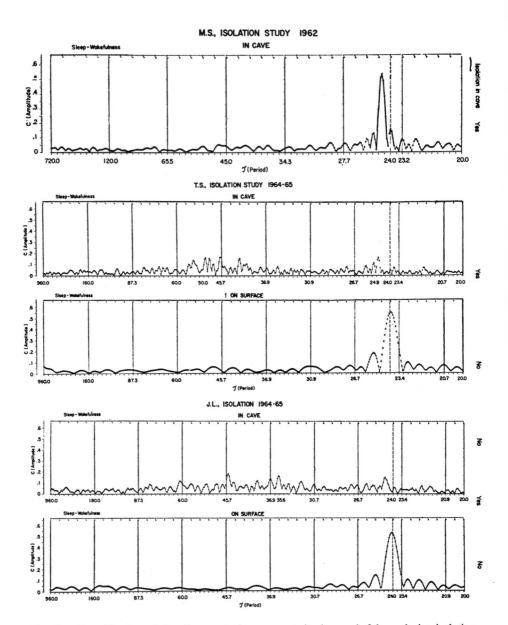

Fig. 6a.   Quantification of circadian spectral components in sleep-wakefulness during isolation and upon return to a 24-hour-synchronized societal routine. Note decrease in amplitude of circadian component and appearance of ill-defined infradian components, described by Siffre *et al.* [24].

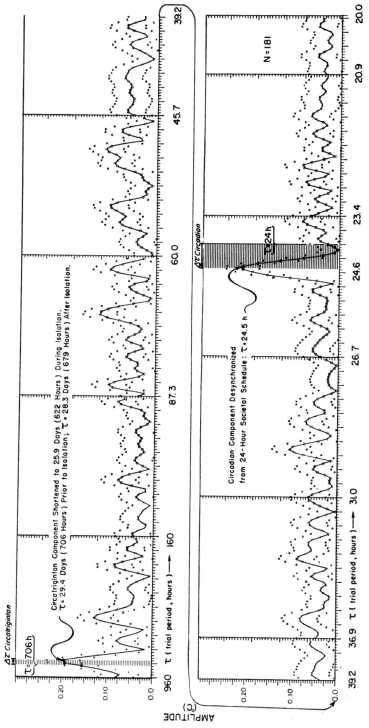

Fig. 6b. Shortening of circatrigintan component (corresponding to menstrual cycle) in rectal temperature series of a healthy woman—concomitantly with a lengthening of circadian component, during isolation in a cave [69]. (See ref. 74–78 for work on gross time relations of the circadian body core temperature rhythm to other variables.)

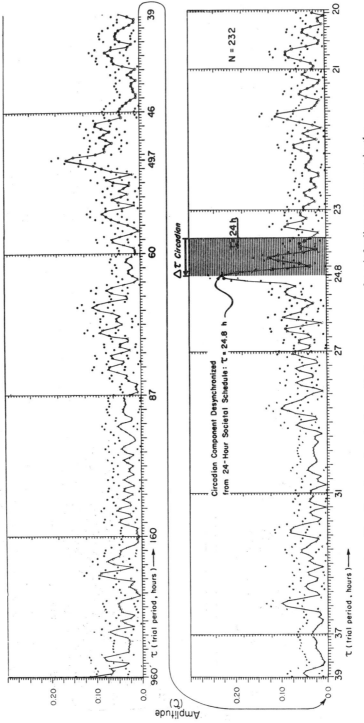

Fig. 6c. Circadian desynchronization with the appearance of an infradian component in the temperature series of a healthy man during isolation from known time cues.

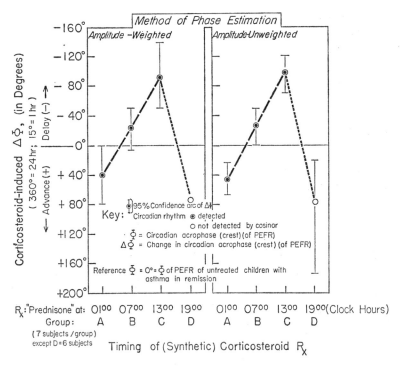

Fig. 7. Drug-induced phase shift of a circadian respiratory rhythm [cf. 25, 26]. Phase shift (*ΔΦ*) of peak expiratory flow rhythm (PEFR) as a function of timing of prolonged corticosteroid therapy in children with severe asthma.

En développant des telles méthodes "microscopiques", on dispose déjà des premiers moyens nécessaires à l'étude des informations relatives aux altérations de rythmes en tant qu'aspect de la chronopathologie. Un tel ensemble de techniques nous permet de donner des résultats plus complets que: "présence ou absence de rythme(s)", tout comme l'histologiste ne se contente pas exclusivement, ou même primordialement, de répondre à la question "présence ou absence de cellules". La signification possible, pour la chronopathologie, d'une constatation telle qu'une transposition de variance de la bande des fréquences circadiennes vers la bande des fréquences ultradiennes, dans le spectre d'une variable physiologique relative à une tumeur pourrait vraisemblablement être similaire à celle de la constatation d'une dédifférenciation dans le tissu cancéreux [11] dans le contexte de l'anatomopathologie microscopique classique.

Pour certaines variations rythmiques de l'Homme, telles que celles du rythme veille-sommeil, la composante ultradienne du spectre pourrait représenter un rythme plus "primitif" par rapport à sa composante circadienne, puisque la première apparaît plus tôt dans le développement néonatal et puisqu'on peut mettre en évidence, au cours du développement, un glissement de variance du domaine ultradien vers le domaine circadien du rythme veille-sommeil dans les premiers mois de la vie. Cependant, il reste à savoir si de telles considérations sont susceptibles d'être généralisées à des fonctions cellulaires tels que les rythmes des mitoses [4] ou, comme l'a rapporté Pierre Passouant de Montpellier, à propos du sommeil "rapide" (REM) des narcoleptiques [12].

## 5.   Le degré de généralité de la structure biologique temporelle

En désignant la totalité des changements réguliers et prévisibles dans le comportement des organismes, ou de leurs subdivisions anatomiques, physiologiques etc. . . . et, en complétant ainsi la structure spatiale classique, la structure temporelle peut-être caractérisée à tous les niveaux d'organisation—l'organisme *in toto*, les systèmes d'organes, les organes, les tissus et les cellules. L'organisation circadienne, discutée jusqu'à maintenant, en est un exemple démonstratif (fig. 17) [2]; de tels rythmes ont même déjà été mis en évidence dans les variations de processus intracellulaires. Il semble donc justifié de considérer la possibilité d'une structure biologique temporelle au niveau des interactions moléculaires et même au niveau des phénomènes atomiques, y compris les phénomènes "microscopiques", dans

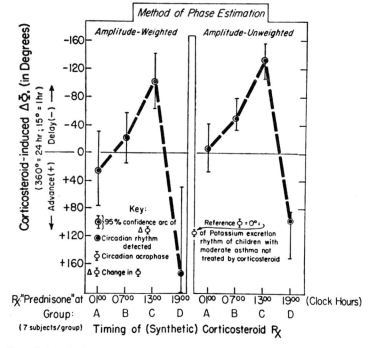

Fig. 8.   Drug-induced phase shift of the circadian rhythm in urinary potassium excretion [cf. 25, 26].

le sens que la physique théorique donne à ce terme. C'est là un des buts des recherches à venir qu'il conviendrait de discuter, lors de cette conférence entre physiciens et biologistes. Mais pour le moment, le biologiste a le devoir de préciser les conditions d'application de la méthode chronobiologique.

Pour qu'un certain aspect temporel du comportement biologique, à tous les niveaux d'organisation, soit accepté comme élément de "structure temporelle", selon la description ci-dessus, il faut que l'on puisse montrer, objectivement et avec une chance d'erreur pouvant être estimée, que ce phénomène n'est pas dû au hasard.

En ne considérant qu'un seul individu, certains aspects de la structure biologique temporelle sont unidirectionnels et non-répétitifs, par exemple le développement, la croissance ou la sénescence de cet individu. D'autres aspects temporels, tels que la naissance et la mort, correspondent à des événements uniques—si de nouveau on se borne à ne considérer que la vie d'un seul organisme.

Si l'on adopte un point de vue plus général, c'est à dire si l'on passe du comportement de l'individu à celui d'une population, les événements auparavant uniques, comme la naissance ou la mort, deviennent, pour la population, des phénomènes rythmés [3]. On peut donc, plus généralement, parler de rythmes comme éléments principaux de la structure temporelle.

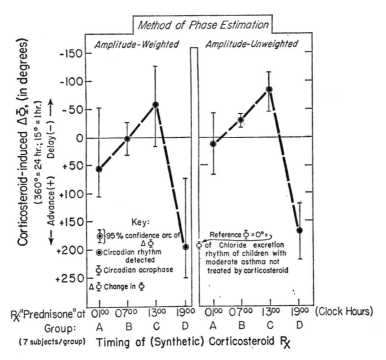

Fig. 9.   Drug-induced phase shift of the circadian rhythm in urinary chloride excretion [cf. 25, 26].

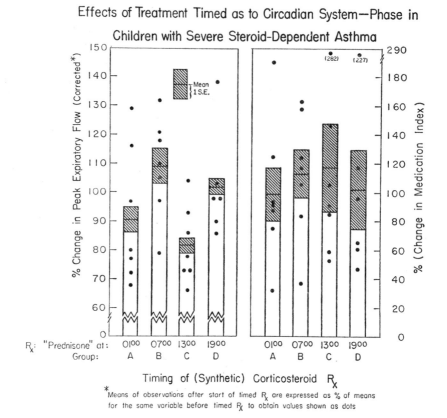

Fig. 10.   Some preliminary chronopharmacologic findings suggesting but not proving the merits of corticosteroid therapy as a function of circadian system phase [cf. 25, 26].

Des orientations spéciales dans des disciplines telles que l'embryologie, la biologie du développement (au sens le plus large), aussi bien que la gériatrie, sont consacrées à l'étude de certains aspects de la structure biologique temporelle. Une grande variété de rythmes de haute fréquence, étudiés en physiologie classique, — *inter alia*, nerveuse, circulatoire, respiratoire — constitue d'autres aspects de cette structure biologique temporelle.

### 6.   Aspects de basses et de moyennes fréquences dans l'étude des rythmes de haute fréquence

Les informations, dans le domaine des hautes fréquences biologiques, peuvent être obtenues à l'aide d'une instrumentation devenue d'usage courant en pratique médicale. Les applications cliniques quotidiennes de l'électrocardiographie et de l'électroencéphalographie sont devenues indispensables au diagnostic et au traite-

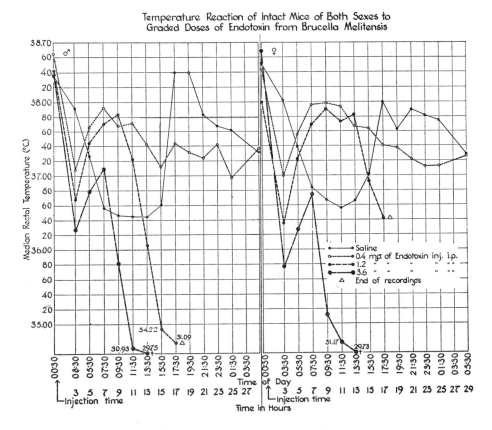

Fig. 11.   Alteration of circadian temperature rhythm of intact ABC mice of two sexes following the administration of graded doses of *Brucella* endotoxin.

ment des malades souffrant d'affections cardiaques ou neurologiques. C'est donc le clinicien qui peut répondre d'une manière relativement satisfaisante à la question "*cui bono*", qui se rapporte à cet aspect de la structure biologique temporelle. Naturellement, il cherche à améliorer l'interprétation de ces rythmes "macroscopiquement" visibles dans un ECG ou EEG et, pour ce faire, il examine actuellement l'utilité potentielle d'analyses additionnelles réalisées à l'aide de programmes spéciaux établis pour les calculatrices électroniques [4, 5]. Si, dans la plupart des cas, de telles analyses sont limitées à des régions spectrales étroites de l'EEG ou de l'ECG usuel, cette limitation peut être levée par l'exploration concomitante de diverses composantes périodiques; c'est le cas des composantes de fréquence circadienne et de fréquence plus basse, qui se trouvent par exemple dans les diverses variations périodiques du pouls, de l'ECG ou de l'EEG superposées aux régions de fréquences ("spécifiques") habituellement examinées. De cette façon, on trouve dans l'EEG, en plus de ces fréquences "classiques" situées dans les régions de 30 à 1,5 cycles par seconde, des composantes additionnelles ultradiennes, circadiennes et autres; (fig. 19) [6, 7].

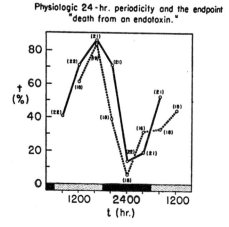

Fig. 12a.   Circadian rhythm in susceptibility to *E. coli* endotoxin and its reproducibility in separate experiments.

Ordinate { % death of group of *standardized* mature C mice
† (%)    { from *E. coli* lipopolysaccharide (Difco, 100 μg./20 gm., i.p.).

Abscissa { times of injection, in 2 experiments (injections begun at 2
t (hr.)  { different time points, during daily light period).

N of mice / group in parentheses.

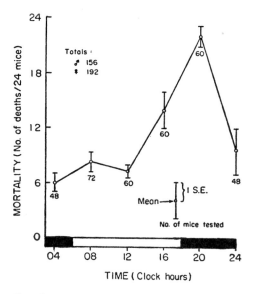

Fig. 12b.   Circadian rhythm in susceptibility to ethanol [cf. also 32].

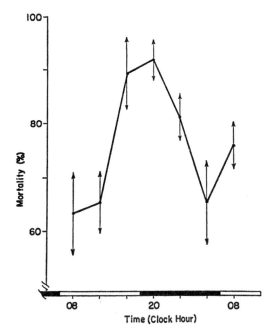

Fig. 12c. Mortality from acetylcholine also is circadian system phase dependent. Circadian susceptibility – resistance cycle to acetylcholine. B$_6$ male mice, about 1 yr. of age. (Pool of results with 3 dose levels (4.0, 4.2 and 4.4 mg/20 gm body wt.) at 4 hr. intervals for 48 hrs. N > 45/time point).

## 7. Détecteurs physiologiques pour l'étude des rythmes de fréquences moyennes et basses

En plus de ces équipements désormais courants, une variété d'instruments complémentaires modernes nous a été fournie par la nécessité d'une recherche biologique scientifique adaptée aux conditions particulières de l'espace extra-terrestre. Ce sont d'abord les "détecteurs physiologiques" que l'on appelle "transensors" aux U.S.A. Ces sondes "sensibles" sont liées à des radio-émetteurs, l'ensemble étant miniaturisé, ce qui facilite beaucoup l'étude de toutes les fonctions rythmées, quelle que soit leur fréquence. Mais pour analyser des rythmes à fréquences moyennes et basses, ces mêmes "détecteurs physiologiques" deviennent presque indispensables. Il en est ainsi pour l'étude dite longitudinale des divers aspects d'une structure biologique temporelle, c'est à dire, pour les collectes des séries de mesures continues ou à intervalles fréquents, pour un individu donné et pour des espaces de temps beaucoup plus longs que les périodes des rythmes évalués. Il convient d'ajouter qu'actuellement la collecte des données expérimentales, par des moyens classiques, demeure encore souvent le procédé de beaucoup le plus sûr et pour cela il est plus raisonnable de l'utiliser. Certaines méthodes classiques de collecte de données physiologiques en série sont aussi "meilleures" que les méthodes de collecte par télémétrie et, souvent, elles sont aussi sensibles et même plus sensibles que ces dernières.

Néanmoins, un avantage très important de ces microsondes sensibles, télémétrant des mesures physiologiques, est que leur utilisation permet souvent de réduire, sinon d'éliminer, l'interférence du procédé de mesure avec les comportements du sujet étudié, notamment les changements de son mode de vie, et les interruptions de son sommeil en particulier. Par exemple, on peut télémétrer les séries de valeurs de la température centrale d'un organisme libre dans ses mouvements, c'est à dire, non perturbé par la présence d'un harnais portant des électrodes et des fils. En outre, les chercheurs eux-mêmes et tous leurs assistants, ne sont plus strictement liés à un laboratoire même si on a besoin de mesures individuelles à des horaires déterminés (fig. 4).

Signalons que le transfert des mesures biologiques télémétrées peut se faire directement de l'instrument récepteur à la calculatrice utilisée pour leur analyse, grâce à l'intervention d'un second type d'instruments et/ou de programmes modernes.

## 8. "Surveillance en marche" (analyses "as you go")

La troisième aide méthodologique moderne repose sur l'emploi d'une série de programmes spéciaux destinée, d'une part, à l'analyse objective et rapide

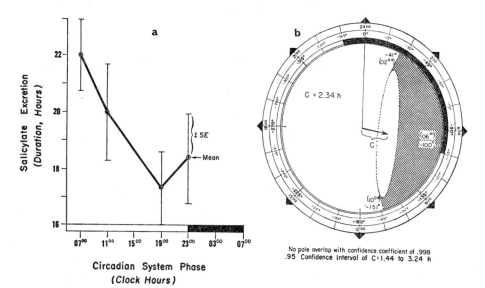

Fig. 13. Even the duration of salicylate excretion depends on the time when the drug is administered. a: Time plot of data (analyzed by cosinor); b: Cosinor summary of circadian rhythm in duration (hours) of salicylate excretion of healthy mature human beings receiving a standard dose of the drug at several different times along the 24-h scale [34].

par des calculatrices électroniques de données télémétrées, et d'autre part, à l'étalement graphique des séries de résultats fournis. Ces programmes représentent un progrès majeur dans l'étude de la structure temporelle. Leur valeur a été comparée d'ailleurs à celle du microscope dans l'étude de structures cellulaires [4]. L'étalement graphique des analyses, sous forme de spectres, de cosinors, etc. . . . peut être réalisé, à partir de mesures effectuées à des intervalles de temps prédéterminés, d'une façon telle que les résultats des analyses constituent un "moniteur automatique" des fonctions physiologiques étudiées. (figs. 4 et 18) [23, 24]. On peut aussi prévoir des analyses destinées à la recherche, aussi bien qu'à une application médicale, telles que les résultats étalés fournissent, en temps utile, des "ordres" correspondant à la nécessité d'un changement du milieu ou à l'administration de certains traitements, etc. . . .

De cette façon et au moment désiré par l'expérimentateur—à certains temps choisis a priori ou qui se sont révélés être intéressants sur la base de critères vérifiés par l'application des programmes destinés à l'analyse séquentielle ("surveillance en marche" (as you go))—on parvient à des interprétations faciles et objectives et aux décisions expérimentales ou thérapeutiques qui sont, pour cette raison, utiles. Cette combinaison d'instruments pour 1) la télémétrie, 2) le transfert automatique des données, 3) les analyses électroniques immédiates et 4) l'étalement séquentiel des résultats correspond à la double nécessité d'avoir un moniteur à décision automatique et une aide diagnostic. Grâce à ces moyens, un progrès méthodologique a pu être réalisé dans l'étude des rythmes de fréquence moyenne, exactement comme on commence déjà à utiliser ces mêmes techniques pour la surveillance des cardiaques menacés de défaillance. Par l'emploi de tels moyens, la détection et la quantification des éléments rythmés, de structure temporelle pour plusieurs fréquences ont déjà été possible, pour plusieurs fonctions physiologiques—résultat qui nous amène à la conception d'un spectre étendu de rythmes qui va, d'une part, de la fréquence de mille cycles par seconde, dans l'activité électrique des gymnotidés, jusqu'à, d'autre part, un cycle couvrant plusieurs années tel qu'on le rencontre dans des populations d'organismes (fig. 19).

## 9. Échantillonnage des séries d'informations temporelles; ergodicité

Actuellement, l'échantillonnage longitudinal n'est pas toujours possible, ou immédiatement réalisable, notamment pour les rythmes de basse fréquence—à part quelques exceptions d'une valeur inestimable (fig. 19b). L'échantillonage longitudinal porte sur les résultats obtenus chez un sujet pendant un espace de temps, T, beaucoup plus long que la période $\tau$ du rythme étudié. L'échantillonage transversal porte sur des résultats obtenus, à partir d'un groupe de sujets aussi homogène que possible, pendant un espace de temps relativement court ($T \leqslant 2\tau$). Ce schéma expérimental, très largement utilisé, peut servir à caractériser un rythme circadien ou un système de rythmes circadiens, pourvu que les sujets soient

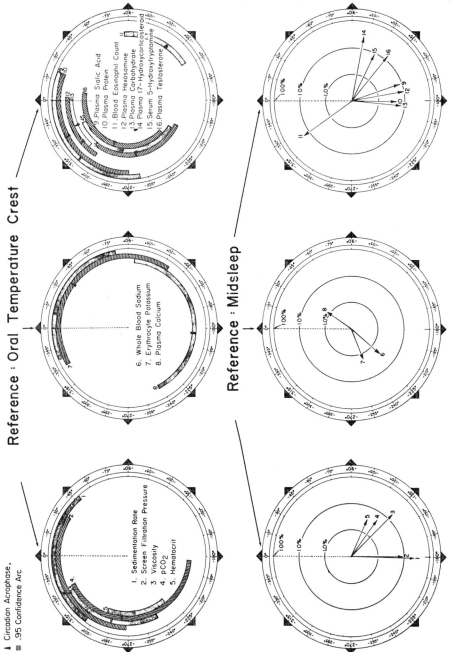

Fig. 14a.  Rhythms in man's "milieu interieur constant". Cosinor summaries of circadian rhythms in human blood.

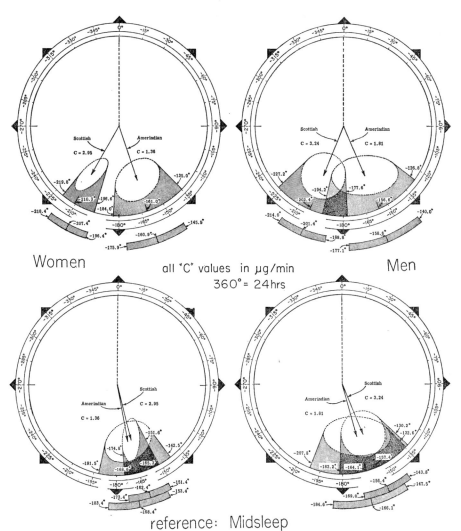

Fig. 14b. Circadian Acrophase (Average Crest Time of Rhythm), $\phi$, of 17-Hydroxy-corticosteriod Excretion by Men and Women from Scotland and South Dutch Guiana (Amerindians); Summaries of Eight Cosinors Providing in Circular Displays an Amplitude (C)-Weighted $\phi$ and of Eight $\phi$-Estimations, Done Without C-Weighting, Shown as Arcs Outside the Circles. Local Midnight, $00^{\circ\circ}$, is Used as Phase Reference for Displays in the Top Half of the Figure; Midsleep Serves as Phase Reference for Displays at the Bottom.

In a cosinor display, $\phi$ is the angle formed by a vector with the phase reference shown as $0°$. The amplitude, C, is usually shown by the length of a vector. In the figure, precise values for C

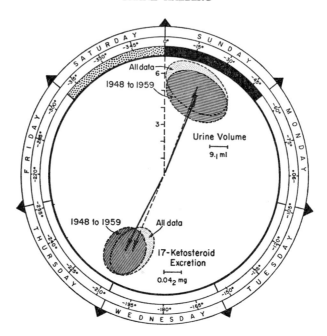

Fig. 14c.  Cosinor summary of the circaseptan rhythms in 17-ketosteroid excretion and urine volume of a healthy man. Rhythms in both variables were synchronized with societal 7-day routine during all of the time span analyzed (urine volume) or during most of that span (17-ketosteroid).

---

Fig. 14b *(continued)*.

are given next to each vector but intergroup differences in C are ignored in order to facilitate a comparison of circadian acrophases. Radii drawn tangent to an error ellipse—representing a 95% confidence region for the C-weighted $\phi$—delineate the 95% confidence arc for the acrophase. Intersects of the same ellipse with the amplitude vector and with a prolongation of this vector provide the confidence interval for amplitude. Outside the circular cosinor displays, C-unweighted $\phi$ estimates also are shown as lines in the center of the corresponding shaded 95% confidence arcs.

When local midnight is used as phase reference, the $\phi$'s of rhythm computed (with C-weighting) by cosinor or without C-weighting for Scottish and Amerindian women are different. The 95% confidence arcs (upper left) do not overlap. The C-unweighted $\phi$ estimates suggest an intergroup difference for groups of men as well as for women from the two populations—so long as local $00^{\circ\circ}$ is used as $\phi$ reference (upper half of figure).

In the lower half of the figure, midsleep is used as phase reference. All acrophases, whether or not computed with C-weighting, are very similar and their 95% confidence arcs overlap. An adjustment in phase reference thus not only eliminates a spurious intergroup difference but also provides an objective, quantitative index of a rhythm's timing in different populations. A statistically significant difference in the distribution of the acrophases from the two populations was not detected by an F-test; this result does not indicate of course that no such difference existed.

effectivement synchronisés de manière similaire. L'échantillonage *hybride* intéresse un nombre inférieur de sujets et une durée d'observation supérieure à ceux d'un échantillonage transversal.

En prenant comme exemple le rythme de la température corporelle de l'Homme, l'estimation de l'acrophase circadienne ($\Phi$) et de ses limites de confiance pour une sécurité de 95 p. 100 (exprimées en degrés, par un angle de phase compté à partir du milieu de l'espace de temps correspondant au sommeil) donne les résultats suivants: –200° (limites de –186° à –216°) dans une étude longitudinale portant sur un même sujet pendant 34 jours; –210° (limites de –188° à –252°) dans une étude hybride: 4 sujets étudiés chacun pendant 3 jours; et –199° (limites de –181° à –220°) dans une étude transversale de 11 sujets étudiés chacun pendant 24 heures.

## 10. Conclusions

Afin de mieux définir les variations physiologiques d'un sujet sain, ou pathologiques d'un malade, on est maintenant en mesure d'apprécier la structure temporelle aussi bien que la structure spatiale à plusieurs niveaux de résolution. Que l'on ait à faire au résultat microscopique d'une structure histologique ou à

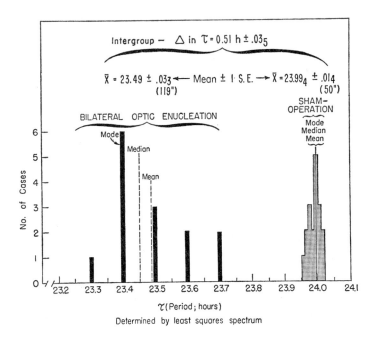

Fig. 15. Summary of circadian period estimates on two groups of mice, originally described at the 4th Conference of the International Society for the Study of Biological Rhythms held in Basel, Switzerland, September 18–19, 1953 [cf. 20].

la résolution par l'ordinateur de la structure spectrale (rythmes), les constatations
"macroscopiques" et "microscopiques" peuvent être intégrées afin d'obtenir des
informations plus complètes. Le fait important demeure que cela peut être réalisé
sans nécessairement mettre sur le même plan ces constatations elles-mêmes,
lorsqu'elles sont faites à deux niveaux différents de résolution. L'utilisation de
l'œil nu répond à une nécessité importante et le recours à la méthode microscopi-

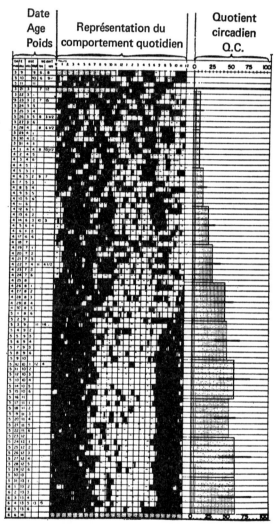

Fig. 16.  Behaviour-Day Chart of a healthy male infant, raised on "self-demand", visualizing
problems encountered in quantifying certain rhythms and near-rhythms. Circadian quotient
gauges in this case the extent of prominence of the circadian rhythm as a function of age.

■ : Sommeil;  □ : Veille;  — : Alimentation;
harcelé : Q.C.;  —— : limites de confiance de Q.C.

que par l'ordinateur à un dessein complémentaire—le résultat "microscopique" étant souvent plus satisfaisant en pathologie morphologique classique spatiale et, peut-être, dans un proche avenir, en chronopathologie.

D'un point de vue historique, l'importance "de la variabilité des conditions organiques" a été reconnue par Claude Bernard [38]:

"... Pour le moment, je veux uniquement appeler l'attention des expérimenta-
"teurs sur l'importance qu'il y a *à préciser les conditions organiques* (sic), parce
"qu'elles sont, ainsi que je l'ai déjà dit, la seule base de la physiologie et de la
"médecine expérimentale. Il me suffira, dans ce qui va suivre, de me borner
"à des indications, car c'est à propos de chaque expérience en particulier qu'il
"s'agira ensuite d'examiner ces conditions, aux trois points de vue physiologique,
"pathologique et thérapeutique" ...

Claude Bernard, qui sur ses vieux jours, décrivait la constance du milieu intérieur, a aussi écrit, en pleine maturité scientifique qu'il y a des conditions physiologiques "dans lesquelles il y a *toujours* du sucre et d'autres conditions dans lesquelles il n'y en a *jamais*" [dans le foie] [37].

Par cette remarque expérimentale, Claude Bernard a souligné l'importance d'une structure temporelle. Les facteurs déterminants de cette variabilité ont fait récemment l'objet des mises au point, des revues et des travaux originaux nombreux [39–63].

## HEURES DE MOINDRE RÉSISTANCE

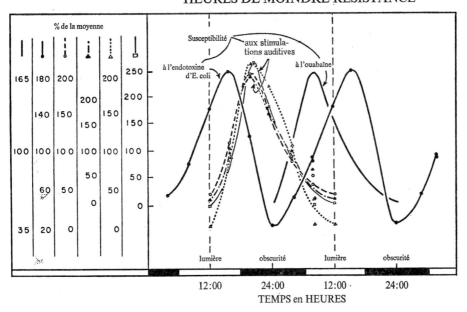

Fig. 17a.  Hours of changing resistance [2].

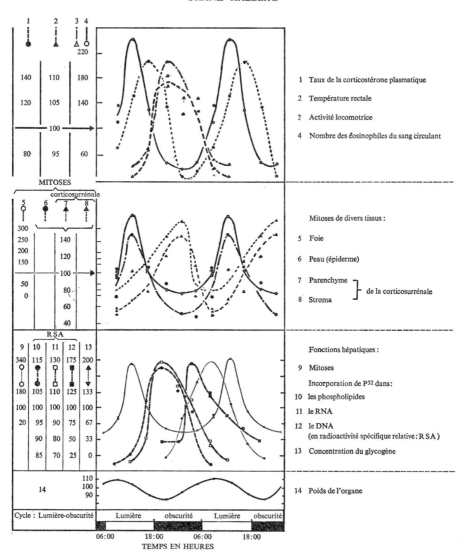

Fig. 17b.  Circadian rhythms at different levels of organization [2].

En ce qui concerne la physiologie humaine, les informations récentes, peut-être les plus intéressantes, ont été apportées par Alain Reinberg *et al.* à propos des "heures de moindre résistance humaine". Par ailleurs, l'équipe composée d'Alain Reinberg, de Jean Ghata, de Michel Siffre et de nous-mêmes a mis en évidence la persistance des rythmes endocriniens, de fréquence moyenne, chez l'Homme adulte sain, pendant l'isolement souterrain prolongé, sans synchroniseurs connus. Ces résultats sur l'Homme peuvent être comparés à ceux d'expériences de désynchronisation externe réalisée chez des Souris, (fig. 20). Chez l'Homme, comme

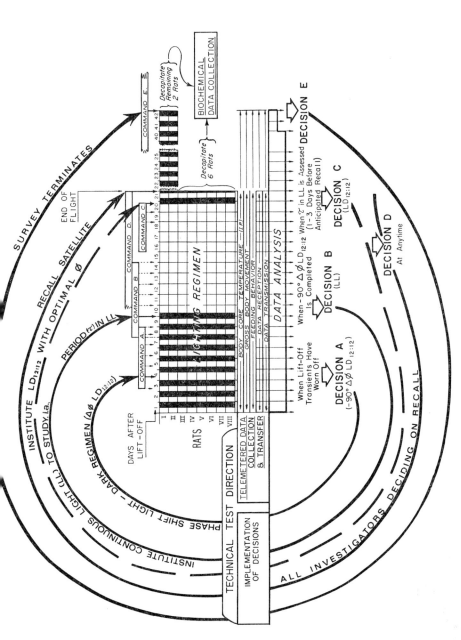

Fig. 18. Sequential decisions originally intended for as-you-go analyses during a Biosatellite survey of rhythms. The same as-you-go analyses may be useful for assessing rhythm characteristics in experimental work on Earth and conceivably in the clinic. Schedule developed in cooperation with Dr. Walter Nelson and Dr. Walter Runge of the Chronobiology Laboratories, Department of Pathology, University of Minnesota.

chez la Souris, on trouve que la distribution des temps internes de certains rythmes désynchronisés correspond à ce qu'on rencontre lors d'un état de synchronisation par un cycle social de 24 heures. (Chez l'Homme, les contraintes horaires imposées par sa vie sociale constituent les éléments du synchroniseur le plus puissant).

D'un autre côté, c'est grâce aux recherches des écoles des Professeurs J. Benoit et I. Assenmacher que le cycle circannuel du diamètre testiculaire du Canard et sa persistance pendant plusieurs années dans l'obscurité complète (sans synchroniseur connu) a pu être mis en évidence. Dans ce climat français de physiologie spectrale, j'espère que j'aurai, quant à moi, dans un avenir prochain comme par le passé, beaucoup de nouvelles choses à apprendre. Je remercie tous ces amis pour l'occasion qui m'a été donnée de coopérer avec eux.

## 11. Spéculations

Certains aspects de la structure biologique temporelle représentent des adaptations évolutives des organismes aux variations cycliques prédominantes de leur milieu (structure temporelle de l'environnement) plutôt que, seulement, des adaptations "personnelles" réacquises (p. ex. "apprises") par chaque individu, à chaque génération.

Dans une perspective de l'évolution des espèces, les variations périodiques de l'environnement, prédominant sur presque toute la surface du globe terrestre, furent de nature géophysique—principalement et essentiellement, parmi ces variations, on doit prendre en considération les changements relativement sûrs des alternances de lumière et d'obscurité suivant différentes échelles de temps: circadienne (environ 24 h.) et circannuelle (environ 1 an). Tandis que la période circannuelle est considérée comme constante, la période circadienne a probablement changé au cours de l'histoire terrestre. Les textes élémentaires d'astronomie [67, 68] font état de l'augmentation lente et graduelle de la durée du jour (ralentissement de la rotation de la Terre sur elle-même) et aussi, à part cela, de variations annuelles, et autres, actuellement non-prévisibles avec précision. Il est tentant de considérer, compte-tenu des données qui précèdent, que la fréquence circadienne la plus élevée enregistrée jusqu'ici par des procédés tels que le spectre de variance et le périodogramme a été trouvée au voisinage d'un cycle par 21 heures—dans le comportement de *l'Escherichia coli* [66]. Nous avons discuté ailleurs [64] la possibilité, actuellement très hypothétique, d'une géochronométrie et d'une biochronométrie de l'évolution par la datation, au moyen de la période des rythmes. La date à laquelle une espèce a évolué pourra être reflétée, en partie, par la période circadienne observée lorsqu'un organisme donné est maintenu dans le cycle lumière-obscurité à la fréquence la plus haute compatible avec une synchronisation rythmique. L'obstacle principal à cette entreprise de datation est qu'une telle hypothèse n'est, au mieux, que partiellement correcte; autrement dit, le $\tau$

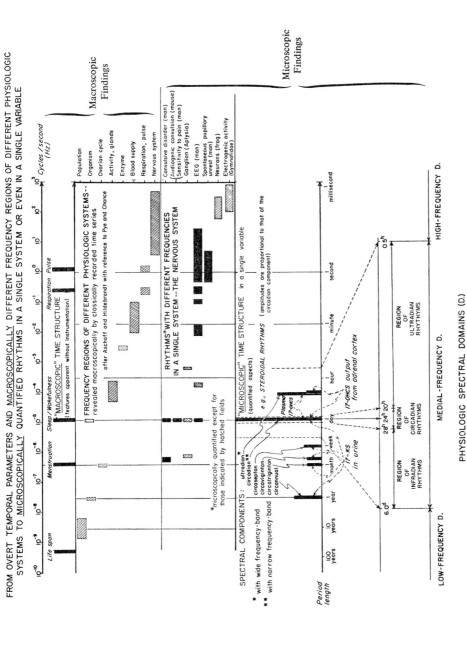

Fig. 19a.   Spectral components of different scope—including those recognized by primitive man (top).

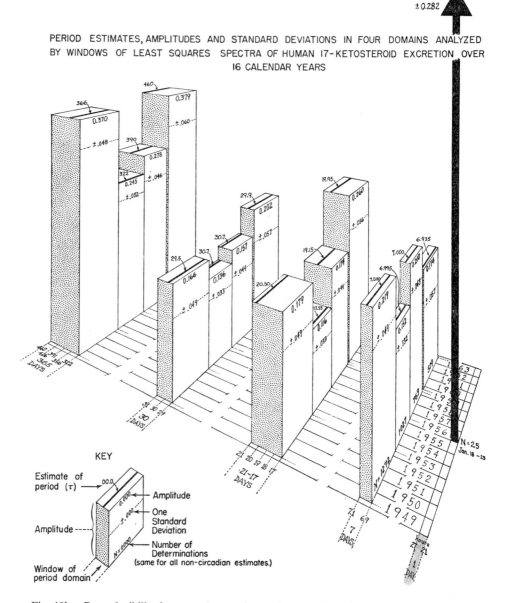

Fig. 19b.  Reproducibility from one data section to the next of certain rhythms with a relatively low frequency, in the 17-ketosteroid excretion of a healthy man. Arrow in heavy print visualizes certain comparable estimates for the much more prominent circadian rhythm of urinary 17-ketosteroid excretion [10; cf. also 72 and 73 for circannual rhythms].

circadien dépend, en partie, de la rotation de la Terre sur elle-même au moment où une espèce terrestre a évolué. Peut-être trouvera-t-on aussi, pour des autres rythmes biologiques, une dépendance similaire, vis à vis de la longueur du jour lunaire, à l'époque du "passage" d'une espèce du milieu aquatique au milieu terrestre—si un tel passage s'est réellement produit [79–81]. De plus, la localisation spectrale d'une bande de fréquence(s) circadienne(s) dépendra de l'histoire du système après son évolution "primordiale"—par exemple, après la transition d'une forme aquatique à une forme terrestre. Au cours de cette histoire, on peut admettre que, par leur reproduction sélective, certaines mutations aléatoires ont persisté parce qu'elles possédaient une valeur d'adaptation positive. On peut donc envisager une meilleure adaptation à des bandes de fréquences circadiennes géophysiques, si les bandes circadiennes des organismes se sont alongées—peut-être plus rapidement que le ralentissement de la Terre elle-même—spéculation qui aurait pour conséquence qu'aujourd'hui les périodes circadiennes peuvent être plus longues que la période terrestre de 24 heures.

## 12.  Conclusions et résumé

Les résultats présentés suggèrent qu'il y a de nombreuses exceptions à la constance présumée des organismes, que nous considérions l'Homme ou les unicellulaires. La structure biologique comme la structure physique est *spatiotemporelle*. La structure temporelle dynamique d'un organisme, à plusieurs niveaux de résolution, est le complément de sa morphologie [82–84] de sa structure spatiale, plus statique; la structure temporelle comprend un spectre de rythmes de différentes fréquences ainsi qu'il apparaît à la fig. 19. Elle montre le début de la métamorphose depuis a) une analyse exclusivement "macroscopique" de la périodicité, qui consistait seulement en une inspection de la représentation, étalée en fonction du temps, de processus biopériodiques, jusqu'à b) la rythmométrie "microscopique", qui est l'étude complémentaire de la structure biologique temporelle et qui permet de représenter une variable physiologique par son amplitude ou sa variance étalée en fonction de sa fréquence, sa période ou sa phase.

La morphologie spatiale est couramment définie par une exploration fructueuse et toujours plus profonde de la cyto-architecture. Les chercheurs en biologie moléculaire, en microscopie électronique et en biochimie joignent leurs efforts dans leurs tentatives de localiser le groupement et l'interaction de différentes molécules en diverses parties d'une cellule donnée. Cependant, toute analyse complète des relations géométriques, dans une structure spatiale, peut avoir (et ce sera un jour une condition indispensable) pour complément une exploration concomitante et, à nouveau, toujours plus profonde des aspects temporels de l'organisation biologique spectrale comme le montrent les figs. 19a et 19b. Une telle entreprise dépend du progrès des techniques de télémétrie en biochimie aussi

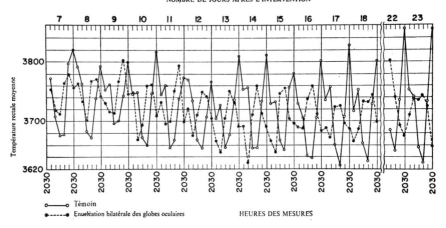

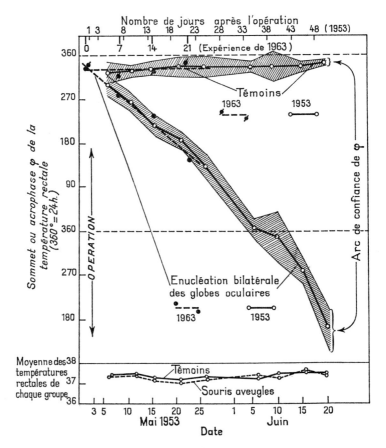

Fig. 20a. Circadian desynchronization following blinding and its reproducibility in studies carried out a decade apart [cf. 2].

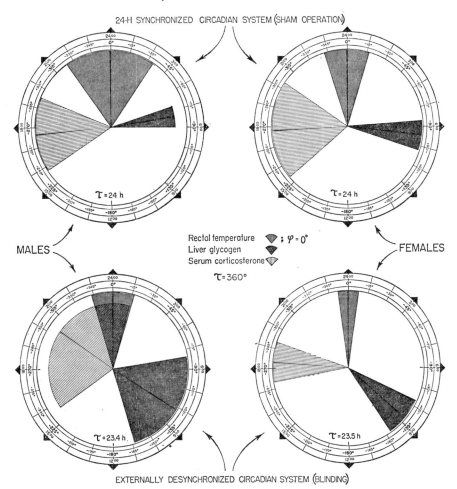

**Fig. 20b.** Lead in circadian acrophase of serum corticosterone in relation to the rectal temperature acrophase—in the 24-hour-synchronized (top) or externally desynchronized (bottom) circadian system of the mouse.

bien qu'en biophysique. Les séries temporelles de mesures collectées grâce à ces techniques, ou suivant des méthodes plus usuelles, peuvent dès maintenant faire l'objet d'analyses numériques spéciales, celles-ci étant devenues réalisables du fait des progrès des calculateurs électroniques. Par ces méthodes, les rythmes sont susceptibles d'être détectés, isolés, définis et développés sous la forme quantitative d'entités statistiques reproductibles.

Chaque rythme, qui à l'origine représentait, peut-être, les résultats de pressions adaptatives et de reproduction sélective en réponse à la localisation géophysique — rythmes circadiens, p. ex. —s'est adapté aussi, sans besoin dès lors de changement de fréquence, aux aspects temporels d'une niche socio-écologique de l'habitat.

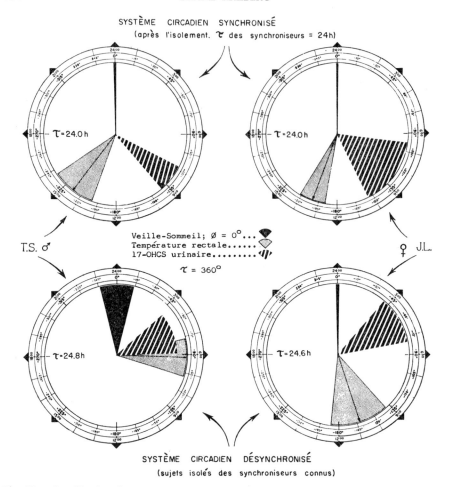

Fig. 20c.   Lead in circadian acrophase of urinary 17-hydroxycorticosteroid excretion in relation to the rectal temperature acrophase—in mature human beings, under conditions of synchronization with a societal routine on top and under conditions of isolation at the bottom.

Pour certaines composantes de la structure spectrale temporelle, nous ne connaissons pas de contre-partie géophysique de même fréquence; tel est le cas des rythmes d'environ une semaine. Cette composante spectrale circaseptidienne a été récemment mise en évidence à partir d'une série longitudinale de l'excrétion quotidienne des 17—cétostéroïdes couvrant 16 années consécutives (fig. 19b).

Qu'un rythme donné ait, ou non, sa contre-partie dans l'environnement, un phénomène prévisible de cette nature représente un des aspects de l'intégration temporelle de l'organisme, en dehors de leur rôle dans l'adaptation à l'environnement. Dans l'un et l'autre des cas—intégration ou adaptation—nous avons à faire à un phénomène qui rend possible une division du travail dans le temps. D'un côté, une division temporelle du travail peut être achevée par l'évolution d'une composante spectrale, avec une fréquence donnée, pour différentes fonctions physiologiques qui y sont reliées; les différences de phase parmi de tels rythmes physiologiquement interdépendants, et ayant une fréquence similaire—la même fréquence, en moyenne—représentent l'ordre temporel dans ce cas. Des différences de phase, entre les rythmes détectés par des méthodes isotopiques de marquage du RNA, du DNA et pour le taux des mitoses hépatiques des souris constituent un exemple d'un tel ordre, c'est à dire d'une telle organisation.

D'un autre côté, les organismes ont acquis, probablement au cours de l'évolution, pour leurs diverses fonctions, des bandes de fréquence spécifiques et différentes en quelques manières: certaines tâches doivent être exécutées plus fréquemment que d'autres. Dans le spectre des fréquences, il y a des grosses différences entre les rythmes d'un cycle par jour et les rythmes ayant une fréquence d'un cycle par environ un mois ou environ un an. Une de ces différences est réduite par la survenue, entre autres, d'une composante spectrale circaseptidienne—peut-être l'intégration temporelle est-elle ainsi facilitée.

Il reste la possibilité de la découverte éventuelle par les géophysiciens, de variations périodiques de l'environnement suivant le spectre des fréquences que le biologiste rencontre—lorsqu'il étudie les aspects chronobiologiques, *inter alia*, de la neurophysiologie ou de la circulation (dans le domaine des fréquences les plus hautes) aussi bien que les aspects métaboliques, reproductifs et autres (dans les domaines des rythmes des basses ou des moyennes fréquences). Les informations à la fois plus approfondies, plus détaillées et plus objectives concernant de tels phénomènes nous conduisent à la *chronobiologie* qui réunit comme spécialisations: (1) la chronophysiologie – Elle répond à l'étude des facteurs physiologiques (nerveux, endocriniens, métaboliques etc.) qui déterminent les caractéristiques temporelles d'un processus biologique; (2) la chronopathologie – Elle répond à l'étude des altérations de ces caractéristiques temporelles déterminant un état pathologique ou résultant de celui-ci (psychose, cancer, endocrinopathie etc.); (3) la chronotoxicologie qui répond à l'étude des variations rythmiques prévisibles d'effets non-désirables ou dangéreux provenant d'agents chimiques, physiques ou autres, y compris les poisons, les substances polluantes et les surdosages thérapeutiques; et l'étude de tels effets non-désirables sur les rythmes biologiques; et (4) la chronopharmacologie – Elle répond à l'étude des effets d'une drogue sur ces caractéristiques temporelles, effets qui varient de manière prévisible et cyclique en fonction du temps.

Les informations plus générales sur l'organisation temporelle constituent le domaine ou se rencontreront dans l'avenir, comme à cette réunion, le physicien, le biologiste et le medecin—pour analyser en commun les principes qui ont conduit à transformer l'adage: *Qui habet tempus habet vitam* en: *Qui habet structuram in tempore habet vitam*, formulation nouvelle qui condense les acquisitions de la physiologie statistique appliquée aux séries de mesures temporelles plutôt qu'aux mesures isolées et non situées dans le temps.

## 11.  Speculations

Certain aspects of biologic time structure may represent evolutionary adaptations of organisms to the predominant cyclic variations of their environment (to the temporal structure of the environment) rather than merely the "personal" adaptations reacquired (e.g., "learned") by each individual in each generation.

Along an evolutionary scale the pertinent environmental periodicities prominent over most of the Earth's surface were geophysical—first and foremost among them the relatively reliable cyclic changes in the alternation of light and darkness, changing along several time scales, circadian and circannual. Whereas the circannual period of the environment is regarded as being on the average more or less constant, the circadian period has probably changed systematically in our terrestrial history. There also is evidence, cited even in elementary astronomy texts [67, 68], to indicate a gradual long-term increase in day length—by virtue of a slow-down of the Earth's rotation around its axis.

It is tempting to consider in this connection that the highest circadian frequency detected thus far in a biologic system not synchronized by a cycle of alternating light and darkness and quantified by procedures such as the variance spectrum and the periodogram has been found to be in the neighborhood of 21 hours. For the behavior of E. coli [66] one might consider the as yet highly speculative possibility of evolutionary biorhythm dating, on the basis of the length of the desynchronized circadian period detected in currently living forms—an endeavor extending the geochronometry and biochronometry that has become possible by comparisons of circadian growth lines in living and fossil marine invertebrates.

The date at which a species evolved might be reflected at least in part by its circadian period observed when the contemporary form of this organism is maintained in a cycle of light and darkness of the highest frequency no longer compatible with the rhythm's synchronization. An obstacle to such an extrapolation made for the rhythm dating of evolution is that such an hypothesis can at best be only partially correct. The circadian frequency of a terrestrial organism can depend only in part upon the speed of the Earth's rotation around its axis at the time when the "ancestors" of this form of life evolved. Conceivably, one may find for certain other organisms originally evolved in an aquatic milieu a similar dependence of their biologic rhythms upon the length of the lunar day at that epoch of the Earth's history when the counterpart of the contemporary form might have "resettled", if indeed it did, on terra firma [79–81].

In any event, the spectral location of a band of circadian frequencies will also depend upon the history of a given biologic system after its early evolution. For example, changes in the circadian period resulting from certain random mutations might have persisted through selective reproduction because they had positive adjustment value. One could indeed envisage a better adaptation of organisms to the prevailing geophysical circadian frequencies via a lengthening of the organism's circadian band.

Conceivably such a lengthening of the organism's circadian period occurred more rapidly than the lengthening of the geophysical day resulting from the slow-down of the Earth's rotation around its axis. For this reason, perhaps, circadian bands that are longer than the terrestrial period, now 24 hours, are encountered in a number of organisms under several conditions. Can it be that an overshoot lengthening of the biologic circadian period occurred because organisms are so structured in time that, according to different lines of evidence, it is easier for several contemporary life forms [71] to delay a rhythm than to advance it?

## 12.  Summary and conclusions

The foregoing evidence suggests that there are many exceptions to the presumed constancy of organisms, whether we consider unicellulars or man. Biologic structure is *spatiotemporal.*

An organism's dynamic structure in time complements at several levels of resolution the more static morphology in space. Temporal structure includes a spectrum of rhythms with different frequencies, as sketched in the abstract fig. 19. This figure conveys the beginning metamorphosis from a "macroscopic" periodicity analysis, consisting of the inspection of bioperiodicity in time displays, to a "microscopic" rhythmometry of the biologic time structure.

Spatial morphology is currently being resolved by the successful and ever deeper probing into cyto-architecture. The molecular biologist joins hands with electron-microscopists and bio-chemists in their attempts to localize the arrangement and/or interaction of different molecules in various parts of a given cell. Yet any analysis of geometric relations in spatial structure can be complemented by a concomitant and again ever deeper probing into the temporal aspects of biologic spectral organization, as indicated in figs. 19a and 19b. This undertaking depends upon the advancement of biochemical as well as biophysical telemetry techniques. Time series collected by such techniques or by more conventional means already can be subjected to special numerical analyses rendered practicable by the advent of electronic computers [2]. By such methods rhythms become amenable to isolation, resolution and display as reproducible statistical entities.

In any event, what originally represented, perhaps, a result of adaptive pressures and selective reproduction in response to the geophysical setting—circadian rhythms, for example—adapted also, with no need thus far for a change in frequency, to temporal aspects of a socio-ecological habitat niche. For certain components of the spectral time structure, we do not know of a geophysical counterpart with the same frequency. A case in point is an about-weekly rhythm. This circaseptan spectral component has recently been documented from a longitudinal series consisting of values for the daily 17-ketosteroid excretion—a series covering about 16 consecutive years (figs. 19a and 19b).

Whether or not a given rhythm has an environmental counterpart, such predictable phenomena represent features of the organism's temporal integration, quite apart from their role in environmental adaptation. In either case—integration or adaptation—we are dealing with a phenomenon which makes possible a division of labor in time. On the one hand, a temporal division of labor can be achieved by the evolution of a spectral component with a given frequency in several related physiologic functions; *differences in phase* among such physiologically interdependent rhythmic phenomena with similar frequency represent the temporal ordering in this case.

On the other hand, organisms also have acquired, presumably, in the course of evolution, somewhat specific and *different frequency bands* for their diverse functions: some tasks have to be done more frequently than others. In this spectrum of frequencies there are sizeable gaps between rhythms with one cycle per day and rhythms with a frequency of one cycle in about one month or in about one year. One such gap is reduced by the occurrence of a circaseptan spectral component—perhaps temporal integration is thus facilitated. The possibility remains that eventually environmental frequencies may be discovered by the geophysicists for those frequencies that we biologists encounter whether we study chronobiologic aspects of neurophysiology in the highest now known frequency domain or whether we evaluate rhythms in the domain of medial or low frequencies.

Such information leads us to the fledgling field of chronobiology, with several subspecialties: a) *Chronophysiology*—study of physiologic factors underlying biologic temporal characteristics, e.g., adrenal, renal and metabolic factors underlying the phase-shifting of rhythms following transmeridian flights; b) *Chronopathology*—study of alterations in biologic temporal characteristics as 1) functions of disease, and, what seems more important, as 2) determinants of disease; c) *Chronotoxicology*—study of undesired or harmful effects from chemical, physical or other agents, including poisons, pollutants and overdoses of drugs 1) upon biologic temporal characteristics, and 2) as a function of biologic timing; d) *Chronopharmacology*—study of drug effects 1) upon biologic temporal characteristics and 2) as a function of biologic timing. Thus, by improving the toxic-therapeutic ratio of a drug, added benefit may be derived from conventional therapy, e.g., a reduction of side effects. More important is the possibility of finding new therapeutic applications for conventional drugs (e.g., ACTH for resynchronization) or of developing new drugs, against the background of any data indicating that rhythm alteration such as circadian desynchronization contributes as a determinant of disease.

## References

[1]  A. Fessard, *Propriétés rythmiques de la matière vivante*, (Hermann, 1936).

[2]  F. Halberg and A. Reinberg, *Rythmes circadiens et rythmes de basses fréquences en physiologie humaine*, J. Physiol. **59** (1967) 117–200.

[3]  I. H. Kaiser and F. Halberg, *Circadian aspects of birth*. Annals New York Academy of Sciences **98** (1962) 1056–1068.

[4]  F. Halberg, *Resolving power of electronic computers in chronopathology, an analogy to microscopy*, Scientia **101** (1966) 412–419.

[5]  F. Halberg, Y. L. Tong and E. A. Johnson, *Circadian system phase, an aspect of temporal morphology; procedures and illustrative examples;* in: The Cellular Aspects of Biorhythms (Springer, 1967) pp. 20–48.

[6]  G. Frank, F. Halberg, R. Harner, J. Matthews, E. Johnson, H. Gravem and V. Andrus, *Circadian periodicity, adrenal corticosteroids, sleep deprivation and the EEG in normal men.* J. Psychiatric Res. **4** (1966) 73–86.

[7]  F. Halberg, *Organisms as circadian systems; temporal analysis of their physiologic and pathologic responses, including injury and death;* in: Walter Reed Army Institute of Research Symposium, Medical Aspects of Stress in the Military Climate, pp. 1–36 (1964).

[8]  F. Halberg, W. Nelson, W. Runge, G. Pitts, D. Smith, J. Tremor, T. M. Edwards, P. Hahn, G. Cook, P. Sebesta and O. Schmitt, *Reproducibility of circadian temperature rhythm in the rat kept in continuous light of 30 lux intensity*, The Physiologist **9** (1966) 196.

[9]  F. Halberg, W. Nelson, W. Runge and O. H. Schmitt, *Delay of circadian rhythm in rat temperature by phase-shift of lighting regimen is faster than advance*, Federation Proc. **26** (1967) 599.

[10]  F. Halberg, M. Engeli, C. Hamburger and D. Hillman, *Spectral resolution of low-frequency, small-amplitude rhythms in excreted ketosteroid; probable androgen-induced circaseptan desynchronization*, Acta Endocrinologica Suppl. 103 (1965) 54 pages.

[11]  M. Garcia Sainz and F. Halberg, *Mitotic rhythms in human cancer, reevaluated by electronic computer programs—evidence for temporal pathology.* J. Nat. Cancer Inst. **37** (1966) 279–292.

[12]  P. Passouant, L. Popoviciu, G. Velok et M. Baldy-Moulinier, *Etude polygraphique des narcolepsies au cours du nycthémère.* Rev. Neurol. **118** (1968) 431–441.

[13]  F. Halberg and H. Simpson, *Circadian acrophases of human 17-hydroxycorticosteroid excretion referred to midsleep rather than midnight*, Human Biology **39** (1967) 405–413.

[14]  M. Sakai, *Diurnal rhythm of 17-ketosteroid and diurnal fluctuation of depressive affect*, Yokohama Med. Bull. **11** (1960) 352–367.

[15]  F. Halberg, P. Vestergaard and M. Sakai, *Rhythmometry on urinary 17-ketosteroid excretion by healthy men and women and patients with chronic schizophrenia; possible chronopathology in depressive illness*, Volume honoring Prof. Jacques Benoit, in press.

[16]  R. P. Doe, J. A. Vennes and E. G. Flink, *Diurnal variation of 17-hydroxycorticosteroids, Na, K, Mg and creatinine in normal subjects and in cases of treated adrenal insufficiency and Cushing's syndrome*, J. Clin. Endocrinol. Metabolism **20** (1960) 253–265.

[17]  A. Lunedei, M. Cagnoni, F. Fantini, B. Tarquini, G. Morace, M. Maiello, A. Panerai, P. T. Scarpelli, R. Toccafondi, *Sindromi Diencefaliche (Problemi in discussione)*, Ed. L Pozzi (Roma, 1967).

[18] M. S. Knapp, P. M. Keane and J. G. Wright, *Circadian rhythm of plasma 11-hydroxy-corticosteroids in depressive illness, congestive heart failure and Cushing's syndrome.* British Med. J. **2** (1967) 27–30.

[19] F. Halberg, *Physiologic considerations underlying rhythmometry, with special reference to emotional illness;* in: *Cycles Biologiques et Psychiatrie,* Ed. J. de Ajuriaguerra (Georg Cie S.A., Genève, 1968) 73–126.

[20] F. Halberg, *Body temperature, circadian rhythms and the eye,* Colloque Int. du C.N.R.S., Montpellier, July 17–22 (1967) in press.

[21] F. Halberg, *Circadian (about 24-hour) rhythms in experimental medicine.* Proc. Roy. Soc. Med. **56** (1963) 253–256.

[22] F. Halberg, M. B. Visscher, E. B. Flink, K. Berge and F. Bock, *Diurnal rhythmic changes in blood eosinophil levels in health and in certain diseases,* The Journal-Lancet **71** (1951) 312–319.

[23] J. Ghata, F. Halberg, A. Reinberg and M. Siffre, *Rythmes circadiens désynchronisés chez deux sujets adultes sains (17-hydroxycorticostéroïdes, température rectale, veille-sommeil).* Ann. d'Endocrinologie, in press.

[24] M. Siffre, A. Reinberg, F. Halberg, J. Ghata, G. Perdriel and R. Slind, *L'isolement souterrain prolongé, Etude de deux sujets adultes sains avant, pendant et après cet isolement.* Presse Médicale **74** (1966) 915–919.

[25] K. Reindl, *Kombinierte Longitudinal- und Transversal-Studien von Zirkadian-Rhythmen bei asthmatischen Kindern,* Verhandlungen der Deutschen Gesellschaft für Innere Medizin, 73 Congress, Wiesbaden. Munich, J. F. Bergmann, pp. 978–982 (1967).

[26] K. Reindl, C. Falliers, F. Halberg, H. Chai, D. Hillman and W. Nelson, *Circadian acrophases in peak expiratory flow rate and urinary electrolyte excretion of asthmatic children: phase-shifting of rhythms by prednisone given in different circadian system phase,* Rassegna Neurol. Vegetativa, in press.

[27] P. Gervais, A. Reinberg, R. Tardif, S. Azabagic, S. Morault, P. Planat and D. Vignaud, *La spirométrie de pointe chez les asthmatiques séjournant en milieu spécifiquement contrôlé (premiers résultats),* Rev. Franc. Allergie 6 (1968) 145–154.

[28] A. Reinberg, *Hours of changing responsiveness in relation to allergy and the circadian adrenal cycle,* in *Circadian Clocks,* Ed. J. Aschoff, (North-Holland, 1965), pp. 214–218.

[29] A. Reinberg, J. Ghata and E. Sidi, *Nocturnal asthma attacks: their relationship to the circadian adrenal cycle.* J. Allergy **34** (1963) 323–330.

[30] F. Halberg and W. W. Spink, *The influence of brucella somatic antigen (endotoxin) upon the temperature rhythm of intact mice.* Laboratory Investigation **5** (1956) 283–294.

[31] F. Halberg, E. A. Johnson, B. W. Brown and J. J. Bittner, *Susceptibility rhythm to E. Coli endotoxin and bioassay,* Proc. Soc. Exp. Biol. Med. **103** (1960) 142–144.

[32] E. Haus and F. Halberg, *24-hour rhythm in susceptibility of C mice to a toxic dose of ethanol.* J. Appl. Physiol. **14** (1959) 878–880.

[33] F. Jones, E. Haus and F. Halberg, *Murine circadian susceptibility-resistance cycle to acetylcholine,* Proc. Minn. Acad. Sci. **31** (1959) 61–62.

[34] A. Reinberg, Z. W. Zagula-Mally, J. Ghata and F. Halberg, *Circadian rhythm in duration of salicylate excretion referred to phase of excretory rhythms and routine.* Proc. Soc. Exp. Med. Biol. **124** (1967) 826–832.

[35] F. Halberg, *Periodicity analysis, a potential tool for biometeorologists,* Int. J. Biometerology **7** (1963) 167–191.

[36] E. Haus, F. Halberg, W. Nelson and D. Hillman, *Shifts and drifts in phase of human circadian system following intercontinental flights and in isolation,* Federation Proc., March-April (1968) p. 224.

[37] F. Halberg, *Claude Bernard, referring to an "extreme variability of the internal milieu";* in: *Claude Bernard and Experimental Medicine,* (Shenkman, 1967), pp. 193–210.

[38] C. Bernard, *De la diversité des animaux soumis à l'expérimentation. De la variabilité des conditions organiques dans lesquelles ils s'offrent à l'expérimentateur.* J. de l'Anatomie et de la Physiologie normales et pathologiques de l'Homme et des Animaux 2 (1865) 497–506.

[39] J. Aschoff, *Response curves in circadian periodicity*, in: *Circadian Clocks*, Ed. J. Aschoff, (North-Holland, 1965) pp. 95–111.

[40] F. A. Brown Jr., *A unified theory for biological rhythms, rhythmic duplicity and the genesis of "circa" periodisms*, in: *Circadian Clocks*, Ed. J. Aschoff, (North-Holland, 1965) pp. 231–261.

[41] K. Pye and B. Chance, *Sustained sinusoidal oscillations of reduced pyridine nucleotide in a cell-free extract of saccharomyces carlsbergensis.* Proc. Nat. Acad. Sci. 55 (1966) 888–894.

[42] J. T. Enright, *Synchronization and ranges of entrainment*, in: *Circadian Clocks*, Ed. J. Aschoff, (North-Holland, 1965) pp. 112–124.

[43] D. S. Farner, *Circadian systems in the photoperiodic responses of vertebrates*, in: *Circadian Clocks*, Ed. J. Aschoff, (North-Holland, 1965) pp. 357–369.

[44] L. Scheving and D. Vedral, *Circadian variation in susceptibility of the rat to several different pharmacological agents*, The Anatomical Record 154 (1966) 417.
L. Scheving and J. Pauly, *Daily mitotic fluctuations in the epidermis of the rat and their relation to variations in spontaneous activity and rectal temperature*, Acta Anat. 43 (1960) 337–345.

[45] A. Reinberg and J. Ghata, *Les rythmes biologiques*, (Presse Univ. Fr. 1964), Sec. ed.

[46] C. S. Pittendrigh, *Circadian rhythms and the circadian organization of living systems*, Cold Spring Harbor Symposia on Quantitative Biology 25 (1960) 159–182.

[47] S. B. Hendricks, *Rates of change of phytochrome as an essential factor determining photoperiodism in plants*, Cold Spring Harbor Symposia on Quantitative Biology 25 (1960) 245–248.

[48] A. Wolfson, *Circadian rhythm and the photoperiodic regulation of the annual reproductive cycle in birds*, in: *Circadian Clocks*, Ed. J. Aschoff, (North-Holland, 1965) pp. 370–378.

[49] C. Kayser and A. Heusner, *Le rythme nycthéméral de la dépense d'énergie.* J. Physiol. 59 (1967) 3–116.

[50] E. Bünning, *Die Physiologische Uhr*, (Springer, 1963).

[51] L. Baillaud, *Les rythmes dans la vie des plantes.* Biol. Med. 53 (1964) 237–265.

[52] J. N. Mills, *Human circadian rhythms*, Physiol. Rev. 46 (1966) 128–171.

[53] A. Lunedei and M. Cagnoni, *Sindromi Diencefaliche*, Relazione al LXVIII Congresso della Societa Italiana di Medicina Interna, Firenze. Rome, (1967) pp. 413; see also p. 175.

[54] H. Simmonet, *Rythmes et cycles biologiques chez les organismes animaux*, Biol. Med. 53 (1964) 266–330.

[55] S. Jerebzoff and E. Lamberg, *Détection par la méthode du spectre de variance de deux rythmes régissant simultanément la sporulation de Myrothecium verrucaria*, C.R. Acad. Sci. Paris 264 (1967) 322–325.

[56] T. Hellbrügge, J. Ehrengut-Lange, J. Rutenfranz and K. Stehr, *Circadian periodicity of physiological functions in different stages of infancy and childhood*, Ann. N.Y. Acad. Sci. 117 (1964) 361–373.

[57] F. Strumwasser, *The demonstration and manipulation of a circadian rhythm in a single neuron*, in: *Circadian Clocks*, Ed. J. Aschoff, (North-Holland, 1965) pp. 442–462.

[58] W. B. Quay, *Regional and circadian differences in cerebral cortical serotonin concentrations*, Life Sci. 4 (1965) 379–384.

[59] I. Prigogine, *Etude Thermodynamique des Phénomènes Irréversibles*, (Desoer, 1947).

[60] I. Prigogine and R. Balescu, *Sur les propriétés différentielles de la production d'entropie*, Acad. Roy. Belg. Classe Sci. Bull. 41 (1955) 917.

[61] I. Prigogine and R. Balescu, *Phénomènes cycliques dans la thermodynamique de processus irréversibles*, Acad. Roy. Belg. Classe Sci. Bull. 42 (1956) 256.

[62] A. Katchalsky and P. F. Curran, *Nonequilibrium Thermodynamics in Biophysics*, (Oxford U. Press, 1966) 248 pages.
[63] M. Reiss, R. E. Hemphill, J. J. Gordon and E. R. Cook, *Regulation of urinary steroid excretion.* Biochemical J. **45** (1949) 574–578.
[64] F. Halberg, W. Nelson, W. Runge, O. Schmitt, G. Pitts, et al., *Tests of circadian rhythm characteristics—design evaluation by results on rodents and men.* In press.
[65] R. Wever, *Über die Beeinflüssung der circadianen Periodik des Menschen durch schwache elektromagnetische Felder.* Z. für Vergl. Physiol. **56** (1967) pp. 111–128.
[66] F. Halberg and R. L. Conner, *Circadian organization and microbiology: variance spectra and a periodogram on behavior of Escherichia coli growing in fluid culture.* Proc. Minn. Acad. Sci. **29** (1961) 227–239.
[67] R. H. Baker, *An Introduction to Astronomy*, (Van Nostrand, 1965), pp. 1–364.
[68] E. A. Fath, *The Elements of Astronomy*, (McGraw-Hill, 1955) pp. 1–369.
[69] A. Reinberg, F. Halberg, J. Ghata and M. Siffre, *Spectre thermique (rythmes de la température rectale) d'une femme adulte saine avant, pendant et après son isolement souterrain de trois mois.* C. R. Acad. Sci. Paris **262** (1966) 782–785.
[70] F. Halberg, J. Reinhardt, F. Bartter, C. Delea, R. Gordon, A. Reinberg, J. Ghata, H. Hofmann, M. Halhuber, R. Günther, E. Knapp, J. C. Pena and M. Garcia Sainz, *Agreement in endpoints from circadian rhythmometry on healthy human beings living on different continents*, Experientia **25** (1969) 107.
[71] C. S. Pittendrigh, V. G. Bruce and P. Kaus, *On the significance of transients in daily rhythms*, Proc. Nat. Acad. Sci. **44** (1958) 965–973.
[72] I. Assenmacher and J. Soule, *Mise en évidence d'un cycle annuel de la fonction cortico-surrénalienne chez le Canard mâle*, C.R. Acad. Sc. Paris **263** (1966) 983–985.
[73] I. Assenmacher, *Influence de la lumière sur certains rythmes physiologiques*, Revue LUX **40** (1966) 3–11.
[74] B. Metz and P. Andlauer, *Rapports de la déperdition latente de chaleur et de la température rectale au cours de la phase vespérale du rythme nycthéméral chez l'Homme*, J. de Physiol. **44** (1952) 293–297.
[75] B. Metz and J. Schwartz, *Etudes des variations de la température rectale du débit urinaire et de l'excrétion urinaire des 17 cétosteroides chez l'homme au cours du nycthémère*, C.R. Soc. Biol. **143** (1949) 1237–1239.
[76] J. Colin, J. Timbal, C. Boutelier, Y. Houdas and M. Siffre, *Rhythm of the rectal temperature during a 6-month free-running experiment*, J. Appl. Physiol. **25** (1968) 170–176.
[77] P. Fraisse, M. Siffre, G. Oleron et N. Zuili, *Le rythme veille-sommeil et l'estimation du temps*, in: *Cycles Biologiques en Psychiatrie*, Symposium Bel-Air III 1967, ed. J. de Ajuriaguerra (Masson, 1968) 257–265.
[78] M. Jouvet, *Perturbation des rythmes circadiens de l'homme et aspects psychosomatiques*, in: *Cycles Biologiques en Psychiatrie*, Symposium Bel-Air III, 1967, ed. J. de Ajuriaguerra (Masson, 1968) 185–203.
[79] D. L. Lamar and P. M. Merifield, *Cambrian fossils and origin of Earth-Moon system*, Geol. Soc. Am. Bull. **78** (1967) 1359.
[80] P. M. Merifield and D. L. Lamar, *Sand waves and early Earth-Moon history*, J. Geophys. Res. **73** (1968) 4767.
[81] D. L. Lamar, J. V. McGann-Lamar and P. M. Merifield, *Age and origin of Earth-Moon system*, Paleogeophysics, in press.
[82] J. Aschoff, *Zeitliche Strukturen biologischer Vorgänge*, Nova Acta Leopoldina **21** (1959) 147.
[83] J. Aschoff, *Survival value of diurnal rhythms*, Symp. Zool. Soc. Lond. No. **13** (1964) 79–98.
[84] F. Halberg, *Chronobiology*, in: Annual Review of Physiology **31** (1969) 675–725.

*Theoretical Physics and Biology*, © 1969, *North-Holland Publ. Co., Amsterdam*

## DISCUSSIONS

B. B. LLOYD: You alluded, Dr. Halberg, to the periodicity of liver glycogen and you also talked about Claude Bernard's earlier statement of liver glycogen being sometimes very low (in fact, I think, sometimes zero) and sometimes very high. Has this periodicity anything to do with the nervous system and is it related to Bernard's work on piqûre diabetes.

F. HALBERG: Thank you kindly, Dr. Lloyd, for providing me with this opportunity to comment on Claude Bernard in relation to the study of rhythms. Let me refer you in this connection to comments made earlier [1]. To the thinker [2], not to mention Bernard the demonstrator of experimental procedure [3], the significance of physiologic timing should have been apparent. Yet if one scrutinizes his writing about the conditions of a "continuous life", one finds only somewhat ambiguous remarks such as the following:

> "The phenomenon of nutrition is accomplished in two times (he probably means two stages), and these two times are always separated one from the other by a period of more or less long duration, which (duration) is a function of a variety of circumstances" (. . . dont la durée est fonction d'une foule de circonstances) [2].

At 52 years of age, the active investigator recognizes explicitly the "milieu intérieur variable" (sic); in 1856, the editors of the Journal de l'Anatomie et de la Physiologie announce [4] that Mr. Claude Bernard will soon publish an introduction to the study of experimental medicine. They do so in a footnote to an article by Claude Bernard himself, written under the title "Of the diversity of animals subjected to experimentation" and—what is more important in our context—"Of the *variability* (italics mine) of organic conditions in which they (the animals) present themselves to the experimenter":

"M. Claude Bernard doit publier prochainement une 'Introduction à l'étude de la médecine expérimentale', 1 vol. in–8 de 400 pages. Nous sommes heureux d'offrir à nos lecteurs un extrait de ce livre, qui est un exposé de doctrines présentant le tableau complet des faits et des idées que le professeur a developpés dans son cours de médecine au Collège de France et dans son cours de physiologie générale à la Faculté des sciences, depuis ses dernières publications de 1859." [4].

This footnote states that they (the editors) are happy to offer to their readers an "extract" of the book prepared by its author. The material in the article then reflects what the active (rather than senescent) Bernard himself regarded as his most important experience. Of primary interest in the same context remains

Bernard's emphasis that "one must keep in mind not only the variations of the cosmic external milieu, *but also the variation of the organic milieu* (italics mine), i.e., that of the actual state of the organism".

Claude Bernard writes further that one might be in great error in assuming that it suffices to experiment on two animals of the same species in order to obtain identical experimental conditions and suggests that *the physiologic conditions of the internal milieu manifest an extreme variability* (italics mine); that, at a given moment, such variability introduces considerable differences into the results from experimentation on animals of the same species that appear to be identical.

He states further that "more than anybody else," he has insisted on the need to study these different physiologic conditions and to have demonstrated that they are the essential basis of experimental physiology. He points out that one must admit in fact that, in a given animal, the vital phenomena vary (sic) only according to precise and determined conditions of the internal milieu.

Students who evaluate rhythms as the elements of an organism's time structure will hasten to agree with such statements much more readily than students of "constancy". From appropriate statistical work one can indeed specify circadian system phases "dans lesquelles il y a *toujours* du sucre et d'autres conditions dans lesquelles il n'y en a *jamais*"—as will become apparent from fig. 1 for the case of glycogen, if not sugar, in mouse liver. Liver glycogen contents of, say, intact ad libitum fed animals continue to be presented by prominent biochemists without a qualification of the sampling time in terms of rhythms. However, the best chemical procedure will yield results that in such instances are physiologically difficult to interpret or actually misleading (fig. 1). One can cite Claude Bernard further in this connection:

"... Pour le moment, je veux uniquement appeler l'attention des expérimentateurs sur l'importance qu'il y a *à préciser les conditions organiques* (sic), parce qu'elles sont, ainsi que je l'ai déjà dit, la seule base de la physiologie et de la médecine expérimentale. Il me suffira, dans ce qui va suivre, de me borner à des indications, car c'est à propos de chaque expérience en particulier qu'il s'agira ensuite d'examiner ces conditions, aux trois points de vue physiologique, pathologique et thérapeutique." [4].

Furthermore, in his *Phenomena*, on page 114, [2] Bernard indicates that constancy presupposes the "self-perfecting" of an organism in such a fashion that the external variations are at each moment *compensated* and *balanced*; the "higher animal", rather than being indifferent to the external world, is in a *close and wise relationship* with it, in such a fashion that animal equilibrium results from a continuous and delicate compensation established by the most sensitive of balances.

The temptation is great to read "rhythm" into "equilibrium", and to attribute to the aged Bernard a hint of the close and wise interactions ("... étroite et

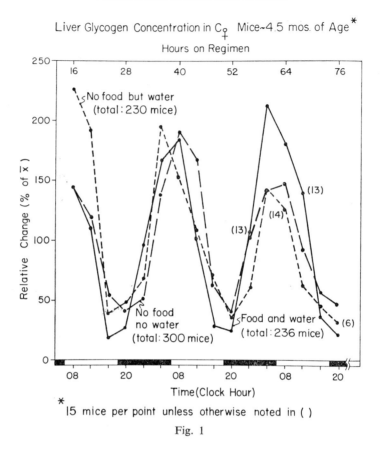

Fig. 1

savante relation . . .") in which the cycles of a terrestrial environment are "anti-cipated" by rhythmic organisms evolved on earth. This would probably be no more than wishful thinking—as far as can be judged from Bernard's writings. It is difficult further to reconcile the remarks on a "close and wise interaction" with the concept of a *shielding* of the organism from the environment by an internal milieu that is constant. Indeed, it is important at this symposium to emphasize that shielding by a basic constancy currently credited to Bernard must not be confused with the view of basic rhythmic physiologic interactions under-lying a superficial constancy or, rather, a limited variability. In the former view, rhythms remain secondary or even trivial considerations as to both interpretation of body function and experimental method. In the latter view, rhythms become primary basic features of temporal integration and adaptation in organisms.

It is probably because Claude Bernard gave precedence to shielding over rhythms that throughout his work on diabetes and glycogenesis there is not a single reference indicating precautions taken to control the rhythm in liver glycogen.

To turn to Dr. Lloyd's question as to whether Bernard's work on the "piqûre diabétique" relates to periodicity, one might suggest that Bernard failed to recognize such a relation. In his *Leçons sur Le Diabète et La Glycogénèse Animale* (Baillière, Paris 1877), Bernard recalls how he arrived at the piqûre after describing its effects; he writes that he had the preconceived idea to augment hepatic secretion and sugar production "en excitant les origines du pneumogastrique, comme j'avais, dans d'autres circonstances, augmenté la sécrétion salivaire en excitant les origines de la cinquième paire." (p. 370). The most prominent periodicity of this variable remains ignored until the explicit statement by Erik Forsgren in 1927 that liver glycogen undergoes about 24-hour periodic changes that are partly independent of nutrition. For a discussion of historical features to Bernard's piqûre and to subsequent work, interested individuals can be referred to a book by Jakob Möllerstrom, *Das Diabetesproblem: Die rhythmischen Stoffwechsel-vorgänge*, (Thieme, 1943).

[1] F. Halberg, Claude Bernard, referring to an "extreme variability of the internal milieu" in *Claude Bernard and Experimental Medicine*, (Shenkman, Cambridge, Mass., 1967) 193–210.
[2] C. Bernard, *Leçons sur les phénomènes de la vie communs aux animaux et aux végétaux* (Baillière, Paris, 1885).
[3] L. Binet, (editor), *Claude Bernard, Introduction à l'étude de la médecine expérimentale* (Paris, Les Chefs-d'Œuvre Classiques et Modernes, 1963).
[4] C. Bernard, De la diversité des animaux soumis à l'expérimentation. De la variabilité des conditions organiques dans lesquelles ils s'offrent à l'expérimentateur. *J. Anat. Physiol. Homme Animaux* **2** (1865) 497–506.

P. DEJOURS: Deux questions à M. Halberg: Pouvez-vous expliquer pourquoi il est plus difficile d'aller d'Ouest en Est que dans la direction opposée? En ce qui concerne les oiseaux migrateurs entre l'hémisphère Nord et l'hémisphère Sud, que devient leur cycle circadien? Existe-t-il un déphasage de 180 degrés lorsqu'ils passent l'équateur?

F. HALBERG: We are searching for the answer to Dr. Dejour's question, which gains greatly in importance in this day and age of global travel not only for passengers but also for the aviator and finally for all of those concerned with work hygiene. Short reflection will lead one to recognize that to adapt following rapid eastward travel, one must advance a rhythm, whereas after a westward flight, one adapts by a delay of rhythm. These two adaptations are distinctly different. It appears that the circadian system has some "polarity", as demonstrated for the rat in my text-fig. 4. The same point is also made by fig. 2 for man. This figure summarizies studies done in cooperation with Dr. Walter Nelson and Dr. Erhard Haus of our laboratory at the University of Minnesota. It shows for the variables and subjects studied a much more rapid delay of rhythm as compared to an advance. Another question concerning the relative ease of adaptations after transmeridian travel relates to the extent of transient frequency and phase desynchronization during such adaptations. From fig. 2 it also appears that while the adaptation following a flight from east to west is faster, the transient desynchronization among certain urinary variables during this relatively short shift time is greater than that occurring in connection with an advance of rhythms.

Such problems of chronophysiology relate perhaps to accident prevention. Certainly the avoidance of performance decrements by manipulating the adaptation of rhythms in the physiologically most favorable fashion compatible with logistic needs can be expected to lower the accident rate following changes in routine.

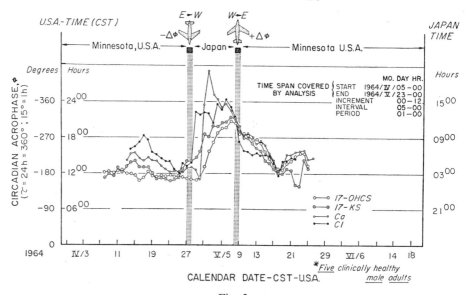

PHASE-SHIFTS OF THE HUMAN CIRCADIAN SYSTEM AS A RESULT
OF 2 INTERCONTINENTAL FLIGHTS, GAUGED BY
URINARY EXCRETION RATES*

Fig. 2

With such problems in mind, Dr. Fessard, Dr. Reinberg and a number of us in other parts of the world intend to assign to the facilitation of rhythm shifts required by transmeridian flight or by odd work routines a prominent place in a chronobiologic project now planned within the framework of the International Biologic Program.

Concerning Dr. Dejour's second question, it would be desirable indeed to telemeter the rhythms of an arctic tern while this animal flies from one Pole to the other. I wish I knew how long this flight takes on the average, and I plead guilty to not having studied classical papers such as those of Bertil Kullenberg

(Über Verbreitung und Wanderungen von vier Sterna-Arten, *Arkiv. Zool.* **38 A** (1946) no. 17 and Finn Salomonsen (Migratory movements of the arctic tern (Sterna Paradisaea Pontoppidan) in the Southern Ocean; Danske Videnskabernes Selskab, *Biol. Meddelelser* **24** (1967) 1.

If the flight from Pole to Pole takes weeks rather than a few days, there may be a slow adjustment to the local setting, as has been reported for the case of sea travel by man. In 1905, Gibson reported an apparent adjustment of the timing of the body temperature rhythm—day by day and coincident with the shifting of the routine—during sea travel from the United States to the Philippine Islands and back. This author, who studied two individuals, describes for both of them a "somewhat limited daily range of variations" in body temperature after arrival in Manila, the Philippines (R. B. Gibson: The effects of transposition of the daily routine on the rhythm of temperature variation; *Amer. J. Med. Sci.* **129** (1905) 1408). During a sea voyage from Melbourne, Australia, Osborne in 1908 found that throughout the journey the evening maxima of the body temperature rhythm followed local time (C. Osborne: Body temperature and periodicity; *J. Physiol.* **36** (1908) 34–41). Conceivably, the arctic terns adjust their circadian rhythms in the case of slow travel across latitudes and longitudes as do human beings. Finn Salomonsen's fig. 2 on page 7 of the above mentioned publication indicates that some terns, at least, follow a transmeridian path during part of their inter-polar journey as champion long-distance migrants, a circumstance kindly indicated to me by Mr. David Cline of the Museum of Natural History of the University of Minnesota.

For information on rhythmicity in arctic birds, two reviews by Donald S. Farner are quite pertinent: Role of extreme changes in photoperiod in the annual cycles of birds and insects, *Federation Proc.* **23** (1964) 1215–1220; and Circadian systems in the photoperiodic responses of vertebrates in *Circadian Clocks*, Proceedings of the Feldafing Summer School, Ed. J. Aschoff (North-Holland, 1965) p. 357–369. Further research may well be aimed at evaluating the time course of any circannual rhythm adaptation during 5 or 6 interpolar round-trips made by an arctic tern in consecutive years.

H. MARGENAU: I have a question or two, Sir. The first one is this. Is anything known about the physiological or chemical mechanism which carries these rhythms? I am particularly interested in those rhythms which one might call active, which originate within the organism itself and are not tied to the surroundings.

The second question which I might voice at the same time concerns the interesting paper by Dr. Lindauer. I would like very much to know whether there is any knowledge or whether there are any conjectures as to how bees become aware of a magnetic field?

M. LINDAUER: In a few years, so we hope, this question can be answered by experimental data. Theoretically a magnetic field can affect the organism (the sensory cells or the neurons):

1) by generation of electromotive force in moving conductors (a honey bee performing the waggle danse is a fairly fast moving conductor);
2) by the force exerted upon moving charge carriers (Hall-effect);
3) by force or torque exerted on para- and diamagnetic particles.

The voltages resulting from the effects can be very small only (in case $1 \approx 0,1$ $\mu V/cm$); however Lissmann and Machin have shown that "electric" fish can sense voltage gradients as low as $0,03$ $\mu V/cm$.

F. HALBERG: In the mammal, adrenocortical steroids represent the mechanism of certain circadian rhythms but not of others. Corticosteroids are, of course, entities that are chemically well defined, recognized physiologically as well and amenable to anatomical localization as to their site of production. Removal of the adrenal gland in man and mouse or Addisons disease—a condition associated with adrenocortical insufficiency—resulted in the obliteration of a circadian rhythm in blood eosinophil cells, at Michigan [1] in Utah [2] as well as at Minnesota [3]. Dr. Azerad, Dr. Reinberg and Dr. Ghata in France, over a decade ago, demonstrated that circadian rhythms in the urinary excretion of certain electrolytes also are obliterated as group phenomena in human adrenocortical insufficiency [4].

Aussi pour le cas de l'incorporation du radiophosphore dans les phospholipides hépatiques le cortex adrénalien semble être un mécanisme critique [5]. Nous avons été un peu troublés lorsque nous avons étudié les rythmes du marquage de l'DNA, il y a dix ans. Nous avons trouvé que les rythmes dans le métabolisme de l'acide désoxynucléique persistent après une adrénalectomie bilatérale. Alors il y a des différents mécanismes endocriniens et cellulaires pour différents rythme circadiens. En passant du mammifère au microorganisme, on signale un travail intéressant qui a été fait par Sweeney et Haxo sur l'acetabularia [6]. Dans ce cas on peut couper les "rhizopodes" de base et enlever ainsi le noyau. Dans ce travail de Sweeney et Haxo et autres, vous avez la continuation d'un rythme circadien photosynthétique, en l'absence des acides nucléiques formes et connus [7, 8]. Finalement, on peut démontrer avec les méthodes modernes de rythmométrie des rythmes mêmes pour *E. coli* [9].

[1]  Mechanisms of diurnal eosinophil rhythm in man, *J. Lab. Clin. Med.* **45** (1955) 247.
[2]  The diurnal variation of blood leucocytes in normal and adrenalectomized mice, *Endocrinology* **58** (1956) 365.
[3]  Diurnal rhythmic changes in blood eosinophil levels in health and in certain diseases, *J. Lancet* **71** (1951) 312; Eosinophil rhythm in mice: Range of occurrence; effects of illumination, feeding and adrenalectomy. *Am. J. Physiol.* **174** (1953) 313–315.
[4]  Disparition du rythme nycthéméral de la diurèse et de la kaliurie dans 8 cas d'insuffisance

surrénale, *Ann. Endocrinol. Paris* **18** (1957) 484–491; See also Rythmes des fonctions cortico-surrénaliennes et systèmes circadiens, in: *Symp. int. sur la Neuro-endocrinologie*, (L'Expansion Scientifique Française, Paris, 1966) 75.

[5]  F. Halberg, H. Vermund, E. Halberg and C. P. Barnum, Adrenal hormones and phospholipid metabolism in liverycytoplasm of adrenalectomized mice, *Endocrinology* **59** (1956) 364.

[6]  B. M. Sweeney and F. T. Haxo, *Science* **134** (1961) 1361.

[7]  E. Schweiger, H. G. Wallraff and H. G. Schweiger, Endogenous circadian rhythm in cytoplasm of acetabularia: Influence of the nucleus, *Science* **146** (1964) 658.

[8]  E. Schweiger, H. G. Wallraff and H. G. Schweiger, Über tagesperiodische Schwankungen der Sauerstoffbilanz kernhaltiger und kernloser *Acetabularia mediterranea. Z.* Naturforsch. **19** (1964) 499.

[9]  F. Halberg, and R. L. Conner, Circadian organization and microbiology: variance spectra and a periodogram on behavior of *Escherichia coli* growing in fluid culture, *Proc. Minn. Acad. Sci.* **29** (1961) 227.

K. MENDELSSOHN: There is of course a rotation of the polarization axis and the thing that is observed is a polarization. It would be a small effect, but biological indicators are sensitive.

H. C. LONGUET-HIGGINS: I have two observations to make. First, an observation on the proposal of Dr. Mendelssohn. He must be thinking of the Faraday effect, because he referred to the rotation of the plane of polarization of light by a magnetic field. This is all very well for radio waves passing through the ionosphere, where you have free electrons with a large mean free path. But for light the effect is quite negligible, and very difficult even to detect. One must then suppose that the earth's magnetic field produces a variation, in the plane of polarization of the light from the sky, of several degrees, because the error that the animals seem to make is of the order of 5 or 10 degrees. This is what your hypothesis seems to imply; but quite possibly I have misunderstood your original suggestion. That is my first point.

The second point is quite different; it concerns the speculations raised by the question which you brought up, Mr. Halberg. I have been trying to think of a reason why it is, apparently, easier to slow down a biological rhythm than to speed it up. These rhythms must arise from metabolic processes, and one may imagine that some of these processes take place with the maximum possible speed, if one thinks of the processes which have been described to us by Dr. Thomas of Denmark. If they do indeed go at maximum speed, it will be very difficult to hurry the rhythm of the overall cycle. One may take as an analogy a person playing a scale on the piano as fast as he can; it will be easy for him to slow down, but quite impossible to go faster.

F. HALBERG: There are examples of a different "polarity" of the circadian system. For instance, Dr. Aschoff found in certain birds that an advance of circadian rhythms occurred faster than a delay [1]. He obtained similar results even in a human being studied in his bunker for possible differences in shift time following advances and delays of schedule [2]. Nonetheless, in our hands a number of human subjects studied after intercontinental flights, and rats, undergo much faster delays of their circadian rhythms investigated thus far, as compared to advances. All I am trying to say is that too broad a generalization may be premature. Furthermore, a number of rhythms have been found to be accelerated in disease. Thus, in human cancer we encounter what might be interpreted as a clear acceleration revealed by variance spectra [3].

This "speeding up" comes to the fore if one takes serial biopsies every two hours from human cancers and analyzes these data thereafter by variance spectra, as was done for the research by Dr. Tähti and Dr. Voutilainen of Finland. From such analyses it becomes apparent that in a number of human cancers a circadian component may no longer be demonstrable. The so-called "ultradian" component predominates, indicating that indeed one can "accelerate" the rate of one cycle in 24 hours; one may do so of course by "napping" as well. Nonetheless, from studies carried out in the isolation of caves on mature healthy men, one might be tempted to agree with the suggestion that we are indeed running with maximal frequency, if we keep in mind that under such conditions sleep-wakefulness at least on occasion changes from a primarily circadian phenomenon to a prominently infradian one. Some such isolated subjects sleep once in 48 hours rather than once in 24 hours, but many more individuals will have to be studied for more variables and under additional conditions before the most interesting polarity commented upon by Dr. Longuet-Higgins can properly be discussed.

Nous sommes de tels poltrons que nous n'osons pas mettre de côté les données sûres que nous avons pour nous lancer un peu dans l'inconnu, mais je suis d'accord avec vous: il est très probable que les opinions de M. Longuet-Higgins sont une bonne hypothèse à l'état actuel des informations.

Par les travaux français de M. Michel Siffre avec Alain Reinberg et Jean Ghata et aussi par les études de John Mills en Grande-Bretagne, on peut démontrer, au moins pour l'état de veille et de sommeil—ce qui n'est pas un bon index pour le métabolisme que dans ces conditions on trouve dans quelques sujets une transposition, au moins partielle, de la variance de la region spectrale circadienne avec des cycles d'environ 24 heures ($\pm$ 4 h) à la région spectrale infradienne (notamment, des fréquences d'environ un cylce en 48 heures).

Ceci s'aligne très bien sur votre proposition: que la veille/sommeil sinon le métabolisme représente un oscillateur qui fonctionne à l'état maximum. Mais comme nous l'avons discuté en haut, pour le cas des cancers humains, on peut trouver un déplacement de fréquence dans la direction opposée, notamment une accélération très claire, révélée par des spectres de variance [3]. Donc, en jugeant sur le cancer, nous constatons que nous pouvons "accélérer", dans une notable proportion, par rapport au rythme d'un cycle en 24 heures.

Ceci nous ramène à des questions de base, qui ont à faire avec l'activité cellulaire comme aussi avec l'organisme global. On trouve donc à tous les niveaux de l'organisme une structure temporelle dont les composantes pourraient être dans une rélation des harmoniques—dans le sens physique plûtot que musical. Ainsi on peut élaborer sur les idées originales de M. Fessard en ce qui concerne la spontanéité des rythmes.

[1]  J. Aschoff, R. Wever, Resynchronisation der Tagesperiodik von Vögeln nach Phasensprung des Zeitgebers, *Z. Vergleich. Physiol.* **46** (1963) 321.
[2]  J. Aschoff, Adaptive cycles: their significance for defining environmental hazards, *Int. J. Biometeorology* **11** (1967) 255.
[3]  M. Garcia Sainz and F. Halberg, Mitotic rhythms in human cancer, reevaluated by electronic computer programs—evidence for temporal pathology, *J. Nat. Cancer Inst.* **37** (1966) 279.

L. Tisza: I would like to ask Dr. Lindauer how long does it take for bees transferred from the Northern hemisphere to the Southern hemisphere to readjust their north sense orientation? Do you know something about it?

M. Lindauer: After the translocation experiment we have tested the bees every second week. Only after the 40th day they had readjusted their orientation to the new situation. The result is different however if we let hatch bees in an incubator and then raise them without sun (the bee colony is put in a cellarroom by artificial illumination). After 4 weeks I took the bees out of the cellar into the field. They were unable to use the sun as compass, they used it on the first 3 days just as it would be a fixed light point on the sky. After the 5th day however (the bees had absolved 500 collecting flights for a goal in 200 m distance) all collectors had *learned*, how fast and in what direction the sun moves across the sky. They had changed from the simple "angle orientation" to the true "compass orientation".

K. Mendelssohn: Since we now seem to have a little time left, allow me to reply shortly to Dr. Longuet-Higgins. I have no wish to be dogmatic about my suggestion that the magnetic rotation of the axis of polarization is the operative mechanism occurring in the bee. It only occurred to me that it may be a possible explanation. Of course, the effect is a small one but the relationship involved is not necessarily a linear one. There are cases where a small external field can, in a suitable substance, affect the direction of a very much larger internal field. This means that the field registered in the sensing mechanism could be of the order of a kilogauss rather than one gauss.

F. Halberg: Si j'ai bien compris votre question, M. Lindauer (nous en avons parlé en aparté): vous avez un renforcement temporel de la mise en condition. Donc, le comportement dépend de la structure temps comme étant un certain nombre des programmes et engrammes de M. Fessard. Pendant très longtemps, la structure temps a été considerée comme étant imprimée par l'extérieur et persistant de l'intérieur. Les études et les données dont on dispose maintenant, au niveau cellulaire, à celui des processus inter-cellulaires, ou même au niveau du comportement de l'organisme global, nous permettraient de supposer que certaines "fréquences" existent dans l'organisme dès la naissance.

Tout comme l'enfant humain, sur auto-demande, aura un déplacement ultradien vers circadien en fonction de son âge pour l'alternance veille-sommeil, les abeilles ont peut-être également un rythme circadien. Il sera bien important d'étudier les abeilles en libre cours. Peut-être est-il possible de mesurer le bruit dans une ruche comme une fonction de temps? Cette mesure parait simple, mais peut-être est elle compliquée par la possibilité qu'un nombre indéterminé abeilles n'est pas synchronisé avec la population entière étudiée. Naturellement il serait bien préférable d'étudier une fonction de l'abeille elle-même, laquelle peut être évaluée par une analyse longitudinale, individuelle. Ceci permettrait de savoir si une mise en condition quelconque—pour la menthe, dont vous avez parlé—est un réflexe conditionné, qui est superposé à une formation temporelle circadienne de base de l'abeille.

Tout ceci se rapproche de la théorie de M. Fessard: il y a un programme, le programme s'écoule. A tel ou tel moment, l'organisme est exposé à une stimulation; il commence à apprendre et ce qu'il va apprendre dépend aussi de sa phase au moment de la stimulation. Puis, ce qu'on a appris doit être reproduit; il faut consulter cette information avant d'agir—comme déterminant d'une action orientée de l'organisme. Les tâches de tenir en réserve ("storage") et de reproduire ("retrieval") l'information sont probablement, au moins en partie, une fonction du système circadien.

The interesting studies described by Dr. Lindauer reemphasize a discussion of long standing as to the extent to which rhythms relate to learning and vice versa (the extent to which learning depends upon an organism's time structure). Some features of a rhythm's synchronization have indeed been compared to conditioned reflexes. Thus Maizelis [1] described as a "cause" of spontaneous changes in motor activity the formation of positive condition reflexes to the time and environment in which muscular work was performed. Few will question the suggestion that conditioning contributes to our time structure, consisting of a number of programs and engrams as discussed by Dr. Fessard. Our "memory traces" when they are being registered also may well be timecoded. This circumstance should not lead one to presume that conditioning usually brings about the rhythm at the outset; instead, the rhythm may well be innate and a determinant of conditioning. Several lines of evidence demonstrate indeed that conditioning can be determined by the stage of a circadian rhythm in which the conditioning procedure is applied—the work of Charles Stroebel at the Institute of Living in New Haven, Connecticut being a case in point [2]. Dr. E. Bünning in Tübingen had once reported that a disturbance at a certain stage of the rhythm, I believe in leaf movement, might reappear in a fashion similar to a "memory trace" on subsequent days with a predictable timing.

To turn back to Dr. Lindauer's most interesting observation, the time may be ripe to study a relatively easily measured yet pertinent variable, the noise of a bee hive or, to avoid confounding from at least a partial lack of inter-bee synchronization, preferably some function that can be measured longitudinally on individual bees, under as constant conditions as possible. If, then, from such work a period desynchronized from both a 24-h solar and a 24.8-h lunar day could be found and the experiment with peppermint and honey described by Dr. Lindauer could then be repeated, one might have at least some tentative information toward the question whether the conditioning described by him represents simply a conditioned reflex or some kind of reinforcement of training, or whether there is indeed a "program" preexisting for the timed conditioning of such stimuli within the organism as suggested at this meeting by Dr. Fessard.

[1] M. R. Maizellis, Time and Conditions of Performance of Muscular Work as Factors of Organization of Diurnal Periodicity, *Bull. Exp. Biol. Med.* **45** (1958) 526.
[2] C. Stroebel, Behavorial aspects of circadian rhythms, *Comp. Psychopathology* (1967) 158.

P. O. LÖWDIN: There is one rythmic phenomenon which I think is very intriging. It's the phenomenon of sleep. Do you care to comment?

F. HALBERG: One of the interesting features of sleep is its multiple frequency structure. At the moment a good deal of work revolves around what is called „fast" sleep or REM (rapid eye movement) sleep. The seven or eight hours of behavioral sleep (diagnosed on the basis of relative quiescence with our eyes closed) are modulated by rhythms of say $\sim 1.7$ hours in a number of functions and perhaps in association with dreaming. It seems pertinent that in this day and age, molecular biology has already aimed at elucidating relations between dreams and molecules, at a symposium held recently at the Massachusetts Institute of Technology. Ultradian frequencies of human sleep become behaviorally overt by day as well as by night in patients with narcolepsy, a problem so rigourously studied by Pierre Passouant of Montpellier.

At the other extreme of the frequency components of spontaneous sleep one finds in some cases of human isolation the infradian component with one cycle in about 48 hours. What seems to be most important with respect to Dr. Löwdin's question may well be the recognition that whereas in human isolation sleep may change from a circadian to an infradian frequency, the adrenal cortical cycle may maintain a primary circadian rhythm. Thus the contention by many classical physiologists that most, if not all, bodily changes along the 24-hour scale are determined by sleep can be ruled out by results from studies by Michel Siffre *et al.* covering several months and allowing the "self-selection" of a different frequency for sleep than for the adrenal cortical cycle [1].

This is not to say, however, that the time relations between these two functions are random. Temporal integration indeed can be achieved on more than one frequency, and such integration is particularly favored by the circumstance that the frequency of sleep-wakefulness "demultiplies" to one-half that of the adrenal cycle—as noted by us in the study on a human subject isolated for several months.

Pendant le sommeil l'enregistrement des mouvements de l'œil, l'électromyogramme, le pouls et la respiration complémentent les études électroencéphalographique du sommeil rapide et il y a déjà des méthodes puissantes pour analyser de telles données si on fait l'enregistrement directement sur bande magnétique [2, 3]. Il sera très intéressant d'enregistrer chez les sujets ambulatoires et sains quelques uns de ces variables pour voir les équivalents d'un composant ultradien pendant la veille aussi bien que pendant le sommeil.

[1] M. Siffre, A. Reinberg, F. Halberg, J. Ghata, G. Perdriel and R. Slind, L'isolement souterrain prolongé, Etude de deux sujets adultes sains avant, pendant et après cet isolement, *Presse Med.* **74** (1966) 915.
[2] D. F. Kripke, C. Clark and J. A. Merrit, A system for automated sleep analyses and physiological data reduction, Document ARL–TR–68–12, August, 1968.
[3] N. Cartwright, D. F. Kripke and P. Cook, Statistical reduction of handstaged sleep analysis. Document ARL–TR–68–5, June, 1968.

G. CARERI: If there exists a periodicity in sleep, then suppose an organism take a sleep out the periodicity. Is there any phenomenon you could detect in this siesta which is different from the phenomenon observing in sleep, because you break the periodicity?

F. HALBERG: Rather than necessarily "breaking" a circadian periodicity by taking a "siesta" one may simply accentuate and/or prolong a physiologic phase of a presumably innate ultradian rhythm (with a frequency much higher than circadian). The ultradian rhythm can be gauged by the telemetered intraperitoneal temperature, among other functions.

Si vous chronométrez la température intra-péritonéale par un détecteur physiologique placé dans l'abdomen d'un rat, et même si l'animal se trouve dans une ambiance sous contrôle aussi absolu que possible (parce que la température d'ambiante est controlée ($\pm$ .5 °C) et les bruits, dans la mesure du possible—on ne peut pas toujours faire les choses parfaitement!) vous trouvez des changements en température corporelle qui couvrent en quelques heures une différence aussi grande que 2 °C que nous ne pouvons absolument pas expliquer dans le sens d'une "réponse" à des facteurs connus. Ce sont des changements ultradiens.

Voici donc un rythme circadien avec une modulation par l'élément ultradien. Et vous arrivez à ceci: la sieste est peut être simplement l'expression du rythme ultradien "approfondi". On trouve des rythmes, ou pararythmes, ultradiens avec des fréquences peu définies qu'on peut démontrer par des spectres de variances aussi dans la concentration du sang en hormones corticosteroïdes, même dans l'effluent de la surrénale cannulée [1]! Je ne sais donc pas ce que la sieste fait à notre physiologie, mais nous faisons beaucoup plus de siestes que vous ne le pensez.

Concerning uncertainties in experimentation, one would have to do so from both theoretical and practical viewpoints. Dr. Fessard's thoughts on the role played by a transducer between stimulus and response included reference to an observer-effect upon the phenomenon being observed, i.e., reference to the equivalent of a biologic uncertainty relation to be considered in the context of Heisenberg's thoughts [2]. In rhythmometry there are in addition many practical points to be considered, including the effect of the interval between consecutive samples upon data analysis on the one hand and upon the subject himself on the other hand. In this connection it is indeed a considerable step forward to dispose of electronic computer programs allowing the analysis of data obtained at unequal intervals whereby certain kinds of interval artifacts are prevented and, what is no less important, the human subject is allowed undisturbed sleep [3]. The only condition for such analyses at unequal intervals is that the density of the data during most of the span be not too drastically different. Of course, for certain tasks other than performance tests, continuous physiologic monitoring—eventually from birth to death—may well be feasible by transducers that, thanks to NASA, already are within the "state of the art".

By achieving such goals, we could obtain for any physical examination at any time of our choice information on the preceding 60 or 80 (or more) circadian cycles in a given monitored variable of interest—body temperature, heart rate or other. Such a desideratum is no more than what we require for the evaluation of high frequency rhythms in current medical practice. More specifically we evaluate heart rates on the basis of 60 or more cardiac cycles. Thus, it seems only fair to require more than a single sample spotcheck and, in some cases, data covering more than a single cardiac cycle, for an assessment of a circadian modulation of the heart rate. However because such monitoring of heart rates and of other pertinent functions such as blood pressure is currently still expensive the student of rhythms will have to demonstrate that such information on certain rhythms with medial or low frequency is worth collecting.

[1] J. H. Galicich, E. Haus, F. Halberg and L. A. French, Variance spectra of corticosteroid in adrenal venous effluent of anesthetized dogs, *Ann. N.Y. Acad. Sci.* **117** (1964) 281.
[2] F. Halberg, Chapter on "Medizin" in Jahrbuch der Internationalen Hochschulwochen des Oesterreichischen College, (Igonta Verlag, Salzburg, 1946) 336–351.
[3] F. Halberg, M. Engeli, C. Hamburger and D. Hillman, Spectral resolution of low-frequency, small-amplitude rhythms in excreted ketosteroid; probable androgen-induced circaseptan desynchronization, *Acta Endocrinologica Suppl.* **103** (1965).

B. B. LLOYD: I have a friend who does experiments on himself from time to time. He was measuring his body temperature day and night about twenty years ago. In the course of one day he was sitting by the gas fire, a primitive sort of heating still used in England, at 6 o'clock in the evening until he sweated profusely. The next evening at 6 o'clock, not sitting by the gas fire, he found that his body temperature went down. I would regard this as an example of a memory and of an anticipation in a homeostasis.

One more point. I think you do see Cheyne-Stokes breathing in a dog if you lengthen the path between the lungs and the carotid body, and fiddle the gaseous atmosphere—a feedback loop showing oscillation rather than being critically damped.

F. HALBERG: Dr. Lloyd considers a feed-back loop in relation to Cheyne-Stokes breathing—indeed in one of the several interesting rhythms of the respiratory system. In discussing such feed-back models for high frequency rhythms, one does not as a rule encounter the dangers and limitations so characteristic of feed-back considerations in the domain of rhythms with medial and low frequencies. The following remarks will be restricted to physiologic phenomena in the latter domain—adrenal physiology, cortical as well as medullary, being a case in point. Much work is being done, in the adrenocortical field in particular, without identifying the stage in which a given "feed-back study" is carried out—despite the demonstration of, for instance, a reproducible circadian rhythm with a large amplitude in adrenocortical reactivity to ACTH, *in vitro* as well as *in vivo*. Also ignored is the added feature that *in vivo*, in the C-mouse, adrenal reactivity to relatively unspecific stimuli such as saline also is circadian rhythmic and that furthermore such rhythms in the reactivity of the adrenal cortex to saline solution on the one hand and to ACTH on the other hand are out of phase with each other. Equally pertinent rhythms in pituitary ACTH content and in the corticotropin releasing factor, CRF, of the hypothalamus also are usually ignored, with the mistaken tacit assumption that whatever one does in terms of a "feed-back study" at one time will be reproduced at any other time. Today, successes in electronics have attracted biologists to an often uncritical transfer of comments out of context, i.e., to an application of control system theory to "biologicals", just as if they were fully inert materials. Thus we repeat the mistakes of a few centuries ago. For instance, so great were Newton's successes and his consequent influence that so distinguished a physiologist as Borelli was prompted to a mechanistic thinking that apparently led him to explain all of digestion by mechanical friction.

The foregoing comments are directed not against a consideration of control mechanisms in physiologic work but rather against the prevailing custom of ignoring rhythms while one does so, whereby mistaken conclusions may be drawn concerning controls. Feed-back considerations then are a useful "scaffold" for physiologists and should prompt him to evaluate rhythms more rigorously and quantitatively and to search for the underlying factors, rather than misusing them as legitimate excuses for studying rhythmic variables without any consideration for time structure.

L. ROSENFELD: I am not sure that I understand your standpoint. It seems to me that scaffolding, as you call it, is not an obstacle to building a house, but on the contrary helps to do it.

F. HALBERG: We need a scaffolding. However, we must not mistake the scaffolding for the house and move into or onto it. For instance, homeostasis has been a scaffolding for some, and unqualified feedbacks or "stress reactions" were a scaffolding for others. I have used feedback models [1] myself but only as a transition to the specification of physiologic phenomena as they relate to chemical compounds in anatomical locations. A simple black box approach ignoring all anatomical, physiological and biochemical rhythms has unfortunately led to the practice of assessing presumed stress reactions or feedbacks at some single time point convenient only to the experimenter. This approach ignores results such as those indicating a rhythmic change as drastic as the difference between death and survival occurring in response to the identical agent and as a function solely of timing [2]. Such results on the hours of changing resistance dramatize the need to, first and foremost, control the stage of a rhythm in work on an organism's frequency structure and, second, use endpoints from rhythmometry as gauges of physiologic "responses".

Pertinent in this connection are not only studies on rodents [2, 3] but also the important French studies of Alain Reinberg on men [4].

[1]  F. Halberg, E. Halberg, C. P. Barnum and J. J. Bittner, Physiologic 24-hour periodicity in human beings and mice, the lighting regimen and daily routine, in *Photoperiodism and Related Phenomena in Plants and Animals*, Ed. Robert B. Withrow, Ed. Publ. No. 55 of Amer. Assoc. Adv. Sci. Washington (1959) 803–878.
[2]  F. Halberg, Organisms as circadian systems; temporal analysis of their physiologic and pathologic responses, including injury and death, in: Walter Reed Army Institute of Research Symposium, Medical Aspects of Stress in the Military Climate, April, 1964, 1–36.
[3]  L. E. Scheving, Circadian rhythms in susceptibility of rodents to nicotine and amphetamine, AAAS, Washington, Resume in *J. Amer. Med. Assn.* **199** (1967) 33.
[4]  A. Reinberg, The hours of changing responsiveness and susceptibility. Pers. Biol. Med. **11** (1967) 111.

H. C. LONGUET-HIGGINS: I cannot see, speaking as an outsider, that there is anything inconsistent between thinking that it is good to investigate the cyclic phenomena in the cell, or in the body, or in the organism, and thinking that it is also good to investigate the non-cyclic processes. Living things are very complicated; one must pay attention to many different things. To stress the value of your own studies is not necessarily to imply that other sorts of studies are not also valuable.

F. HALBERG: Dr. Longuet-Higgins emphasizes of course that there are many approaches to any one biologic problem. Let me hasten to agree and apologize if I gave the impression that rhythmometry is a panacea for any and all problems. Nonetheless, it is difficult to accept "homeostatic" work on rhythmic variables when, as is customary, the identification of the stage of a rhythm in the variable used for sampling is altogether ignored. Knowledge of the stage of the circadian rhythm may be as desirable for a variable such as corticosteroid as is the statement on whether a blood pressure measurement is systolic or diastolic. Our interpretation of a blood pressure determination of, say, 100 mm Hg will be drastically different as a function of whether it is diastolic or systolic; accordingly one may distinguish a hypertensive patient from a hypotensive subject. Quite clearly the Reverend Stephen Hale must have been aware of the difference between diastole and systole when he measured blood pressure directly by attaching a glass tube to the artery of a horse and observing how a column of blood rose and fell with the heartbeat. According to Dr. Leonard Wilson, Borelli had already discussed the change in pressure occurring with each contraction of the heart and had reported that with the systole the aorta stretched and subsequently relaxed. Circadian changes in serum corticosteroid or liver glycogen are no less drastic. Furthermore, the research work on yeast by Kendall Pye and Britton Chance, who regretfully could not be present, shows "ultra-ultradian" trains of oscillations with relatively very little damping once trehalose, a carbohydrate, is added to the medium. [1]

Thus, many aspects of metabolism are rhythmic and yet biochemical journals continue to publish papers by molecular biologists, endocrinologists and others, who ignore any and all periodicity in variables previously shown to undergo large amplitude rhythms amenable to rigorous standardization (e.g., murine liver glycogen). However to remedy this situation is hardly the major task for students of rhythm. Many of us are lead to study a given rhythm because we encounter it as a source of variation that must be controlled, yet in my opinion the primary tasks of the chronobiologist in statu nascendi lie on an entirely different plane.

We can raise new questions as to body function in health and disease—whether our interest relates to 1) transmeridian dyschronism after an intercontinental flight, to 2) the hygiene of shift workers, to 3) the failure of rhythmic integration in a psychotic patient, or to 4) the rhythm alteration found in a cancer patient. Such rhythm alterations could constitute determinants of a performance decrement, a resistance deficit or of an actual disease process. If indeed rhythms should prove to play an important role in such processes, sound biophysical theory for the ubiquitous and nontrivial spectral structure of organisms will become yet more desirable. Meetings such as the present one may contribute

to better human performance and health by fostering such theoretical developments—that can be anticipated from the interaction between physicists and
biologists and that can hardly fail to have eventual applied value.

[1]   K. Pye and B. Chance: Sustained sinusoidal oscillations of reduced pyridine nucleotide
      in a cell-free extract of *Saccharomyces carlsbergensis*, *Proc. Nat. Acad. Sci.* **55** (1966) 888.

A. REINBERG: Dans son analyse, portant sur les phénomènes rythmiques des
nerfs et des muscles, le M. Fessard a dit que l'activité rythmique apparaît comme
une des propriétés fondamentales de la matière vivante. Nos connaissances
actuelles permettent de généraliser cette notion à tous les systèmes vivants. Il
apparaît en effet que des processus rythmiques peuvent être mis en évidence à
tous les niveaux d'organisation:

1) des organismes uni-cellulaires jusqu'à l'homme y compris,

2) à tous les niveaux d'organisation chez un être vivant, l'homme par exemple
ou les mammifères supérieurs. Qu'il s'agisse de fonctions globales physiologiques
ou biologiques, de groupe d'organes, de tissus, de cellules et même de fonctions
sub-cellulaires, l'existence de rythmes a été prouvée objectivement. Par exemple:
rythmes de l'incorporation du $P^{32}$ dans le D.N.A. et le R.N.A. des cellules
de divers tissus.

3) l'analyse objective de ces rythmes, réalisée suivant des processus de calculs
électroniques mis au point par M. Halberg et ses collaborateurs de l'Université
du Minnesota, permet de donner une estimation des différents paramètres et de
caractériser ces variations périodiques. Il s'agit en particulier de la période ou de
la fréquence des phénomènes, de leur amplitude et de leur relation de phase.
C'est ainsi que, pour une même fonction physiologique (le rythme cardiaque,
l'excrétion urinaire des 17 cétostéroïdes, la température par exemple), des analyses
spectrales ont pu être réalisées permettant d'estimer les différentes périodes
suivant lesquelles chacune de ces fonctions peut se manifester. L'étude des rythmes
biologiques introduit donc une nouvelle dimension dans l'analyse des phénomènes
vivants: le temps. Les résultats acquis permettent déjà de discuter les phénomènes
sous leur aspect chrono-biologique et chrono-physiologique. Ces recherches
ouvrent de nouveaux chapitres qui sont: la chronopharmacologie, la chronopathologie et la chronothérapeutique.

# DISCUSSIONS GENERALE ET CONCLUSIONS

*3ème séance*

## PRÉSIDENT A. LICHNEROWICZ

A. LICHNEROWICZ: Nous sommes au commencement d'une séance importante, parce qu'elle est la dernière. Je voudrais vous proposer de la partager en deux parties. Nous avons fait le cycle, la boucle si je puis dire. Des considérations théoriques générales nous ont amenés, à travers des phénomènes biologiques extrêmement intéressants pour nous tous, à de belles séances, comme celle de ce matin.

Un certain nombre de nos collègues, physiciens ou théoriciens, soit par les hasards de la discussion, soit par une trop grande discrétion naturelle, n'ont pas pris suffisamment la parole. J'aimerais donc que, dans une première partie, nous bouclions la boucle avec une espèce de retour théorique, en donnant la parole à un certain nombre d'entre vous, de manière aussi libre qu'ils le veulent.

Je souhaite entendre tous nos collègues, plus particulièrement le Pr. Onsager, que nous avons l'honneur d'avoir parmi nous.

*Theoretical Physics and Biology,* © 1969, *North-Holland Publ. Co., Amsterdam*

## GENERAL DISCUSSIONS AND CONCLUSIONS

L. ONSAGER: At a conference like this, few if any questions are answered. Rather, when biologists and physicists with varied experience in different fields come together, each can learn the lore and the problems of the others.

The clarification of pertinent theories and principles will help the biologists to organize and interpret their physical observations. Conversely, while we have cause to believe that biological systems obey the general laws of physics, they do present such unique situations that the proper adaptation of physical theories becomes a formidable task. This week the biologists have alone much to help us define the problems involved, and we shall need more such help in the future.

K. MENDELSSOHN: Listening to the biologists at this conference I was, like Dr. Onsager, most impressed by the progress which has been made in their field, particularly when we look at the almost fantastic success of the molecular biologists. One therefore wonders, as a physicist, whether and where we can make a useful contribution. We must remember that, unlike physics, biology is operating with concepts which are purposely designed to fit the study of the properties of life.

Probably what the physicists need is the formulation of one or more new concepts to deal with a subject which is so new to them. This will be a major task for the completion of which one cannot hope in the framework of a conference but perhaps – and this may be similar to what Dr. Onsager hopes – we may learn to ask the right sort of question. This in itself is extremely difficult.

What, for instance, is in the terms used by the physicist the salient feature of life? In private discussion with one of the eminent biologists at this conference I was told that: "Life has neither purpose nor meaning". This seems fair enough, except that physicists occasionally have the habit of referring to "meaningful" explanations and while they are unlikely to talk about the purpose of a physical system, they are forever interested in its future state. In early thermodynamics a forecast of the future state of the system was often arrived at by noting its "tendency" at the time of observation. Say, you have two containers connected by a tap, and one should be empty while the other is filled with air. As the tap is opened, the air flows into the empty container until the pressure in both containers is equal. This "tendency" was noted and led to the second law of thermodynamics well before the kinetic explanation of the phenomenon was provided.

As a physicist one may equally ask: "What is the tendency of life?" One of the answers is clearly that it tends to survive. It does so at very considerable odds against the random statistics usually employed by physicists or, as Thornton Wilder says in the title of his play, we live "by the skin of our teeth". How life manages to survive is an intriguing question about which the biologists know a lot but, I suspect, that this is not one of the primary questions on tendencies.

Another clear tendency which we observe is that life develops. The biologists have amassed a great deal of information about the fact that life develops under the environmental constraint, part of which is, of course, provided by life itself. However, in the process of development the structures of life become increasingly complex. It is by no means clear whether this tendency towards complexity is a feature which is forced on life or whether it must be treated as a separate basic phenomenon. Possibly, in order to survive against each other, the various forms of life have to become more complex and then this tendency towards complexity would be a necessary corollary of survival, requiring no separate explanation.

Physicists, dealing with a world of random processes connect complexity by statistical treatment to improbability. According to this, life is extremely improbable. Thermodynamically speaking, a state of low probability is one of low entropy. However, this statement by itself is not very helpful since it seems unlikely to lead to a meaningful description of life. The statement of complexity may be insufficient if it turns out that only a certain type of complexity is capable of showing the phenomena of life. Earlier in this conference Dr. Bergmann used the term of "specificity" when discussing living structure and we must ask ourselves whether there exists a specificity principle without which life cannot be described.

Possibly, here the right question to ask might run as follows: "Can we arrive by random processes at many stable structures which each are as complex as the simplest structures of life?" If this were possible—and we don't know whether it is—then more is required than the high degree of complexity which we can possibly postulate without further basic assumptions beyond the tendency to survive. In this case we would have to specify a certain type of complexity which is distinguished from other equally complex structures by the fact that it confers the property of life on only one of them. This is a question which, even at this stage, can perhaps be decided.

The great complexity of life in itself should not deter us—irrespective of whether or not a specificity principle is required. One somehow would suspect that complexity, too, and even specific complexity might be described and handled by simple methods. As Plato already suspected, God is a mathematician, a fact that has now been borne out by two millenia of observation. However, one can probably go further and suspect that God is a simple mathematician in his grand design and that it should be capable of simple solution. Somehow it seems that he tends to leave the complex solutions to his research students, the archangels, but that these complex solutions are not needed for comprehending the basic principles of the creation.

Physics is full of examples of this kind. Going back to the behaviour of a gas, the motion of its individual molecules, changing direction and speed after each collision, presents an aspect of formidable complexity. The solution of the Newtonian equations of motion for even only 10 or 20 molecules becomes quite hopeless. Nevertheless, a simple solution exists which has become the stock in trade of any first year student. All that was required, was to look at averages rather than at individual events and borrowing the mathematical methods developed for the problems of state craft.

We therefore have no reason to assume that because of their complexity the phenomena of life must resist solution by simple methods. That so far we have failed to find the right method is no argument to the contrary. In fact, we have hardly tried. Admittedly the assumption that the basic features of the physical world are revealed to our mind in a form which we regard as "simple" is an article of faith which can never be proved. On the other hand, all our scientific endeavours are, in any case, based on another article of faith, i.e. that of an integral creation. I am sure that somewhere the correct approach, leading to a simple treatment of the problems of life, lies already hidden in our minds. To find it must be our main task.


M. MATTHIAS: Just one very short remark to my friend Mendelssohn. We had that discussion three years ago in Colgate when I tried to tell him that in order to understand, we have to believe because the other way round, it's a hopeless affair. In my opinion, today, if a theoretical physicist talks to a biologist what he really needs is an interpreter. The interpreter should be the experimental physicist... I will really refrain from saying anything more but one thing.

I was baffled at the discussion about memory and computers and their connections, and the classic assumption that this must go one way or the other as a flip-flop mechanism like in a computer. If you consider that most of us cannot function with the speed of life, it's baffling to assume that a memory or anything that we perceive should go like an electric computer with the speed of light. Completely apart from the fact that the access to a computer of that kind would be experimentally a very difficult problem, I was amazed that nobody seems to object to the fact that it may be unrealistic to use these entirely physico-technical gadgets to treat something as complicated as a biological problem.


L. ROSENFELD: I am tempted to contradict the last objection, simply because it is very difficult to speak in general terms without defining exactly what one means by such things as "complexity", for instance. One of the aims of science at any rate is just to make it possible for us to cope with complex situations by creating a sort of shorthand symbolism which helps us to characterize them. In investigating biological phenomena, there is a challenge for us to find such simple schemes which may help us to orient ourselves in their complexity, and, to begin with, to classify the various orders of organization that we encounter in biology.

So I am more optimistic than you.

# REMARK ON THE PARAMETRIZATION
# OF THE STATES OF COMPLEY SYSTEMS

IRVING SEGAL

*M.I.T., Boston, Massachusetts, U.S.A.*

## 1. Introductory

While the mathematical approach appears to have a great deal to offer towards the solution of some of the problems discussed in this conference, both technically and in spirit, there are visible few places in the interface between theoretical physics and biology where the more sophisticated, contemporary type of mathematics appears effectively applicable at this time. Certainly non-linear partial differential equations should be applicable to transport and other relevant phenomena, but specific results seem unlikely to become accessible until some simple empirical laws have been attained concerning the fundamental interactions. Digital computer models are also qualitatively quite appealing, but suffer from similar limitations.

These limitations may be illustrated by a simple qualitative model for a phenomenon on which Professor Fröhlich remarked extensively, the self-limiting nature of the growth of a cell or other organism. Let us suppose that the state of the organism at a given time $t$ can be represented by the Cauchy data for a second-order hyperbolic equation, admitting a positive definite temporal invariant; if at an early time the solution of the equation is small while its first temporal derivative is large, then at a late time, the solution is limited in size by the conservation law and can attain its maximum only when its temporal derivative vanishes. These considerations are exemplified by the equations of the form

$$(*) \qquad \Box \, \emptyset = m^2 \emptyset + g \emptyset^p \qquad (g > 0, \ p = \text{odd integer}),$$

in which case it has even been mathematically established that there is appropriate limiting behavior for the solutions as $t \to \pm \infty$, for general choices of $m$ and $p$ corresponding to a growth to maturity. But there is no empirical basis for any asymptotic equations parallel to the asymptotic equation

$$\Box \, \emptyset = m^2 \emptyset$$

in the case of equation (*), still less for the crucial interaction term $g \emptyset^p$, in any biological situation known to us.

The fundamental difficulty in the mathematical treatment of complex systems such as biological ones is that there is no simple parametrization, a priori or otherwise, for their state spaces; and the experience with biological systems in recent decades has not been notably encouraging in this respect. Naturally, one can hardly begin to treat the foundations of the kinematics or dynamics of systems whose phenomenology is theoretically so complicated. My feeling is however that there are mathematical possibilities in the direction of a simplified phenome-nology which have been overlooked, perhaps because of their relative sophistication, and are worthy of investigation.

## 2. Phenomenological generalities

Proceeding conservatively, — as well as in line with interesting models which have been proposed, — we consider a quantum-mechanical biological system. Advances in theoretical phenomenology make it easy to cover all possibilities, including those based on classical mechanics, by the definition of a "system" as an abstract operator (Jordan) algebra, whose (hermitian) elements represent the observables of the system; states of the system are then determined by the expectation values $E(X)$ of the observables $X$ in the state $E$ in question, and it is postulated that $E$ is a positive normalized linear functional on the algebra $\mathscr{A}$ of all observables; this means that

$$E(aA+bB)=aE(A)+bE(B), \quad E(1)=1, \quad E(A^2) \geqslant 0$$

for any observables $A$ and $B$, and real numbers $a$ and $b$. The state space of the system is then represented by a convex set $\Sigma$ consisting of linear functionals on $\mathscr{A}$. The kinematics and dynamics of the system are given by groups or families of automorphisms of the algebra, in terms of which the concepts of stationary, equilibrium, and ground states may be defined and treated, in appropriate cases.

The problem of approximating to $\mathscr{A}$ or $\Sigma$ by systems of substantially lower dimension is amenable to a number of general considerations. One could give a precise mathematical definition for the best approximation $\Sigma'$ of a given dimen-sion to the state space $\Sigma$, based on considerations of minimization of deviations in expected values of observables and maximization of the degree of temporal invariance of $\Sigma'$. In practice, tactical considerations modify such strategic ones, which are nevertheless useful for avoiding or making explicit preconceived notions concerning the state space, and for suggesting alternatives to the conventional models. These matters may be illustrated by the problem of the parametrization of

## 3. Molecular states

The Schrödinger wave function provides an excellent description, so far as is known, of the state of a biologically interesting molecule, and its usefulness is too well known to require any elaboration. On the other hand, even for moderately simple molecules, the wave function is an enormously complex object, whose mathematical determination within a preassigned accuracy is just beginning to become practical. In addition, one commonly does not require all the information contained in the wave function, but is concerned with a number of summary (or "collective") aspects. Especially in the case of larger molecules, some simple mathematical alternative to the Schrödinger wave function seems highly desirable.

Since the Schrödinger wave function is in any event an approximation to the theoretical state of a molecule, one need feel no great hesitation in exploring qualitatively rather different lines. While there seems presently no indication of a significant quantitative deviation of classical non-relativistic quantum mechanics from observations in the area of molecular physics which appears relevant to biological phenomena, there is no certainty that relativistic and quantum-field-theoretic effects may not later be found to play a rôle. The phenomenon of biological amplification over long periods of time, and the delicacy of the control mechanisms already discovered for insects, birds, etc., indicate that this possibility should be born in mind. There is a possible parallel with the introduction of the spin into non-relativistic quantum mechanics; although of marginal importance energetically, it was of great importance for the understanding of selection rules and the classification of the observed states. On the other hand, for a purely classical type of analysis the Schrödinger wave function is likewise not directly applicable.

Consider therefore the more directly phenomenological generating function for the state of the molecule, defined as follows. Let $P_1, Q_1; P_2, Q_2; \ldots; P_n, Q_n$ denote the momentum and position vectors for the $n$ particles of which the molecule is constituted; then the expectation value of

$$T = \exp \left\{ i \sum_{j=1}^{n} (u_j \cdot P_j + v_j \cdot Q_j) \right\},$$

as a function $F(u_1, \ldots, u_n; v_1, \ldots, v_n)$ of the arbitrary fixed vectors in question is this generating function, which has been considered in quantum mechanics by Wigner, Moyal, and many others. For a quantum-mechanical system in the state given by the Schrödinger wave function $\Psi$, the generating function $F = \langle \Psi, T\Psi \rangle$; it completely determines $\Psi$, within an irrelevant constant phase factor; it has the advantage of being closer to the physics than the Schrödinger function, and of being applicable to pure and mixed states, and to classical and quantum-mechanical systems, alike; it is however no easier to determine than the Schrödinger function and it has the disadvantage, that it is not easy to determine when a given function is the generating function of a state, and for this reason is not readily approximated by simpler generating functions.

A convenient mode of approximation may however be based on the following feature of generating functions. For simplicity, let us write simply $z$ for the $6n$-vector $u_1, u_2, \ldots, v_1, v_2, \ldots$. Then $z$ is an element of a linear vector space, and the notion of a "positive definite" function of $z$ is well-defined, for example as the Fourier transform of a positive mass distribution. We may now state the

*Theorem: If $F(z)$ is the generating function of a state, and if $G(z)$ is any continuous positive definite function, then $F(z)G(z)$ is again the generating function of a state.*

Now a particular type of positive definite function, as is well known, is one of the form

$$G(z) = e^{if(z)-Q(z)},$$

where $f(z)$ is a linear function of $z$ and $Q(z)$ is a positive definite quadratic form in $z$. On the other hand, the general second approximation to a given generating functional $F(z)$ can be put in the form

$$F'(z) = F(z)\, e^{if'(z)-Q'(z)}$$

for some linear and quadratic functions $f'(z)$ and $Q'(z)$, where however $Q'(z)$ is not necessarily positive definite. Experience in these matters indicates that the positive definiteness of $Q'$ corresponds to the generating functional $F'(z)$ being at a higher level than $F(z)$. In other terms, the generating functional for two (or more) systems in interaction (with a positive interaction energy) should differ from the product of the respective generating functionals, to within second order, by a factor of the indicated type with a positive definite quadratic form; so that *this second approximation is again the generating functional of some state.*

Consider for example the case of two systems $S_1$ and $S_2$ which are spatially separated and in states represented by the generating functionals $F_1(z_1)$ and $F_2(z_2)$; suppose that they interact as they come closer, producing a compound system with generating functional $F(z_1, z_2)$. Now a generating functional transforms in a very simple way under euclidean spatial displacements, or more generally, under any linear inhomogeneous transformation in phase space which is canonical (i.e. preserves the fundamental commutation relations between the $P$'s and the $Q$'s; in mathematical terms, such a transformation is "symplectic"). An approximation to $F(z_1, z_2)$ which should give somewhat more than the geometrical disposition of the systems $S_1$ and $S_2$ within the compound system, corresponding to such a transformation, is then of the form

$$F^{(1)}(z_1, z_2) = F_1(T_1 z_1)\, F_2(T_2 z_2)\, e^{if_1(z_1)+if_2(z_2)},$$

where $T_1$ and $T_2$ are the linear transformations corresponding to the basic linear transformations in 3-dimensional configuration (or more generally, 6-dimensional phase) space, and $f_1$ and $f_2$ correspond to the vector displacements in these spaces. The next approximation to $F(z_1, z_2)$ is obtained by multiplication with $\exp\{-iQ(z_1, z_2)\}$, where $Q$ is a positive definite quadratic form, chosen to minimize the energy variance as determined from the generating functional $F^{(1)}(z_1, z_2)\, \exp\{-iQ(z_1,z_2)\}$; this approximation $F^{(2)}(z_1, z_2)$ should give, hopefully, a good first-order account of bond strength, among other information. Any number of systems, such as the atomic constituents of a molecule, or monomers in a polymer, may be treated in the same way.

This should be a relatively simple approximation scheme to explore, either analytically or numerically. The number of free parameters is easily adjustable, according to the number of parameters on which the quadratic from Q is permitted to depend, and on whether one goes beyond the usual displacements in configuration space to the admissible ones in phase space. Its main qualitative advantage is that the approximating generating functional is always the generating functional of some bona fide state; this means essentially that all probabilities computed through the use of the functional will be non-negative, and in the case of mutually exclusive compatible events will add up to unity as they should. Additional qualitative advantages are:

(1) Symmetry is never lost; the approximating generating functional can easily be made to be invariant or transform appropriately in the same fashion as the exact generating functional, under the compact or finite groups relevant to molecules;

(2) Additional information concerning the actual generating functional can be incorporated into the initial approximation, before correction by the exponential quadratic factor, providing a more accurate approximation;

(3) The generating functional is relatively convenient for Fourier analysis, computation of moments, and adaptation to higher approximations, such as may be obtained by replacing the exponential quadratic factor by the Fourier transform of a Gaussian function multiplied by a non-negative polynomial of given degree (the degree O corresponding to the case here discussed);

(4) The present approximation scheme provides a relatively uniform and consistent method and formalism, adaptable to a wide range of physical systems, including relativistic and quantum-field-theoretic ones. In particular, it is adaptable to the treatment of states which do not contain an exact number of particles; a special class of such states, introduced mathematically in 1960, has since been found useful for the study of optical coherence, and their fermion analogue, recently developed by Shale and Stinespring, may prove relevant to electron phenomena (including possibly that reported by Dr. Szent–Gyorgi).

On the other hand, it may well be that some difficulties may arise upon further investigation of the present scheme, or in its application to molecules of biological interest. Our purpose has been simply to attempt to start the ball rolling, so to speak, by giving a theoretical indication of the possible existence of interesting mathematical alternatives to the Schrödinger wave function for the approximate description of complex molecules. While the accuracy of the indicated description may be significantly less than that which is theoretically possible with the use of fully delocalized molecular orbitals, it should be practically attainable relatively conveniently, and may yet be sufficiently precise for some interesting quantum biological purposes. The mathematical questions which arise, — for example, that of the choice of the $T$'s, $f$'s, and of $Q$, for simultaneous minimization of $E(H^2) - E(H)^2$, with supplementary side conditions (where $H$ represents the energy operator and $E$ the expectation value functional corresponding to these choices), — should be well within the range of contemporary mathematical techniques.

S. L. Sobolev: Je ne suis pas d'accord sur le fait que les mathématiciens n'ont rien fait au niveau des conceptions et au niveau des cas techniques. Et je veux donner quelques exemples qui se rattachent aux problèmes biologiques.

Le premier exemple: qu'est-ce que l'organisme humain? ou, peut-être, l'organisme animal? C'est un automate, qui a un état, dans lequel il se trouve, et selon lequel il peut réagir d'une façon ou d'une autre sur chaque action qu'il extériorise. Il y a quelque chose qui n'est pas tellement pauvre dans la théorie des automates. Nous n'avons pas traité, dans ce colloque, de cette théorie. Car je doute, maintenant, que ce soient des résultats tellement développés qu'ils puissent être appliqués immédiatement à la biologie. Mais je ne doute pas pour autant que la technique et la théorie des automates, comme leurs concepts, soient inapplicables. Je suis sûr, cependant, que nous ne pouvons pas trouver la solution de nos problèmes dans cette théorie. Mais nous pouvons en tirer des choses très utiles pour le développement de la biologie.

Il y a aussi d'autres questions. Il y a un très intéressant paradoxe. D'un ordinateur pour "curlatives" électroniques (je ne sais pas comment mieux dire en français), nous ne connaissons rien sur la façon dont il marche. Ce problème "devinez comment il marche" est beaucoup plus compliqué que de construire le nouvel automate, car la question surpasse mille fois et davantage toutes les possibilités de ce même automate, de ce même ordinateur. Ce que je dis est un théorème déjà bien démontré: c'est un fait dont il faut tenir compte.

Nous ne pouvons pas, maintenant, construire un automate qui fonctionne comme le cerveau humain, ou même comme le petit organisme de quelque microbe très primitif. C'est dire qu'il est très difficile de penser que nous puissions déchiffrer, maintenant, tout ce qui se fait dans la biologie. Mais nous avons déjà un très bel exemple au niveau de la physique classique, qui a créé la théorie des probabilités et la physique statistique, qui n'a rien de commun avec le mouvement des particules singulières et vous donne les méthodes pour décrire et résoudre tous les problèmes se rattachant aux grands ensembles. La position est beaucoup plus difficile, maintenant. Car ni la théorie des grands ensembles de la physique statistique, etc. . . . , ni la théorie des petits ensembles, qui contient seulement les opérations en nombres finis, ne sont applicables. L'une est trop générale, quoique pourvue de quelques propriétés très importantes, l'autre, trop pauvre, ne peut opérer avec tel nombre d'opérations qu'on peut effectuer par une calculatrice moderne. Il s'agit toujours de nombre d'opérations qui se perdent dans l'espace vu le nombre de molécules dont le globe terrestre est construit.

Alors, il faut espérer — j'espère: je suis très optimiste quand même, bien que j'aie émis quelques paradoxes sur ces théories — que le développement de quelques nouvelles branches de la mathématique: la théorie des automates, la logique mathématique, peut-être quelques nouvelles branches qui sont entre la logique mathématique, et l'analyse classique, et l'analyse fonctionnelle, qui sont déjà bien développées, bien qu'on ne puisse résoudre quelques problèmes difficiles —

j'espère, je crois que le développement de ces branches, encore insuffisamment
en progrès, nous donnera enfin la possibilité de résoudre non pas sans doute
tous les problèmes biologiques, mais ceux qui, d'entre eux, sont les plus importants.

G. CARERI: I would like to continue perhaps on the same way as Mendelssohn;
I am also an experimental physicist. I find some difference between the way we
see things and the way theoretical physicists do.

There is a question I will try to put; I am afraid nobody can answer. Is it
true that what exists inside a living cell is a kind of organization which can also
operate between many cells in an organ, and between many organs in the individual, and between many individuals in a society and so on? Then there is a
common pattern perhaps. I agree with you Sobolev, the analytical way we should
describe this pattern, this hierarchic order between subunits and units, is perhaps
what is missing today. Hierarchy of interactions this is what we are looking for.
This is what attract some physicists, which are used to think in this way, in low
temperature and in solid state physics. That is essentially why I am here. I believe
that in this search for things that happen inside a cell and that are similar to
the major ones that happen even in a human society, we must be open minded,
we must use also our private moral life, in the sense that we should learn something
so general to be applied in all life that surrounds us. I do not expect everybody
to agree with me, of course. I see here people who do not. But I would like to
use a phrase by the painter Georges Braque: "I do not aim to convince you,
all I want is to let you think about it".

Concerning the extreme improbability for existence of life that Mendelssohn
was saying, let me quote that in Florence five hundred years ago, a committee
of six people joined around a table. One was Michelangelo, the other was Leonardo
da Vinci, and among others I remember Botticelli, Perugino, Filippo Lippi. Now
how could these very improbable events happen? There is perhaps one possible
reason: these people were very much influencing each other. Whenever life
operates, we see strongly interacting members! And I think this is something
that should be explored, and deeply, in all possible systems, from the cell to
men and to the society.

W. M. ELSASSER: Ma remarque sera très courte. Je reviendra de cette conférence
avec une nouvelle conception. Tout le monde sait que l'organisme est une structure. Mais, de toutes les indications du M. Fessard et du M. Halberg, j'ai déduit
ce que je savais déjà d'une façon générale mais pas du tout de façon précise:
que l'organisme, ayant une structure quadridimensionelle, ne peut être considéré
comme une structure à trois dimensions. Si j'ai raison de généraliser les données
de ces messieurs, c'est parce que nous savons que par exemple dans l'hydrodynamique, il y a une différence radicale entre les dimensions de l'espace, selon
qu'on en ait deux ou trois. On peut prouver, d'une manière mathématique, qu'il

n'y a pas de turbulence dans un espace de deux dimensions. Mais tout le monde sait qu'il y a de la turbulence dans trois dimensions. De sorte que la différence entre une structure à trois dimensions et une structure à quatre dimensions pourrait bien être une chose fondamentale. C'est donc aux mathématiciens et à nous, les physiciens mathématiques, d'élaborer cette différence. Sans cela, l'on tirerait des conclusions, à partir des structures à trois dimensions, et l'on commettrait des erreurs qui sembleraient ridicules dans cinquante ans.

H. FRÖHLICH: Dr. Pullman in his talk has told us that there are two types of approach to this kind of physics. One was that of quantum chemistry whereby you specify detailed problems which he and others have been able to solve so nicely and to make very much progress. But there is also a physics concerned with general features. In the preceding discussions a number of people have come up to ask, are there some general principles? In other discussions some statements were made which one might call very vague. In physics, however, one speaks about general principles in a very exact way. Both Dr. Prigogine and I tried to do this in a certain way on the first day. I turned out when we met that we both had used overlapping concepts as if we had actually prepared our relative contribution— which was not the case. I think that many of the biologists and also of the physicists have not been really aware of what we have said because they were not all up to date in what has happened during the last five years.

What Prigogine said was something very general but nevertheless exact. Namely that big systems have an equilibrium state, and if we take it slightly out from this equilibrium state, they tent to fall back into it. This is on what physics had previously mainly been concentrated. He has found in a very general way without being specific that if we take such a system very far from equilibrium then another way of stabilisation can arise. But this is a stabilization which requires always a flux of energy through the system. That is what he thought resembles the state of some biological systems.

I have started with the concepts we are trying to derive in a more intuitive way. When I started I said to you that in my opinion biological systems are large systems, where a few degrees of behaviour are far from equilibrium and stabilized in a nearly stationary way. Now each big system consists of many atoms; the single atom does not see really a large size organization and it practically oscillates with very minute deviations as if it were in an equilibrium system. This is the reason for considering a few degrees of functions.

Now I tried to link up this general concept with some developments that have taken place in recent years in physics. They start out from super-conductivity, superfluidity where we found that a few modes of motion have exceptional behaviour. But these are systems in equilibrium. Later it was found that in lasers for instance something occurs which is related to this macroscopic order of very subtle nature. It is closely connected with certain phase correlations and we found that for instance lasers show this correlation as do superfluids and supercon-ductors.

We had a discussion whether this was a classical or quantum concept. It is both! while it builds up it's quantal, but later on the quantal aspects are suppressed, I then tried to see whether biology seen in a naive way, offers such a possibility, namely, to excite few modes of a certain behaviour in a way in which it can be stabilized. Here I think arises a most striking possibility out of the phantastic dipolar properties of biological molecules. So I made an estimate, I introduced a concept that in physics is extremely well known though very little use is made of it namely that of longitudinal electric modes. I assumed, that they be strongly excited and asked whether they can be stabilized. The answer is that there is the possibility of stabilization if we introduce deformations of a general nature, and so go from the usual linear behaviour into the nonlinear region. Then we went down a step further and we found that the estimate I made of the frequencies are in the region which according to Careri may also be expected to obtain resonances of biological molecules.

Je veux donner une idée du raisonnement que suit le physicien. Nous nous épuisons sur des principes établis et définis, et, à moment-là, nous fabriquons un modèle. Le modèle a été une invention, quelque chose qui présente une corré-lation avec une réalité bien plus complexe. Et puis, nous essayons de traiter ce modèle, aussi exactement que possible. Puis, nous repassons à l'objet réel.

Je crois que les concepts de la physique quantique vont se vérifier dans la biologie. Il n'y a pas de doute. Mais alors il faut faire des modèles, pour étudier certaines suggestions expérimentales. J'ai la joie de pouvoir vous dire que quelques laboratoires vont s'attaquer à vérifier ces suggestions. Je voudrais terminer en disant quelques mots sur la nature générale de ces discussions.

D'une part, il y a un domaine de la physique qui s'attache au détail et qui se donne pour tâche d'élucider certains phénomènes explicites, qui ont été découverts (ce qui a été très long à étayer . . .). Il y a autre chose: il faut le courage de faire des prédictions, de faire quelque chose qui ne soit pas tout à fait orthodoxe. Il y a, par exemple, la superconductibilité, dont on a tant parlé ici, pendant de nombreuses heures. On a dit que c'était un système très obscur. Les gens ne le comprenaient pas très bien. Il y avait un principe sur lequel tout le monde était d'accord: inversant la tendance, on ouvrirait la voie au progrès; donc, il ne faut pas faire de prédictions en partant d'hypothèses, mais garder toute la précision nécessaire sur le plan mathématique; il faut respecter tous les principes généraux de la physique. Mais, ceci assuré, il ne faut pas hésiter à faire des prédictions, en demandant à l'expérimentateur de les vérifier, bien que les résultats ne viennent pas toujours confirmer les prédictions. Cela peut toujours être un moyen d'ouvrir la voie à des progrès. C'est ce que le M. Prigogine et moi-même avons essayé de vous suggérer de faire, en vous proposant des façons nouvelles d'aborder ces problèmes.

Cela rejoint ce qu'a suggéré M. Mendelssohn et ce qu'a suggéré M. Careri. Mais il faut d'abord essayer de faire des prédictions et les étudier avec beaucoup de précision et de rigueur.

H. C. LONGUET-HIGGINS: I hope Dr. Fröhlich didn't take me to say that one should never make predictions. But there is a difference between predicting things that *might* happen and things that *should* happen. May I speak in a lighter vein for a minute or two?

The virtues of physics as a mental discipline are very great. So, at one time, were thought to be the virtues of a study of the Classics. The classical education opened all doors; one was enabled to think logically ,and to distinguish between the good and the bad argument.

There is a story which dates from forty or so years ago, of a dinner party in Oxford. The wife of a distinguished classical scholar was introduced to her neighbour, and asked him "And what do you do?" He said "Madam, I am a physicist." The lady replied: "Oh, indeed. My husband says that anyone with a classical education can get up physics in a fortnight."

M. MARGENAU: Although I am grateful to be called upon for comments in this final session, I speak with a sense of embarrassment since I do not have sufficient knowledge in the field of biology to contribute positively to the solution of the problems that have been raised. I have found myself mainly on the receiving end during the exchange of information in our interesting discussions. But I shall make bold once more to speak briefly as a philosopher of science with basic concerns about the integration of physics and biology.

As I view past efforts at establishing union between disparate disciplines, I am impressed by the circumstance that such attempts have never succeeded in the plane of observational facts. The manifest phenomena in any given science are rarely similar to those in another and attempts at comparison and classification almost never provide common ground. Integrative qualities lie in the texture below the surface of facts, in the realm of principles and theories in terms of which the facts are explained. Only these theoretical roots, the axiomatic foundations of different disciplines are likely to unite them, and I am happy to observe that in most of the discussions during the last few days, basic theories like quantum mechanics and thermodynamics, not specific facts, have been called upon to provide a synthesis. The deeper you go into the theoretical bases of biology and physics, the better will be your prospect at unification. And here, perhaps, even the philosopher can play a role. For the deepest roots of all science are found within certain philosophical premises.

I shall not try to confirm or contradict any of the things some of my colleagues have said, even though I have been at times prompted to do so. If I were to level criticism it would be directed against the cocksureness I perceived in the attitude of some of the theoretical physicists at these meetings.

Interesting epistemological problems have been raised. I was especially impressed by the paper presented by Dr. Fessard who outlined for us one of the most basic problems of the theory of knowledge in very definite scientific terms. When he discussed the connection between the unobserved responses of a subject and the external stimuli that occasion them, when he spoke of the merely conscious modes of awareness of a subject in contradistinction to the observer's observation or knowledge of them. I drew from his comments the conclusion that you can not avoid facing the serious problem of the meaning and the role of consciousness in biological processes. Nothing has been said about it at these meetings; yet I think it ought to be in the minds of physicists who are interested in the problems of life. Specifically, I am raising the question as to the feasability of the pretended union so long as the phenomenon of consciousness is neglected.

What I wish to say next may seem heretic and out of place. Practically all discussions this week have been based on the premise of mechanistic causation. I use this phrase in a rather general and generous sense, including the statistical form of causality that is peculiar to quantum mechanics. Perhaps it is an insult to this audience to inject the word teleology on this occasion. While I do not want to talk like an Aristotlian, referring to vague notions of purposiveness or to the even vaguer ideas of du Nouy who coined the phrase telefinality, I am anxious to propose something a little more specific and articulate. What I have in mind is the kind of teleology which inheres in the exclusion principle of Pauli. Let me explain. If physicists had limited their studies to the laws controlling the behavior of individual particles they would never have discovered the Pauli principle or its antithesis, the principle of symmetry for bosons. These principles of symmetry have no relevance to individual particles but become significant and coercive in the interplay between particles of the same kind. As is well known, they are sources of organization in the non-living world, providing explanations for all the so-called cooperative phenomena. The lesson I wish to draw is this. Just as the Pauli principle regulates occurrences involving many particles and could never have been discovered on the plane of individuals, it may well be that biological organization is tied to principles which can not be discovered within physics because they have no applicability to phenomena in the inorganic (perhaps I should say unconscious) world. What these principles are is difficult to predict, but I would guess that they are laws of invariance, laws of symmetry of a novel kind. However, in the sense which I have tried to outline in earlier remarks they must not be in conflict with the laws of physics, just as the Pauli principle is not in conflict with Schrödinger's equation.

Thus, I plead for open mindedness and for the maintenance of an expectation that the unravelling of problems in biological science may well expose principles of physics which the physicist, even the expert in quantum mechanics and statistical mechanics, could not derive from his present knowledge.

I am prompted to conclude with this memento. Let us not try to force physics and biology together. The natural symbiosis of these two disciplines can not help but be fruitful. But if you take over ideas from physics too literally and install them in biology you may contract the same difficulties which confronted people who, some years ago, formed a group of eager disciples of a discipline called social physics. It took concepts like force, pressure, temperature, entropy and tried to formulate in terms of them a theory of social behavior. I believe this movement has failed because it did not recognize that concepts are not completely interchangeable between different disciplines. It would be most astounding to me if a science like biology would not ultimately have to develop concepts of its own, concepts with no counterparts in physics. On the other hand, this open-mindedness, this liberal expectation should never prevent us from staging the sort of fruitful intercourse which to me has made the present occasion unusually memorable.

R. Kubo: Since I have not spoken yet in this conference, I would like now to make a very brief remark from a physicist's point of view, although I doubt a little if I should talk after the last remark of Dr. Margenau.

The point I like to make is concerned with a general theorem about the relationship between response of a system to outside stimulus or disturbance. The concept of response to stimulus is very often talked about in physics and also in biology. Now, there is a well-established theorem [1] in statistical physics which has been proved extremely useful in recent years in analysis of many body problems, which are typically complex problems in modern theoretical physics. If the external stimulus is an external force $F$ which is defined by the additional term, $-AF$, in the Hamiltonian (energy) of the system and if the response of the system is observed in a physical quantity $B$, the relationship between the stimulus and the response is represented by such an equation as

$$B(t) = \int_{-\infty}^{t} \varphi_{BA}(t-t')\, F(t')\, dt',$$

provided that the response is linear in the stimulus. Here $\varphi_{BA}(t-t')$ is the response at the time $t$ to a preceding pulse stimulus at $t'$. It is a function of $t-t'$ if the system is in a stationary state. If the force is periodic, then the linear response is described by an admittance (susceptibility) which is defined by

$$\chi_{BA}(\omega) = \int_{v}^{\infty} \varphi_{BA}(t)\, e^{-i\omega t}\, dt.$$

This is just a generalization of admittance function which is familiar in electric net work theories.

A general theorem now states that the response is very closely related to the fluctuation within the system in the absence of such external disturbances. This is the *fluctuation-dissipation* theorem. In fact, it is generally shown that the response function $\varphi_{BA}(t)$ is given in terms of a correlation of two fluctuating quantities $A$ and $B$ at different time points; namely

$$\varphi_{BA}(t) = \langle A(t_0)\, B(t_0 + t)\rangle / kT,$$

where $k$ is the Boltzmann constant and $T$ the temperature and $\langle\,\rangle$ means a statistical average. This fluctuation-dissipation theorem is very fundamental in the non-equilibrium statistical mechanics. It should be emphasized particularly that the theorem should apply more generally to systems in stationary states, not necessarily in thermal equilibrium [2]. Then the response function should be determined by fluctuation in a stationary state of the observed system. This observation is important for biological applications of the fluctuation-dissipation theorem if such should be possible, since a biological system can be regarded stationary, though not in thermal equilibrium.

[1]  R. Kubo, *J. Phys. Soc. Japan* **12** (1957) 570.
[2]  R. Kubo, Rep. on Progress in Physics **24** Part I, 255 (1966); Tokyo Summer Lectures in Theoretical Physics 1965, Part I. Many-Body Theory, Ed. R. Kubo (Shokabo, Tokyo, and W. A. Benjamin, New York).

A. LICHNEROWICZ: Nous avons retouché, si j'ose dire, des choses relativement concrètes, bien qu'elles soient encore abstraites.

W. REICHARDT: In relation to Dr. Fessard's paper some remarks were made concerning the repetitive structure of retina units, neuronal units in the first optical ganglion (lamina) of the housefly Musca and the nervous projection of the retina onto the lamina. In this connection it was shown in our laboratory that the dioptrics of the Musca ommatidium acts as an inverting lens system. The distal endings of the rhabdomers at the basis of the dioptric apparatus are separated and arranged in a typical asymmetric pattern. The optical axes of the individual rhabdomers of one ommatidium are the geometrical projection of the distal rhabdomer endings into the environment, inverted by 180° by the dioptric apparatus. The divergence angles between optical axes of the rhabdomers of one ommatidium correspond to divergence angles between the appropriate set of ommatidia in such a way, that seven rhabdomers of seven ommatidia are looking at one point in the environment (in the intermediate region between dorsal and ventral part of the eye: 8 to 9 rhabdomers of a set of 8 to 9 ommatidia). These facts were established from sections of living eyes and confirmed by using special optical methods in the intact animal.

The pattern of decussation of individual retinula cell axons between retina and lamina was predicted adopting the hypothesis that the fibers of retinula cells No. 1 to 6, whose rhabdomers are looking at one point in the environment, project into a single "cartridge" in the lamina. These predicted connections were confirmed by Braitenberg. We have therefore a one to one correspondence between a lattice of points in the environment and the lattice of "cartridges" in the lamina.

It was shown that the unfused rhabdomer structure of the Musca ommatidium increases the effective entrance pupil of the eye by a factor of seven (respective 8 to 9 in the intermediate region between dorsal and ventral part of the eye) compared to the classical apposition eye. – The Musca compound eye can be regarded as a "neural superposition eye".


P. O. LÖWDIN: I have been asked to try to make some general comments from theoretical points of view on the relation between physics and biology. Let me start by mentioning the great importance of biology for physics by reminding you of the little frog which, in 1791, led Galvani to the discovery of the electric element – an event which has certainly been of fundamental importance for the development of the entire theory of electric phenomena.

In physics, one deals with phenomena in nature which are as simple as ever possible, whereas, in biology, one studies systems of extreme complexity. It seems clear, however, that the basic phenomena in biology follow the laws of physics and chemistry. In science, one should remember that "theory" is nothing which exists by itself, and that "theory" is essentially the quintessence of experimental experience expressed in mathematical form. The last hundred years of experiments in physics and chemistry have led to the formulation of the fundamental laws which are now expressed in concise form by quantum theory.

From the time of Newton, the laws of nature have usually been expressed in differential calculus, but there is now a trend to shift over to "modern mathematics" and set theory. The experimental data are collected in "sets", and correlation between various data are then explained in terms of "mappings" of sets on each other. Of particular importance are the linear sets (or linear spaces) and their mappings in terms of linear operators. The modern language is essentially geometrical and, starting out from the ordinary 3-dimensional space, one constructs $n$-dimensional spaces or $\infty$-dimensional spaces of the type introduced by Hilbert. A child crawling around on a floor is living in a 2-dimensional space, when it rises on its feet, it enters the 3-dimensional world, and when it becomes a scientist, it may use all these experiences to enter the multi-dimensional spaces of mathematics and modern physics. The properties of a space may be described by means of a specific reference system, but, since the laws of nature must be independent of any specific choice of this reference system, one has a "general principle of relativity" which is even more general than the principles formulated by Einstein.

In connection with classical physics and modern quantum theory, one should remember that they both require "initial conditions" which are usually only prescribed by the experimental set-up. If the physical situation at the initial time $t_0$ is given by the wave function $\Psi_0$ or the density matrix $\Gamma_0$, the situation at the time $t$ is described by the wave function $\Psi$ or the density matrix $\Gamma$, respectively, according to the laws

$$\Psi = U\Psi_0, \qquad \Gamma = U\Gamma_0 U^\dagger,$$

where $U = U(t, t_0)$ is the evolution operator for the system. It should be observed that this operator $U$ is, in principle, determined by the theoretical laws of physics, whereas the information about $\Psi_0$ or $\Gamma_0$ has to come from the experimental situation under consideration. In biology, the initial conditions and the boundary conditions must come from biology, and theory is hence essentially a tool for handling biological data according to laws found under simpler conditions. The application of theoretical physics to biology may thus also be considered as an interesting extrapolation procedure from simple to extremely complex systems.

This procedure may be more complex than one anticipates. For example, let us consider a master skier gowing down a slalom hill on skis and ask for an explanation of his motion on the basis of the laws of nature. One should study the motion of the center of mass, the motion of the skier with respect to this center, and the physiological processes underlying what is going on. The interesting point is that one has considerable difficulties in explaining even the pure mechanical aspects of the motion, since it is hard to get hold of sufficient experimental information about what is really going on. Even if the entire motion is divided into small parts, it is hard to get hold of the "initial conditions" for each part — even if the skier believes that he knows how to "initiate" various turns and swings. If one puts a theoretical physicist who knows everything about the laws of mechanics (but nothing about skiing) on a pair of skis and asks him to go down the slope, the result will be rather startling, but the interesting point is that, even if he later learns to ski, he may not know what he is doing mechanically. Even an analysis by means of slow-motion picture has not yet revealed the full details of all the dynamical processes involved, and still skiing is a very slow process in comparison to the physiological processes in the body.

I have chosen this example to indicate that one should not expect too much of theory itself in connection with biological phenomena; it is essentially a tool for handling biological data, and it may provide a useful conceptual framework for discussing the elementary processes going on.

In three excellent lectures by Fröhlich, Prigogine, and Pullman, we have heard about various theoretical aspects on biology referring to different levels of "molecular sophistication", which are supplementing each other in a most interesting way.

Transport phenomena are of fundamental importance in biology, and one has to understand energy and momentum transfer as well as charge transfer on a macroscopic as well as a microscopic level. Prigogine has chosen to discuss transport phenomena and basic properties of certain systems far away from equilibrium, and our general understanding of these things are of great importance also in biology. Fröhlich has pointed out the possible importance in biology of the so-called "intermediate region" in the electro-magnetic theory – the region in which usually both the longitudinal and transverse components of the field strengths are of importance, and he has emphasized the effects of the existence of "longitudinal waves". It should perhaps be observed that he is not suggesting the introduction of any strange new phenomena, but a careful analysis of the consequences of Maxwell's equations for the intermediate region which may lead to results of direct importance in cellular biology. By means of the Hückel-method, Pullman has finally given a first analysis of the over-all properties of the conjugated systems, which are planar molecules of essential importance in biochemistry due to the occurence of mobile $\pi$-electrons.

The main point is perhaps that, as time goes on, biologists are making better observations and finer measurements and they have now reached a level where the fundamental particles come in. The transport phenomena are now dealing with electrons and protons, and Szent Györgyi speaks here of electron transfer as one of the most important physiological processes.

To me, it is particularly fascinating that the memory unit in biochemistry is connected with a hydrogen bond, and that it seems to consist of an electron lone-pair: which is either empty or has caught a proton H and formed the structure: H. This leads to a flip-flop mechanism with two possibilities which correspond to the symbols 0 and 1 in a digital electronic computer.

In the genetic code, according to the Watson-Crick model, the information is contained in the four base pairs AT, TA, GC, and CG, which are nitrogen bases joined by hydrogen bonds according to the diagrams:

An analysis of the formulas shows that the genetic template consists of patterns of electron lone-pairs and protons, and that the normal nitrogen basis are characterized by the following combinations

$$
A \left\{ \begin{matrix} : H \\ : \end{matrix} \quad \begin{matrix} : \\ H : \end{matrix} \right\} T
$$

$$
G \left\{ \begin{matrix} : \\ : H \end{matrix} \quad \begin{matrix} H : \\ : \end{matrix} \right\} C
$$

If a proton is misplaced and one gets a "tautomeric" form, the genetic code is also changed, and one may obtain a mutation.

Each hydrogen bond is a proton shared between two electron lone-pairs, and, since each electron-pair offers the proton an attraction represented by a potential well, the proton in the hydrogen bond is actually situated in a double-well potential with two minima separated by a barrier. The proton may move from one minimum to the other above the barrier or through the barrier by means of the quantum-mechanical "tunnel effect", and this process leads to a deterioration of the genetic code [1]. The tunnel-effect depends on the fact that the proton is a de Broglie wave-packet, which may penetrate classically forbidden regions.

In Uppsala, we have recently calculated the shape of the energy potential surfaces for the mobile protons in the genetic code under the assumption that they simultaneously polarize the electronic clouds of the base pairs involved, and the results indicate that the minima are so asymmetric that the protons stay where they are supposed to be in the genetic code with a thermal error due to the tunnel effect which, at $T = 310$ K, is of the order of magnitude $10^{-10}$ for the GC-pair and $10^{-11}$ for the AT-pair [2]. Since the biological mutation rates per base pair and generation are in the region $10^{-8}$ to $10^{-12}$, the theoretical figures obtained are almost "too good to be true", but it would certainly be of some interest if one could predict the spontaneous mutation rates in biology on the basis of the laws of physics and biology.

The main conclusion seems to be that, when the biologists refine their observation and measurements to the extent that they approach the level of the fundamental particles, they are bound to find quantum effects, and personally I believe that the field of "quantum biology" is here to stay — irrespective of whether this extension of biology is really desired or not. In such a case, theoretical physics is going to be of even greater importance for biology in the future.

It is always difficult for new ideas to come into science, and even cross-fertilization is a very slow process. Max Planck discovered quantum mechanics around 1900, and it is said that in the next decade the new approach conquered physics completely — however, the truth seems to be that Planck's older opponents died one after each other, leaving space for the new ideas to develop. Something similar happened to Einstein and his theory of relativity.

I think that the main importance of a conference of this type is to diminish the inevitable friction which occurs when two scientific fields approach each other, particularly when they are as different as theoretical physics and biology. Both the biologists and the theoreticians are sincere scientists which may have completely different or even controversial aspects on one and the same problem, but both fields may still gain immensely by a closer collaboration which is based on some form of compromise and understanding and not on the death of one of the parts involved. We have hence all reasons to be grateful to the organizers of this conference for the contact they have established.

[1]  P. O. Löwdin, *Advances in Quantum Chemistry* (Ed. P. O. Löwdin, Academic Press, 1965): 213.
[2]  P. O. Löwdin, Proc. Study Week on Intermolecular Forces at Pontifical Academy, Rome (1966); Pontif. Acad. Scie. Scripta Varia No. 31.


Mrs. GRENE: Dr. Lloyd mentioned about one of his experiments that it worked better if the subject did not know what the experiment was about. Perhaps it's better that you didn't know that the philosophers among you were using you as their subject so that we could try to see how you go on and what your methods of thinking and of handling problems are. In this sense, philosophical problems have been implicit in much of the conference. At the same time there were also explicit philosophical problems that arose on occasion, and a good many have come out this afternoon. Perhaps I could just summarize what seem to me to be the principal philosophical problems that ought to be dealt with, not necessarily at such a conference as this — since I gather that most of you prefer to deal with the detailed problems of interaction between the disciplines — but certainly by philosophers who are thinking about more general problems.

First, Dr. Margenau mentioned what I should certainly want to mention also at the head of my list, the very beautiful model of epistemology which Dr. Fessard gave in his talk: Dr. Fessard spoke of the problem that arises when one considers the observer of any piece of animal behavior; this can be generalized to include the behavior of theoretical physicists or biologists or any other knowers. In all these cases there is not only the "input" from the external world in the subject and his output; there is also an input in the observer and his output, and in between there is what happens in himself, which he has to use as a kind of model for what he takes to go on in his subject. He can't get inside his subject, but he has to use the structure of his own experience to model the experience of his subject. Dr. Fessard described this situation as the physiologist confronts it; insofar as the epistemologist is (as Polanyi calls him) an "ultrabiologist", he is faced with exactly the same situation. He is interpreting a piece of human behavior — knowing physics or biology or what you will — and he can do so only by using the transition from his input to his output to model what he imagines takes place in his subject.

Not only does Dr. Fessard's model shed light, moreover, on the central situation

of the epistemologist; it also gives the philosopher something to think about
in connection with the question of the relation, philosophically speaking, between
physics and chemistry on the one hand and the biological sciences on the other.
Several people, Dr. Careri for instance, and Dr. Margenau, have mentioned the
distinction between levels of organisation in the organic vs. the inorganic world.
There seem to be hierarchies of structure and function not only as between the
non-living and the living, but in the living world itself. It seems to me that Dr.
Fessard's little diagram showed very clearly how this may be, especially for the
study of behavior. There seems to be a fundamental difference here in the way in
which biological knowledge is organised in relation to its subject from the way
in which physics is related to its subject. If you are talking about protons and
electrons, for example, you do not have the double relation that Dr. Fessard
described, for you are not dealing with a subject which is itself responding to
its environment as you are also doing. The reference to the responding organism
seems to introduce into biology a new logical level which is missing in physics
and chemistry. How is one to talk about this problem adequately? Does the
proliferation of levels arise already in physics, or is there something different
about it in biology? I would have gathered from the remarks both of Dr. Margenau
a few days ago and of Dr. Careri today that both of them think that there is a
continuity here. You just start with the Pauli principle, and go on multiplying
complexities, all along the same line. On the other hand, Dr. Fessard's example
and others that I have dealt with on other occasions lead me to believe that
perhaps there is a different kind of complication of levels when one comes to
organic beings, that there is a qualitative change in the kind of hierarchy one is
dealing with. The problem here is one which no one has yet, so far as I know,
definitively solved, but which is a matter of lively controversy at the moment.

And this leads on to the question whether one can give a proof of nonreducibility
for living things, that is, a proof that there is a greater logical richness in biological
than in non-biological subject-matter. I was impressed by Dr. Löwdin's way of
putting this: that is, that when you are dealing with biological material you have
a vast number of starting points and that in this way biology is more complex
than physics. I think there is a job for logicians here, to try to demonstrate what
kind of logic is entailed here. Again, this may be done perhaps in terms of some
kind of mathematics, as Dr. Sobolev mentioned a couple of times. It might be
done perhaps in terms of the logic of parts and wholes, which again logicians
and philosophers of science have dealt with very inadequately. At any rate there
is clearly an important field here in which philosophers and mathematicians
should collaborate.

Another philosophical question that, again, was touched on only today in
connection with the relation between physics and biology is the question of the
temporal organisation of living things. Maybe the physical world, viewed rela-
tivistically, is itself four-dimensional, but certainly the four-dimensional character

of living beings is something pretty complicated and perhaps it has to be dealt with, as Dr. Halberg suggested strikingly this morning, with different tools. Here again there is a whole nest of philosophical problems.

Finally, there is another question which was touched on this afternoon by Dr. Careri. That is the question of the structure and function of communities, not only the complexity of a single organism and of its morphogenesis and life history, but the question of the way in which an organism enters into and is controlled by its community. I hoped that we would hear something of this from Dr. Lindauer. It's certainly a subject that people in insect behavior and other ethological fields are concerned with. This again is certainly one of the places where biology ought to touch on philosophical questions and I was very glad that Dr. Careri mentioned it.

A. LICHNEROWICZ: Il nous est revenu que vous désiriez savoir s'il pourrait y avoir une suite à ce colloque. Ce colloque a eu le mérite de réunir des gens que la vie ne faisait pas se rencontrer. Ils avaient tort, nous le savons tous aujourd'hui, de ne pas se rencontrer. Certains d'entre vous ont proposé le principe d'une nouvelle conférence, dans environ deux ans. La spécificité de notre actuelle Conférence est qu'elle a été vraiment pluridisciplinaire tout en couvrant un large éventail, peut-être trop large. La nouvelle conférence pourrait continer à s'intituler, en précisant davantage, peut-être: "De la Physique Théorique à la Biologie". Bien entendu, pour l'organiser l'actuel Comité serait élargi à toute l'Europe d'une part, et, d'autre part, à l'Amérique et à l'Asie.
(*Assentiment général, se traduisant par un oui unanime*)
Il en sera donc ainsi décidé.

F. HALBERG: Je vote une ovation au Professeur Marois, pour l'accueil qu'il nous a réservé.

A. LICHNEROWICZ: Je voudrais dire à chacun d'entre vous, au nom du Comité d'Organisation, notre gratitude pour être venus ici, pour avoir créé cette atmosphère à la fois de très haut niveau scientifique et particulièrement amicale. Au nom de ce Comité, merci. C'est vous qui avez assuré ce que l'on peut appeler le succès de cette Conférence. Je laisse maintenant la parole à l'Institut de la Vie, sans lequel rien n'aurait pu avoir lieu.

F. DE CLERMONT-TONNERRE: C'est par l'amitié du Pr Marois et la confiance du Comité d'organisation que le Vice-Président, co-fondateur de l'Institut de la Vie, a l'insigne honneur de parler devant vous.

Permettez-moi, au nom du Conseil d'administration de l'Institut de la Vie, au nom du grand Conseil et au nom du Comité de Patronage de notre Institut, de vous remercier tous! ... Tous les participants à ce Colloque, les uns arrivés de tellement loin! Et vous êtes venus pour nous. Croyez que nous en sommes profondément reconnaissants et émus.

Reconnaissants, particulièrement, de la confiance que vous nous témoignez par votre présence. Et d'avoir répondu, comme vous l'avez fait, à l'invitation de notre jeune Institut; d'avoir pendant six jours travaillé en son nom, avec autant d'ardeur! C'est pour nous une responsabilité lourde, tant dans l'organisation actuelle que pour l'avenir.

Votre confiance, je n'en veux qu'un témoignage. Lorsque M. Lichnerowicz, tout à l'heure, évoquait le lieu où se tiendrait le futur Colloque, j'ai remarqué, dans cette salle, ce que plusieurs murmures et suggestions laissaient entendre: que vous ne seriez pas fâchés de revenir à Versailles. Nous ne savons pas quelle sera la décision du Conseil élargi, mais je suis sûr que cette décision vous l'apprécierez tous, car je tiens à vous dire ceci: où que se tienne ce Colloque, l'Institut de la Vie sera fier de vous y voir participer. Puis-je ajouter que la caractéristique essentielle qui me paraît, du point de vue de l'Institut, s'être dégagée de notre confrontation, pourrait être son étonnant esprit de liberté intimement associé à la plus grande rigueur. Comment ne pas penser que les conséquences en seront fructueuses!

Permettez-moi, pour conclure, d'ajouter un mot au nom de la discipline que je représente, puisque, aussi bien, c'est la seule occasion que j'aurai d'en parler devant vous.

... L'Histoire aussi est une science de la vie. Et, de plus en plus, elle se tourne vers vous, vers votre aide. D'abord, pour augmenter ses possibilités d'investigation. Pour développer encore davantage son esprit de critique, son ambition est, sous votre égide, à son tour de connaître la rigueur des lois et d'arriver à des résultats certains.

Merci, au nom de cette discipline, de l'aide que vous lui apportez tous les jours! Croyez que, pour un historien, avoir assisté à vos débats est l'expérience la plus enrichissante qu'il puisse avoir. Ainsi sommes-nous devenus des amis, au sein et par-dessus nos disciplines respectives. Cette amitié anime nos travaux futurs et garde vivant le souvenir des heures exaltantes que nous avons passées ensemble.

Ce n'est qu'un au revoir que nous vous disons, Amis de l'Institut de la Vie!

# LIST OF PARTICIPANTS

M. AGENO,
Istituto Superiore di Sanità,
Laboratori di Fisica,
Viale Regina Elena 299,
ROME (Italie).

P. AUGER,
Professeur à la Faculté des Sciences de
Paris, Président de la Commission des
Sciences de l'U.N.E.S.C.O.
12, rue Emile Faguet,
75 – PARIS 14e (France).

S. BENNETT,
University of Chicago,
Laboratories for Cell Biology,
939 East – 57th street,
CHICAGO – Illinois 60637 (U.S.A.).

E. D. BERGMANN,
Laboratoire de Chimie Organique,
Université de Jérusalem,
JERUSALEM (Israël).

M. A. BOUMAN,
National Defense Research Organiza-
tion TNO,
Institute for Perception RVO-TNO,
Kampweg 5,
SOESTERBERG (Pays-Bas).

S. E. BRESLER,
Institute of High Molecular CPDS,
Academy of Sciences,
LENINGRAD B. 164 (U.R.S.S.).

G. CARERI,
Università degli Studi – Roma,
Istituto di Fisica "Guglielmo Marconi",
Piazzale delle Scienze 5,
ROME (Italie).

B. CHANCE,
Johnson Research Foundation,
Department of Biophysics and Physical
Biochemistry,
School of Medicine,
University of Pennsylvania,
PHILADELPHIA – Penn. (U.S.A.).

F. DE CLERMONT-TONNERRE,
Vice-Président du Conseil d'Administration de l'Institut de la Vie,
14, rue du Conseiller Collignon,
PARIS – (France).

E. G. D. COHEN,
The Rockefeller University,
NEW-YORK – N.Y. 10021 (U.S.A.).

M. H. COHEN,
Institute for the Study of Metals,
University of Chicago,
5640 Ellis Avenue,
CHICAGO – Illinois 60637 (U.S.A.).

A. COURNAND, Prix Nobel,
College of Physicians and Surgeons,
Columbia University,
Department of Medicine,
630 West 168th Street,
NEW YORK – N.Y. 10032 (U.S.A.).

J. D. COWAN,
University of Chicago, Chairman,
Committee on Mathematical Biology,
937 E. 57th Street,
CHICAGO – Illinois 60637 (U.S.A.).

D. M. CROTHERS,
Yale University,
Department of Chemistry,
Sterling Chemistry Laboratory,
225, Prospect Street,
NEW-HAVEN – Connecticut 06520
(U.S.A.).

C. Crussard,
Directeur Scientifique de la Société
Péchiney,
23, rue Balzac,
75 – PARIS 8e (France).

A. Dalcq,
Secrétaire Perpétuel de l'Académie
Royale de Médecine de Belgique,
Palais des Académies,
1, rue Ducale,
BRUXELLES (Belgique).

K. H. Degenhardt,
Institut für Humangenetik und Ver-
gleichende Erbpathologie
der Universität Frankfurt,
Paul-Ehrlich-strasse 41,
6 – FRANKFURT (R.F.A.).

P. Dejours,
Faculté de Médecine de Paris,
Laboratoire de Physiologie,
45, rue des Saints-Pères,
75 – PARIS 6e (France).

Rév. Père A. Dou, S.J.,
Catedra de Ecuaciones Diferenciales,
Facultad de Ciencias – C.U.,
Departamento de Ecuaciones Funcio-
nales,
MADRID – 3. (Espagne).

J. Duchesne,
Université de Liège,
Département de Physique Atomique et
Moléculaire,
Institut d'Astrophysique,
COINTE-SCLESSIN (Belgique).

M. Eigen, Prix Nobel,
Max-Planck-Institut für Physikalische
Chemie,
Bunsenstrasse 10,
3400 – GÖTTINGEN (R.F.A.).

W. M. Elsasser,
University of Maryland,
Institute for fluid Dynamics and Applied
Mathematics,
MARYLAND – 20740
(U.S.A.).

A. Fessard, de l'Académie des Sciences,
Collège de France,
Institut Marey,
Laboratoire de Neurophysiologie Générale,
4, avenue Gordon-Bennett,
75 – PARIS 16e (France).

H. Fröhlich, F.R.S.,
University of Liverpool,
Chadwick Laboratory,
LIVERPOOL 3 (Grande-Bretagne).

P. Glansdorff,
Université Libre de Bruxelles,
Faculté des Sciences,
Pool de Physique,
50, avenue F. D. Roosevelt,
BRUXELLES – 5. (Belgique.)

P. P. Grassé,
Président de l'Académie des Sciences,
Laboratoire d'Evolution des Etres Organisés,
Faculté des Sciences,
105, boulevard Raspail,
75 – PARIS 6e (France).

Mme. M. Grene,
University of California,
Department of Philosophy,
DAVIS – Cal. 95616 (U.S.A.)

F. Gros,
Institut de Biologie Physico-Chimique,
Fondation Edmond de Rothschild,
13, rue Pierre Curie,
75 – PARIS 5e (France).

H. Haken,
Institut für Theoretische Physik der Technischen Hochschule Stuttgart,
Azenbergstrasse 12,
7 – STUTTGART 1. (R.F.A.).

F. Halberg,
University of Minnesota,
Medical School,
Chronobiology Laboratories,
Department of Pathology,
MINNEAPOLIS – Minnesota 55455 (U.S.A.).

T. L. Hill,
University of Oregon,
Department of Chemistry,
College of Liberal Arts,
EUGENE – Oregon 97403 (U.S.A.).

A. Katchalsky,
Weizmann Institute for Sciences,
REHOVOTH (Israël).

O. Klein,
Ringen 21,
MORBY-STOCKSUND (Suède).

M. Kotani,
Osaka University,
Faculty of Engineering Science,
TOYONAKA – OSAKA (Japan).

R. Kubo,
University of Tokyo,
Department of Physics,
Faculty of Science,
BUNKYO-KU – TOKYO (Japan).

A. D. McLachlan,
University Chemical Laboratory,
Lensfield Road,
CAMBRIDGE – (Grande-Bretagne).

A. Lichnerowicz,
de l'Académie des Sciences,
Collège de France,

Chaire de Physique mathématique,
Place Marcellin Berthelot,
75 – PARIS 5e (France).

M. Lindauer,
Zoologisches Institut der Universität,
Siesmayerstrasse 70,
6 – FRANKFURT a.M. (R.F.A.).

B. B. Lloyd,
University Laboratory of Physiology,
OXFORD (Grande-Bretagne).

H. C. Longuet-Higgins, F.R.S.,
Department of Machine Intelligence and Perception,
University of Edinburgh,
Forrest Hill,
EDINBURGH – 1. (Grande-Bretagne).

P. O. Löwdin,
Quantum Chemistry Group for Research in Atomic, Molecular, and Solid-State Theory,
University of Uppsala,
UPPSALA (Suède).

F. Lynen, Prix Nobel,
Max-Planck-Institut für Zellchemie,
Karlstrasse 23–25,
8 – MÜNCHEN 2. (R.F.A.).

O. Maaløe,
Det Mikrobiologiske Institut,
Øster Farimagsgade 2 A,
COPENHAGUE (Danemark).

M. Magat,
Laboratoire de Physicochimie des Rayonnements,
Faculté des Sciences,
91 – ORSAY (France).

H. Margenau,
Yale University,
Physics Department,
217 Prospect Street,
NEW-HAVEN – Conn. 06520 (U.S.A.).

M. MAROIS,
Professeur Agrégé à la Faculté de
Médecine de Paris,
Président du Conseil d'Administration
de l'Institut de la Vie,
89, boulevard Saint-Michel,
75 – PARIS 5e (France).

B. MATTHIAS,
University of California, San Diego,
Department of Physics,
Revelle College,
P.O. Box 109,
LA JOLLA – California 92038 (U.S.A.).

P. MAZUR,
Instituut Lorentz voor Theoretische
Natuurkunde,
Nieuwsteeg 18,
LEIDEN (Pays-Bas).

K. MENDELSSOHN, F.R.S.,
University of Oxford,
Department of Physics,
Clarendon Laboratory,
Parks Road,
OXFORD (Grande-Bretagne).

J. MONOD, Prix Nobel,
Institut Pasteur,
25, rue du Docteur Roux,
75 – PARIS 15e (France).

R. S. MULLIKEN, Prix Nobel,
University of Chicago,
Laboratory of Molecular Structure and
Spectra,
Department of Physics,
1100 East – 58th Street,
CHICAGO – Illinois 60637 (U.S.A.).

S. OCHOA, Prix Nobel,
New-York University Medical Center,
Department of Biochemistry,
550 First Avenue,
NEW-YORK – N.Y. 10016 (U.S.A.).

L. ONSAGER, Prix Nobel,
Yale University,
Department of Chemistry,
Sterling Chemistry Laboratory,
225, Prospect Street,
NEW-HAVEN – Connecticut (U.S.A.).

J. POLONSKY,
Compagnie Générale de Télégraphie
sans Fil,
Directeur du Département Télévision,
132, avenue de Clamart,
92 – ISSY-les-MOULINEAUX
(France).

I. PRIGOGINE,
Université Libre de Bruxelles,
Faculté des Sciences,
Service Chimie-Physique II,
50, avenue F. D. Roosevelt,
BRUXELLES – 5. (Belgique).

B. PULLMAN,
Professeur à la Faculté des Sciences de
Paris,
Administrateur de l'Institut de Biologie
Physico-Chimique,
Université de Paris,
Faculté des Sciences,
Laboratoire de Chimie-Quantique,
13, rue Pierre Curie,
75 – PARIS 5e (France).

W. REICHARDT,
Max-Planck-Institut für Biologie,
Spemannstrasse 34,
TÜBINGEN (R.F.A.).

A. REINBERG,
Laboratoire de Physiologie,
Fondation Adolphe de Rothschild,
29, rue Manin,
75 – PARIS 19e (France).

S. A. RICE,
Director of the Institute for the Study
of Metals,
University of Chicago,
5640 – Ellis Avenue,
CHICAGO – Illinois 60637 (U.S.A.).

L. ROSENFELD,
Nordisk Institut for Teoretisk Atom-
fysik,
Blegdamsvej 17,
COPENHAGUE Ø (Danemark).

M. P. SCHÜTZENBERGER,
Institut Blaise Pascal,
23, rue du Maroc,
75 – PARIS 19e (France).

I. SEGAL,
Massachusetts Institute of Technology,
Department of Mathematics,
CAMBRIDGE – Mass. 02139 (U.S.A.).

S. L. SOBOLEV,
Académie des Sciences de l'U.R.S.S.,
NOVOSIBIRSK (U.R.S.S.).

A. SZENT-GYORGYI, MD. PH. D., Prix
Nobel,
Laboratory of the Institute for Muscle
Research at the Marine Biological
Laboratory,
WOODS-HOLE – Massachusetts
(U.S.A.).

A. TAVARES DE SOUZA,
Instituto de Histologia e Embriologia
da Faculdade de Medicina,
Universidade de Coïmbra,
COIMBRA (Portugal).

L. TISZA,
Massachusetts Institute of Technology,
Department of Physics,
CAMBRIDGE – Massachusetts 02139,
(U.S.A.).

Mme TONNELAT,
Faculté des Sciences,
Institut Henri Poincaré,
11, rue Pierre Curie,
75 – PARIS 5e (France).

H. H. USSING,
Institute of Biological Chemistry,
University of Copenhagen,
2 A, Øster Farimagsgade,
COPENHAGUE K. (Danemark).

TH. VOGEL,
Directeur du Centre de Recherches
Physiques,
Centre National de la Recherche
Scientifique,
31, chemin Joseph-Aiguier,
13 – MARSEILLE 9e (France).

E. WOLFF,
de l'Académie des Sciences,
Administrateur du Collège de France,
Laboratoire d'Embryologie Expérimentale,
49 bis, avenue de la Belle-Gabrielle,
94 – NOGENT-sur-MARNE (France).

R. WURMSER,
de l'Académie des Sciences,
Institut de Biologie Physico-Chimique,
Fondation Edmond de Rothschild,
13, rue Pierre Curie,
75 – PARIS 5e (France).